PLANT COLLECTING IN ANOTHER PLANET

E. H. WILSON IN AUSTRALIA 1920-21

First published in 2025 by
UWA Publishing
Crawley, Western Australia 6009
www.uwap.uwa.edu.au
UWAP is an imprint of UWA Publishing,
a division of The University of Western Australia.

This book is copyright. Apart from any fair dealing for the purpose of private study, research, criticism or review, as permitted under the Copyright Act 1968, no part may be reproduced by any process without written permission. Enquiries should be made to the publisher.

Copyright Margaret Grose © 2025
The moral right of the author has been asserted.

ISBN: 978-1-76080-308-7
Design by Upside Creative
Printed by Lightning Source

UWAP acknowledge we are situated on Nyoongar land, and that Nyoongar people remain the spiritual and cultural custodians of their land, and continue to practice their values, languages, beliefs and knowledge. We pay our respects to the traditional owners of the lands on which we live and work across Western Australia and Australia.

PLANT COLLECTING IN ANOTHER PLANET

E. H. WILSON IN AUSTRALIA 1920-21

MARGARET J. GROSE

FOR KON

WHO HAS RETURNED TO
THE LAND OF TUART,
JARRAH, MARRI, SALMON
GUM AND GIMLET, AND
THEIR COMPANIONS

*Nature has
manifold ways
for rearing trees*

VIRGIL,
GEORGICS, BOOK II, LINE 9

*In the matter of trees,
Australia is rich,
but decidedly strange*

ERNEST WILSON,
BOSTON EVENING STANDARD,
30 SEPTEMBER 1922

CONTENTS

| ACKNOWLEDGEMENTS | viii |
| PREFACE | x |

CHAPTER 1	An expedition to a new world of botany	1
CHAPTER 2	A puzzle frames the Expedition	13
CHAPTER 3	Into the other planet's *'veritable botanic garden'*	23
CHAPTER 4	*'The price ... exacted'* in Western Australia's Wheatbelt	73
CHAPTER 5	Into southern forests to see *'the Adonis and Hercules of the tree world'*	141
CHAPTER 6	Arid, with woodlands	187
CHAPTER 7	*'Stop making havoc of your heritage of natural forests'*	233
CHAPTER 8	Wilson mortified by Australia's 'habit of destructiveness'	282
CHAPTER 9	Extended journeys	283

| EPILOGUE | 309 |

APPENDIX 1	Selected images of Harvard Herbaria specimens	313
APPENDIX 2	Wilson's transcribed diaries 1920–1, Australia	316
APPENDIX 3	Plant specimens from Wilson's Australasian Expedition in the Harvard Herbaria	349
APPENDIX 4	A note on Wilson's plant specimens held in Australian herbaria	365
SELECTED BIBLIOGRAPHY	382
INDEX	387

OPPOSITE: Ernest Wilson in Queensland, 1921, detail from Wilson Y-514 AA.

FRONTISPIECE: *Eucalyptus guilfolyei*, the yellow tingle, collected by Wilson as 'Euc. Guilfolyei Maiden. Tree up to 100 ft. abounds Frankland River S.W. Australia', 10 November 1920, no. 388.

ACKNOWLEDGEMENTS

I thank the Arnold Arboretum of Harvard University for a Sargent Award that enabled me to spend time at the arboretum's wonderful facilities in 2018. Sincere thanks to the members of the Arnold Arboretum. I thank especially William (Ned) Friedman, Director of the Arnold Arboretum, Jonathan Damery, editor of the journal *Arnoldia*, and Lisa Pearson, head of archives, for their enthusiasm and support of this project, which grew and grew. Thanks to Peter Del Tredici, urban ecologist and senior research fellow at the Arnold Arboretum, for alerting me to Wilson's neglected collection, and to Fay Rosin for sorting out the administration involved in my visit.

 This project could never have been completed without the help given to me by Anthony Brach of the Harvard Herbaria, who attended to my every inquiry, gave extraordinary assistance by imaging all the plant specimens found that were from Wilson's Expedition to Australasia, and then did further searches of the collection when I had returned to Australia. I thank the Harvard Herbaria's allowance of time, access, and space.

 For the great working environment in the Arboretum's guest house in Roslindale in 2018, I thank Carolina Osuna-Mascaró, who was awarded her PhD in October 2020.

 Throughout the work on this book, I appreciated what Wilson called 'that friendly feeling among scientists'. In Australia, Alex George, one of Australia's foremost taxonomic botanists, and Stephen Hopper, Professor of Biodiversity at the University of Western Australia, examined some of Wilson's images to determine the species, and I thank them for their knowledge and clarity. I thank John Dargavel (d. 2024), forest historian from the Australian National University in Canberra, who gave me valuable information about Australian forestry one hundred years ago and insights into early 1920s Australia.

 Working and travelling in the Western Australian Wheatbelt and Goldfields, where Wilson travelled in 1920, was sheer pleasure thanks to the friendship, hospitality, and knowledge of the Wheatbelt of Pauline and James (d. 2023) Scott of Kellerberrin. I also thank Malcolm French for his deep knowledge of eucalypts, Wheatbelt landscapes and revegetation. I thank the Kalgoorlie members of the

Goldfields Speciality Timber Group, and to Neil and Suellen Turner in Bunbury, who all provided insights into the woods of the trees of the regions Wilson visited in Western Australia. I thank Ian Kealley, a Goldfields' forester with great knowledge of this region, for discussions on trees of the Great Western Woodland and the Goldfields.

I thank the Western Australia Herbarium for their assistance with Wilson's specimens held in Perth.

Emma Russell from History at Work did the very difficult job of the first cut on transcribing Wilson's handwritten diaries, which are in pencil and exceedingly difficult to read. Wilson was a master of both the scrawl and the incomplete scribble for a word, and some lines of text are faded and smudged. Any blanks in the text remain only after many forensic attempts to decipher words. Tricia Dearborn copy-edited the first draft of the book, checked my data and species names, standardised my footnotes, and suggested helpful changes to my text. I also thank three very helpful reviewers of the book. Samantha Trafford from University of Western Australia Publishing did a wonderful job of editing, and Max McMaster of indexing, and I also thank Kate Pickard for her advice and guidance.

I thank the University of Melbourne and the Melbourne School of Design, where I have worked since 2007 in the landscape architecture group, for their support of this book.

Finally, I thank my son Konstantine Panegyres for his support of this work. All errors, of which there must be some, are mine.

Margaret Grose
Kellerberrin
Western Australia
2024

PREFACE

It was early November when I arrived in Boston from Australia on my botanical quest. I was to investigate the remains of an expedition made to Australia by Ernest Wilson, a renowned English plant explorer based at the Arnold Arboretum in Harvard University in the early twentieth century. Most of Wilson's plant collecting had been done in China between 1899 and 1911, and it was for his travels and collections in eastern Asia that he had found fame, with thousands of plants and seeds collected and many introductions of Chinese species to North America and elsewhere. While Wilson was well-known for his Chinese expeditions, his last major expedition had been to Australasia, India and Africa from 1920 to 1922, but little was known about this trip at Harvard of either landscapes or herbarium specimens. No one in Australia appeared to know about this trip at all; it was as if it had vanished from sight. I was curious to know what had been collected, where Wilson had gone during his trip and why this collection had never been examined. Perhaps, when he had returned, he had put the plant collection aside, for later. Perhaps, having outlined the trip in horticulturally based articles called *Plant Hunting* in 1927, he felt no need to do further. Perhaps the answer, I thought, was because Wilson had been killed in a car accident in October 1930[1] and the collected plant specimens and photos had been left to languish.

    I had heard about Wilson's trip to Australia by chance on my first visit to the United States in 2016, while working at the Massachusetts Institute of Technology. Peter Del Tredici, an urban ecologist and botanist and a senior research scientist from the Arnold Arboretum, had given me a tour of the arboretum. Peter had pointed out that some collections of Ernest Wilson's were in the arboretum's archives, and included some notes, small diaries, and a collection of images from an Australian trip in the 1920s. He thought it likely that some plant specimens were held at the Harvard Herbaria, but no one knew for sure, and possibly none would be found. Wilson's images from this expedition were not labelled. 'This is definitely Australia, isn't it?' someone asked of an image of Kirstenbosch Botanic Garden in Cape Town. It was the correct ancient continental grouping, but the wrong new continent. Clearly, there was work to be done, and I went back to Boston towards the end of 2018 to examine Wilson's collection of Australian material from his expedition to Australasia, India, and Africa.

It is a long trip from Australia to the east coast of the United States of America. Los Angeles to Boston is 4200 kilometres (2611 miles), for the USA is wide. West–east across Australia, from Perth to Sydney, is 3290 kilometres (2044 miles). Even though I had lived in England for many years, the change of landscape from Australia to the more northern reaches of the Northern Hemisphere always gives an immediate impression of difference, with a complete change of plant species, and contrasting colours to Australia's vegetation.

It was November, and the university year had just ended in Australia. Having come from a part of the world where summer's heat had already been knocking on the door, I found Boston cold. Even in the dark night as a taxi took me to the Arnold Arboretum's house on Walter Street in Roslindale in Boston's southern suburbs, I could see that the Northern flora was turning leafless. When I turned off the light that night and pulled up the venetian blinds to let in the evening, I gave a small gasp in delight. There, in moonlight, was an enormous American beech, *Fagus grandifolia*, still in glorious autumn colours. How remarkable this tree and its companions seemed to me, because I come from what Wilson called "another planet" in its botany, and I stared out at the beech in wonder at colour, form, and the deciduous response to winter that is extremely rare in Australia, which is an evergreen world. During the night it snowed, something that had not happened in my part of the world since the Last Glacial Maximum, about 18–22,000 years ago, if then.

The morning dawned sunny, white, and apricot-leaved. I set out to meet people at the arboretum's laboratories at Weld Hill, cautiously nudging slippery ice with my boot and poking at snow on my way to find what remained of Wilson's collection. Squirrels darting in the arboretum grounds reminded me again of the differences between my part of the planet and the rest, because squirrels are found on every continent except Australia due to Australia's extraordinary isolation for millions of years. This was a theme that I would come back to again and again in my journey to see what Ernest Wilson had found, seen, collected, and thought about during his expedition to Australia.

NOTES

1   Obituary of Dr EH Wilson in *Nature*, 1 November 1930, 126:693–4.

CHAPTER 1

AN EXPEDITION TO A NEW WORLD OF BOTANY

In 1920, the great plant collector Ernest Wilson travelled by steamer to Australia to survey Australia's unique flora; in particular, its trees. This expedition was Wilson's last long trip overseas. It incorporated sojourns in Australia and New Zealand, then travel through Malaysia, India, and eastern and southern Africa on his slow route back to America in 1922.

Wilson had never been to Australia when he arrived at the age of forty-four. He had already established a worldwide reputation as one of the foremost plant collectors of the previous century. His fame was due to his long expeditions to China, Formosa (now Taiwan), Japan, and Korea, which were funded by the Arnold Arboretum of Harvard University, where he was based. He had brought back to the United States hundreds of plants from eastern Asia, and many, such as the handkerchief tree (*Davidia involucrata*), had already been planted extensively in European and North American gardens by 1920. He had become known as 'Chinese' Wilson, a term that he appeared to relish, though he never boasted about his accomplishments.[1] He wrote of his travels in well-written, popular, readily accessible books of great interest, such as the two-volume *A Naturalist in Western China with Vasculum, Camera, and Gun*, published in 1913, which addressed the nearly eleven years he spent in China.[2]

His Australian trip was described in his *Plant Hunting*, again published in two volumes.[3] *Plant Hunting Volume 1* was fundamentally an armchair travel book for plant lovers, charting his travel from Australia to New Zealand, Malaysia,

India, and Ceylon, to east Africa and southern Africa. It included information primarily about the plants he met along the way, with some geography and historical perspectives. It was re-published after Wilson's death as *Smoke that Thunders*.[4] Early in his trip, and while still in Australia, Wilson wrote that 'Australia is a new world to one familiar only with the flora of the Northern Hemisphere. Everything is different, all the species, excepting certain aliens and naturalized weeds, most of the genera and many of the families'.[5]

This was to be a very different trip from those Wilson had made previously in Asia for the Arboretum, where his main purpose had been collecting in wild places for new plant introductions that were largely unknown to the rest of the world. However, in Australia, Wilson was shown known plants by botanical experts – English speakers who were senior players in the big issues of the day concerning forests and botanical plant collecting – and he was not collecting for plant introductions for the Arnold Arboretum, because changed US quarantine regulations disallowed that possibility.

There seem to be many reasons for the expedition to Australasia. When asked by newspaper men in Australia the purpose of the expedition, Wilson always framed it as a census of the world's trees. To them, he said:

> I am Assistant Director of the Arnold Arboretum of Harvard University, and am working on a scheme we have to try and take a census of the world's trees and see what they are, before they disappear. We have already done America, and I have spent 21 years in the Far East on the same work. I was in Korea, China and Japan, Manchuria, and through parts of Siberia. We want to get a census of the world's trees because they are disappearing so rapidly. If we do not get such records of them in the shape of photographs and specimens, a hundred years hence many will have disappeared entirely. Our work is pure research, purely scientific; we have nothing whatever to do with the commercial aspect. My institution is established solely for the advancement of the study of trees and pure research work.[6]

He noted later in Australia that 'We at Harvard deplore the wanton destruction of trees the world over, and are filled with alarm, for if this continues much longer we shall have no standing trees left, and will pay to go to a museum to see what a tree was like'.[7]

Wilson pointed out that it might be conceived that he was interested in commercial aspects simply because this study would give rise to information on 'how common a tree is and how rare it is, and whether it is good for fire-wood or good for market, and so forth. But every tree is interesting for its own sake'.[8] Other reasons for the expedition were liaising with botanists and arboretums in Australia and other Commonwealth countries, and gathering specimens as examples for herbarium material. Live plants or seeds could no longer be sent back to Boston due to changes in United States customs regulations since the time of Wilson's Chinese trips.

On Wilson's return to the United States in 1922, the *Boston Evening Transcript* asked its readers:

> For what was this formidable journey undertaken? Not this time' they said, for the purposes of collection. Mr Wilson, though he is undoubtedly the greatest living botanical collector, went this time for another main purpose, which was to get in touch with the correspondents of the Arnold Arboretum in these regions; to establish means of mutual help with collectors and authorities; to effect exchanges with institutions and collections, to the end that the herbarium of the Arboretum shall possess specimens of every woody thing on earth.[9] Wilson reiterated this idea many times while in Australia. To the *Ballarat Star*, in 1921, he said 'that his object was to ascertain how Australia could help America and how America could help Australia in regard to forestry, and to help to promote that friendly feeling among scientists which should exist if our civilization was to be saved'.[10] Wilson also said that his reports and those of his colleagues would form a library of the trees of the world.[11]

New customs regulations aside, Charles Sprague Sargent, Director of the Arnold Arboretum, was very aware of the unlikelihood of finding plants to extend the Arboretum's living collection from Australian sources. He noted in a report of the Arboretum's first fifty years that 'No plants from the southern hemisphere, not even from the high Andes, southern Chile or the higher mountains of New Zealand have proved hardy in the Arboretum'.[12] Many species from other climates have trouble growing in Boston due to the difficult combination of snowy winters combined with warm, humid summers.

The hardiness issue with tree seeds was common. For example, Sargent had been offered tree seeds from southern India. He replied with the comment: 'I am sorry to say that there is no hope that any of the trees of southern India will ever grow in this climate', referring to Boston's severe winter.[13] Sargent was, however, keen and 'anxious'[14] to obtain herbarium specimens from southern India because he described the Arnold Arboretum as a 'general one' and the aim was to 'contain specimens of the woody vegetation of all parts of the world'.[15] In a letter to Captain R. N. Parker, Conservator of Forests in India, Sargent commented that 'American herbaria are very poor in Indian plants and this is particularly important to us here in view of our studies of the Chinese silva'.[16] Despite that limitation of climate, Wilson was interested in recognising plants in Australia that might be suitable for California's warmer climate because of funding for the expedition received from Henry Huntington, who created his now famous gardens in San Marino, near Los Angeles.[17]

Sargent believed that:

> If the Arboretum is to become a great institution for gathering and spreading information about trees and allied plants, specimens and a series of photographs of every species of tree in the world should be found in its herbarium ... It is work that this Arboretum should begin and steadily push forward.[18]

But Sargent knew that he needed good men to go out and plant-hunt. 'Such information can be obtained only in small part of correspondence and the information which the Arboretum needs can only be successfully obtained by agents sent out to obtain it.'[19]

Wilson, as the Arnold Arboretum's most successful 'agent', would have been expected to take many photographs during his Australasian Expedition. Sargent noted that in the Arboretum in 1922 'The two thousand eight hundred picture [*sic*] made by Wilson in eastern Asia and Australasia form the most valuable and interesting part'[20] of the 9,600 photographs made and held at the Arboretum. At that time, the Australasian Expedition images had not been collated or annotated and were only examined nearly one hundred years later.

Wilson did not speak Chinese; neither did he seem to have attempted to learn the language,[21] though he rendered many plant names in their Chinese version as well as the scientific one.[22] In Australia and elsewhere on his new trip, he was travelling about the British Commonwealth, among English speakers. Not only was he to meet and readily converse with many scientists, foresters in the field, and politicians, but in Australia at least, he was able to make many critical comments about government and forestry policy and management practices. A distinguishing feature of this trip was that newspaper reporters followed him, interviewed him, quoted him at length, and reported his entire expedition within Australia. This had not occurred in China, Japan, Formosa, or eastern Siberia or, at least, it has not been reported and recorded. In those earlier travels to eastern Asia, his diaries and letters provided most of the material. For this book, I had scant diary material and have had to rely very strongly on newspaper reports.

Letters that reveal the organisation behind the expedition to Australasia are almost non-existent. Wilson wrote: 'I am having quite a little work over my passage to Australia. The latest is the ship that was to sail has be [*sic*] taken off— another is to take the place & sail about the same time. Of course everything will come out all right but it means that I have to keep in constant touch with the agents.'[23]

Letters to Sargent track his path:

> '3rd September 1920: My boat sails from Toulon on the 11th and I leave England on the morning of the 9th & I am quite glad to be moving on.'

> 'Sept 11th. Grand Hotel Toulon I am now informed that it will not sail before 6pm tomorrow. It is an ex-German liner that has been tied up for more than [*indecipherable*]. It is a bit of a nuisance for there is nothing to do here except enjoying the brilliant sunshine.'

He finally sailed from France on 12 September 1920 and travelled via Suez and Ceylon. His voyage was slow, dirty, and uncomfortable; he did not travel first class but steerage; the ship was dingy, and hot through the tropics.[24]

On 17 September 1920, aboard the SS *Königin Luise*, near Port Said, he wrote:

> We are due at Port Said tomorrow so I am sending this note to let you know that all is well. We left Toulon on the afternoon of the 12th & touched at Naples for an hour on the 14th but I did not go on shore. This ship is an ex German liner and very ill-found. She can only make twelve knots under the most favourable conditions & God knows when we shall reach Fremantle. There are three of us in a lower deck cabin with one port hole. The ship is crowded and the bath and lavatory accommodation disgraceful. There is no laundry on board so we shall have a nice lot of linen by the time our destination is reached … I am longing to get to Australia & to work, for I am utterly bored![25]

> Dear Professor Sargent,

> We are due at Colombo tomorrow morning & I am sending this note to let you know that all is well. In the Red Sea & Indian Ocean it was frightfully hot. We had faltering winds & not a breath of air in the cabins. Sleep except on deck was out of the question & even there difficult. I feel as limp as a wet rag but otherwise am all right. We are promised a full day on shore at Colombo & if possible I shall [ ][26] up to Peradeniya. A breath of cool air & the sight of something green will be very very welcome. From Colombo to Fremantle is said to be twelve days so I shall be in Australia before this reaches you.

Peradeniya is the Royal Botanic Gardens of Sri Lanka, about an hour from Colombo by bus today; it is now a suburb of Kandy.

Wilson wrote on the SS *Königin Luise*, off Fremantle, on 13 October 1920: 'We are due to arrive at Fremantle tomorrow morning and I shall be very glad to bid farewell to this good ship.' Fremantle is the port for Perth, the capital of Western Australia.

Wilson's fame preceded him, and he was met by important foresters and botanists in every Australian state that he visited. The press introduced him well and usually with a flourish. For example, the *World's News* from Sydney in July 1921 said this of him:

> Mr Wilson … is the Livingstone of horticulture. He is the most redoubtable, audacious, and persevering plant-hunter the world has ever seen. There are no flowers or shrubs in civilisation with which he is not acquainted … No one else knows so many varieties of native and foreign flowers, shrubs, and trees, and no other explorer has transported from their habitats the countless strange plants that this unusual man has to his permanent credit.[27]

They went on to describe Wilson's exploits in China and other parts of eastern Asia. 'The vocation of plant hunter is not so genteel as the name might suggest. It is a man's job, charged with red-blooded adventure. For example, the natives of Formosa are largely head-hunters, and, as Mr Wilson travelled throughout the island, his experiences were many and dangerous.' They outlined the discovery of the 'famous incandescent lily in the hinterland of Tibet', and the story of how Wilson broke his leg gaining the regal lily (*Lilium regale*), that he 'collected ten thousand bulbs from acres of beautiful blooms and shipped them by man, mule, and steamer to America'.[28] Of China and Wilson, they said:

> there is no man outside of Western China, for but one instance, who has a greater knowledge of that country. He has spent eleven years in the interior, and brought back 1400 specimens of exotic plant life, which he found in the course of his trip of two thousand miles in the extreme border. And he travels comparatively simply, with luggage, guns, and a few coolies.[29]

Though China was his focus and the source of his foremost finds, they noted that Wilson 'is equally enthusiastic over Korea, where there is a delicious climate and other natural attractions. It is occupied largely by the summer homes of mandarins. The natives, he says, excel in agriculture and porcelain-making'.[30] Wilson's previous trip had been to Formosa (Taiwan) in 1918–19. Far from the imaginative description by the press, Wilson wrote of Formosa as 'a rich and beautiful island & its forest wealth is very great. To have visited the island is a privilege I greatly appreciate & I shall carry away with me none but the pleasantest of recollections'.[31] The Australian press increased the glamour, romance, and danger, perhaps influenced by Conan Doyle's *The Lost World* of 1912, or Edgar Rice Burroughs's *Tarzan* of the same year.[32]

His Australian journey was described as one to the Southern Hemisphere

> which he anticipated will require two years … The fruits of this journey will be mostly autobiographical. He will investigate rare pamphlets and other literature on horticulture and arboriculture which are hidden in botanical museums in Australia, and in Calcutta, Bombay, and other cities. Whatever flowers and plants he may be so fortunate as to come across must needs go to England, for the American Federal Horticulture Board has prohibited virtually all such importations.[33]

There were many and varied reasons that Wilson, Sargent and other interested parties saw or gave as the purpose for this visit. All were likely true. The 'mostly autobiographical' explanation given by the Australian press is a mystery, except that, as in his previous travels, this trip was to give rise to many personal anecdotes and responses as he travelled through landscapes that were new to him. In *Smoke that Thunders*, his trip to Australia took up fourteen of the twenty-nine chapters the book contained. Wilson's style was to include descriptions of the landscape as well as the events and experiences of his journeys, including many references to geological history. In keeping with this style, I have tried to add some substance and stories about the landscapes through which Wilson travelled in Australia. In doing so, I have noted some things that might be obvious to Australian readers but not to American readers. I have also tried to move between the world of learning and scientific knowledge and the world of everyday conversation. The book is a reflective on Wilson's journey to Australia and is not an exposé on the environmental movement in any part of Australia in the 1920s or preceding it; I make some comments on what happened to the landscapes that Wilson saw and noted as of concern to him.

Despite there being some information in the archives, little was known at Harvard about the Australian component of this trip. While diaries did exist, they had never been examined. There were images, but none had been identified as to a place visited or tree species. It was not known if any plant specimens were housed in the Harvard Herbaria or elsewhere. When I arrived, no one had any idea what would be found, if anything. I was drawn to examine Wilson's trip because of the sheer absence of knowledge at Harvard – and also in Australia, where botanists seemed astounded that the famous Wilson had once visited. It was for me a voyage of discovery, to find what was held in Harvard's herbaria and arboretum without Australia's knowledge, and to tell that story. I was also keen to see if Wilson's images and trip could shed light on Australia's treatment of its landscapes 100 years earlier. Where had Wilson travelled? What had he seen and thought? What might Wilson the photographer have preserved for us? Had he said what he thought of this new world of botany, so different to the rest of the world and strikingly different to the Northern Hemisphere?

To approach these questions, I first examined all the details held in the archives of the Arnold Arboretum. These included some newspaper articles and Wilson's diaries; his scrawled pencil entries were very incomplete and difficult to decipher. Then, in sessions in the Harvard Herbaria, I went through the Australian flora and noted and imaged specimen sheets that were labelled as being from Wilson's Expedition to Australasia, India, and Africa. On my return to Australia, I examined newspaper records from across the continent to discover whether and how Wilson's trip had been recorded; this proved a major source of information, with often lengthy articles about Wilson's trip.

Finally, I travelled along some of the route that Wilson took that was best recorded in his diaries – his route through Western Australia. As Wilson did in his own books, I have tried to give some flavour of this country to the reader, with notes on geography, history, botany, and palaeohistory, to emphasise the great difference in this landscape that Wilson could see so clearly. I focused on Western Australia because Wilson's journeys in eastern Australia are unclear, bar a few sites mentioned in newspaper reports. In doing so, I hope my own shadow in the story improves the understanding of place and people rather than intrudes.[34]

During the research and writing of this book, I have had one foot in the scholarship of Wilson's notes and plant collection and his comments to newspaper men, and the other foot in the landscape – both as seen by him and the landscape of today. The landscapes Wilson saw were on the cusp of staggering changes. Some of my comments are descriptions of actions taken more than a century ago and the outcomes that resulted, now seen in the context of today's greater understanding of lands that quickly proved so different to European and North American sources of agricultural knowledge that Europeans had brought with them. They faced their challenges with well-supported beliefs, erasing trees by the thousands.

The equilibrium of millennia was destroyed, and one can still feel it like a sorrow across much of Australia.

Sometimes, thinking of Wilson at a site as I followed his exploration of what he came to call the botany of the 'other planet', I felt able to contract the century between us and conjure up the past, but that moment was often fleeting. Though I strived to see the bush, the trees, the forests, the opened-up agricultural land with the eyes, ideas, and feelings of newness of the men and women of the 1920s, I was aware that I knew enough of this country that I could never see these landscapes with a stranger's eye, particularly the western lands.

The record of Wilson's visit is an incomplete one, as no one can reconstruct the long nights he had alone, the dusty travel, his companions and their chatter and discussion of policy and trees, the endless and boring travel arrangements that were so often frustrated by changes outside of Wilson's control, the isolation from home, the intrusion and demands of the press, the expectations of the Arnold Arboretum's Director, the pressure to collect and record accurately – often while besieged by heat and flies – and the sheer learning that Wilson did during his months in Australia. Much must be left out because it is unknown. But who pressed the specimens, who climbed the tree to gain a good specimen of fruits high in a eucalyptus canopy, who organised the water bottles for these men in suits but with dusty boots and sweat on their brows?

NOTES

1. R Briggs, *'Chinese' Wilson: A life of Ernest H Wilson 1876–1930*, HMSO, London, 1993, p. 6.
2. EH Wilson, *A Naturalist in Western China with Vasculum, Camera, and Gun: Being some account of 11 years' travel, exploration, and observation in the more remote parts of the flowery kingdom*, 2 volumes, Doubleday, New York, 1913.
3. EH Wilson, *Plant Hunting*, volumes 1 and 2, Stratford, Boston, Mass., 1927.
4. EH Wilson, *Smoke that Thunders*, Waterstone, London, 1985.
5. EH Wilson, 'Notes from Australasia', no. 1. *Journal of the Arnold Arboretum*, 1921, 2(3):160.
6. 'A study of trees. Professor Ernest H. Wilson's visit. Our wonderful flora', *Western Mail* (Perth), 25 November 1920, p. 9. This was published under the subheading 'His mission'. Wilson had been made assistant director in 1919.
7. *West Australian* (Perth), 21 October 1920, p. 7.
8. *Western Mail*, 25 November 1920, p. 9.
9. *Boston Evening Transcript*, 30 September 1922, magazine section, pp. 1, 4.
10. 'Our timber reserves: Australia's greatest asset. American expert's views', *Ballarat Star*, 21 April 1921.
11. *Western Mail*, 25 November 1920.
12. CS Sargent, 'The first fifty years of the Arnold Arboretum', *Journal of the Arnold Arboretum*, 1922, 3(3):127–71. The comment is from p. 137.
13. Letter from Sargent to Father L Anglade, Sacred Heart College, Shembagur, Madura District, India, 21 October [year uncertain], catalogue no. 464.
14. Letter from Sargent to Dr LL Uhl, Kodaikamal, Madura District, India, 21 October [year uncertain], catalogue no. 463.
15. Ibid.
16. Letter from Sargent to Captain RN Parker, Conservator of Forests, based in Chamba, NW India, catalogue no. 420.
17. Huntington's contribution of $1,000 was noted by Wilson in a letter to Sargent from Perth, 17 November 1920, AA Archives, W.XIV:A, box 21, folder 11.
18. Sargent, 1923, p. 170.
19. Ibid, p. 170.
20. CS Sargent, 'The first fifty years of the Arnold Arboretum', 1922. The comment is from p. 167.
21. This contrasts with William Purdom, who arrived in China in 1909 and 'picked up the language very rapidly'. F Gordon, *Will Purdom: Agitator, Plant-hunter, Forester*, Royal Botanic Gardens, Edinburgh, 2021, p. 62.
22. As, for example, in his book *A Naturalist in Western China*.
23. Letter to Sargent, from Kew (just back from Edinburgh), 23 August 1920.
24. This contrasts with the comment that he travelled first class after this fame in China, in TM Holway, 'History or romance? Ernest Wilson and plant collecting in China', *Garden History*, 2018, 46(1):3–26.
25. Letter to Sargent, AA Archives, W.XIV:A, box 2, folder 11.
26. This format is used throughout the book to denote a word that could not be transcribed from Wilson's notes or diaries—his handwriting can be hard to decipher.
27. *World's News* (Sydney), 1 July 1921, p. 13.
28. See Michael S Dosmann, the current keeper of the living collections at the Arnold Arboretum, in his article on Wilson's collection of the regal lily, 'A lily from the valley', *Arnoldia*, 77(3):14–25.
29. *World's News* (Sydney), 1 July 1921, p. 13.
30. Ibid.
31. Wilson's letter to Sargent, from his diary notes dated 14 April 1918. Reprinted as 'Taiwan Dispatches', with selected letters, in *Arnoldia*, 76(3):28–33.
32. Wilson wrote about cannibals in Formosa elsewhere.
33. From *World's News*, Sydney, 1 July 1921, p. 13.
34. I have followed somewhat the style of the Arnold Arboretum's journal *Arnoldia*, which blends scholarship with storytelling.

CHAPTER 2

A PUZZLE FRAMES THE EXPEDITION

At Harvard, the remnants of the Expedition to Australasia, India and Africa of 1920–2 were scattered. There were images, diaries, and a small box of ephemera including newspaper clippings held as physical entities at the old Hunnewell Building of the Arboretum in Jamaica Plain. An unknown quantity of plant specimens were held in the Harvard Herbaria on Divinity Avenue, Cambridge, Massachusetts.

At the Arnold Arboretum were Wilson's glass-plate negatives of the Expedition, which had been digitised in 2018 by brothers Scott and Michael Dietrich, who were working on Wilson's photographic legacy at Harvard. The Dietrichs had digitised Wilson's Australasian images to such a high standard that the zoom could be used to determine the tree species depicted in many of the images, with assistance from taxonomic botanists in Australia. In addition, I zoomed in to examine not only the trees, but the men in the photos.[1] Immediately, I could see that Wilson's images showed landscapes and trees from Western Australia. Many trees were readily recognisable, as most of the major species are distinctive from one another, while the identification of others needed a closer look at the fruits and leaves using the zoom capacity.

In the collection of glass-plate negatives, I could see little of eastern Australia in either trees or landscapes. At the beginning, that absence did not make much of an impact on me, but my curiosity about that absence soon grew into a nagging feeling. There were a few images of tropical rainforest in Queensland, but little of the south-east of the continent, except some that were possibly of Tasmania. Why so few images of eastern Australia? He had gone there – to Melbourne, to Sydney, and into the hinterland from those cities – in 1921.

As I commenced a search for plant specimens in the Harvard Herbaria, I began to sense something odd in the same vein. Anthony Brach, from the herbarium's curatorial team, looked after me and introduced me to the catalogue. Anthony said, 'Well! Where do you want to start?', and I suggested the Myrtaceae, the great family of Australia that includes the eucalypts. He had chuckled at me starting on the 'big one'.

Anthony showed me the colour system. Australian specimens were housed in blue folders, which were easy to see among the swathes of folders of green, orange, pale orange, yellow, pale yellow, beige, white, off white, and hues of pinks and reds, in a collection dominated by specimens from North and South America. If the folders were blue, the contents needed to be examined. The work was done on a long, high table, standing, peering, and writing into my notebook with the yellow pencils that I had been given, as well as using new little paper envelopes to collect and contain broken-off bits of specimens, and fresh new clear plastic clips to replace the rusty metal clips of one hundred years ago.

I hunted through the blue files to find labels with Wilson's name and handwriting. 'Near a saltlake', he noted on one. 'In a group of trees', he wrote on others. 'Centipede bush', he wrote of a now rare little beautiful shrub, *Daviesia euphorbioides* (#32), collected at Hynes Hill, a tiny settlement;[2] 'Black toothbrushes', *Grevillea apiciloba* (#82) from Westonia;[3] and 'Geraldton Wax', *Chamelaucium uncinatum* (#209), on another showy shrub that was common in gardens even in Wilson's time. I enjoyed the intimacy of the found material, the smell of the once-living eucalypts, the touch of someone else's work and care from so long ago, of plants collected in the Great South Land, as Australia used to be called by seamen, geographers, and cartographers.

The collection of plant specimens is contained within rolling cabinets, called compactors, and I noted every day the written directive that 'compactors can be opened by releasing the locking bar and smoothly rotating the handle of the appropriate bay'. They glided. The instructions also suggested to 'move one row at a time to prevent strain on the system'. There is great pleasure in working systematically through a collection, and the closure of the compactors, the turning off of the lights, the shutting of the door, all signalled a clear end of the day's work at the herbarium, which closed at 5 pm promptly. If you overstayed, you might be locked in for the night.

As piles from the Wilson collection gathered, Anthony would kindly take them away and make a detailed digital image of each one at high resolution. We had no prior idea as to how many specimens we might find from this Expedition. The number grew and grew. By the end, Anthony had imaged nearly six hundred plant specimens, and yet I had not completed the entire herbarium search by the end of my stay in late 2018. There were some genera housed in a special part of the herbarium, all wrapped in plastic to prevent insect damage, and these have mostly not been examined.

I started in on *Eucalyptus* because I presumed that Wilson, as an arborist, would have been deeply interested in this very large genus that almost defines the Australian continent. There are now more than eight hundred species of *Eucalyptus*, and this number usually surprises overseas visitors to Australia, who often think that there are only a few 'gum trees'. The eucalypts include the world's tallest flowering plant, *E. regnans*, known as the mountain ash, a hardwood that reaches heights of 100 metres (328 feet). Timber records in the nineteenth century reported an *E. regnans* as the tallest tree in the world, with a fallen tree in Victoria in 1872 reported at 133 metres (436 feet). This tree's top had been lost, perhaps in a storm. The intact tree was estimated to have once been over 152 metres (nearly 500 feet). However, debate ensued as to the accuracy of the measurements – possibly a conflict between surveyors and lumber men – and the tallest tree by known survey was measured as 115 metres (377 feet) in 1880, also in the Victorian forests[4] after it had been felled!

Wilson wrote about tall eucalypts in *Smoke that Thunders*:

> The height of these trees has been much exaggerated, as much as 525 feet having been stated. The tallest tree authentically measured was 375 feet, which leaves a good margin in its favor over the 340-foot Redwood (Sequoia sempervirens) measure[d] as it lay on the ground by Professor Sargent near Scotia on the Eel River, California, in September, 1896.[5]

The loss of tall timber underscores one of the reasons Wilson was keen to see the Australian forests first-hand – these eucalypts rivalled the heights of those of much greater fame in the United States.[6]

Eucalypts are not all tall: some are small and gnarled; many are low shrubs and suitable for home gardens; some have brilliant flowers; some have glabrous pale-grey leaves that are suitable for the floricultural trade; some have huge bud caps, the feature that gave the genus its name – *eu* (well) and *kalyptos* (covered); some are single-trunked; others have a mallee form, an Aboriginal term referring to plants with multiple trunks that emerge from underground lignotubers.

Looking at Wilson's diaries at the Hunnewell Building, I noted that he was captivated by the variety of the eucalypts and the colours and forms of their trunks. But as I continued at the Harvard Herbarium, I noticed again the strange absence of eastern Australian flora. In an herbarium, a genus is stored, as would be imagined, with folders containing species stored in alphabetical order. Thus, with *Eucalyptus* I went through *E. accedens, E. astringens*, and on and on … but I stopped suddenly – I had gone through sheets of *E. regnans*, the mountain ash, and none of the specimens housed in the herbarium had been collected by Wilson. I double-checked. Where was Wilson's *E. regnans*? As one of the tallest trees in the world and a fine timber and plantation tree, it would have been a species that his Australian hosts would have taken him to see. Surely. Yet there was no specimen from Wilson's Expedition in the Harvard Herbaria. I was dumbfounded and curious. I began to notice other unexpected absences, and I checked for other species.

The oddly absent trees from Wilson's collection in the herbarium were all from eastern Australia. This was a significant realisation in the first few days of my stay at Harvard. What had happened to them?

Wilson's diaries, though sketchy, showed that time had been spent collecting in the bush and forests of eastern Australia. For example, visiting rural areas out of Sydney in early January 1921, he filled in three pages of eucalypts collected near Picton, Moss Vale, and Gundagai in New South Wales: 'En route gathered the following species of Eucalypts', he wrote.[7] At least two dozen eucalypt species were in his list, with plant-specimen sheet numbers given, including *E. creba* ('Narrow-leaved ironbark', #661), *E. maculosa* ('White Gum', #662 and # 664), *E. piperita* ('Sydney peppermint', #663), *E. dives* ('Broad leaf Messmate', #665), *E. radiata*

('Narrow leafed Messmate', #666), *E. punctata* ('Grey gum', #667), *E. longifolia* ('Wooly butt', #668), *E. paniculata* ('Grey Ironbark', #669), *E. sieberiana* ('Coast ash', #671), the famous *E. haemastoma* ('Scribbly Gum', #672), *E. leptospermum* (#673), *E. tereticornis* ('Red Gum', #674), *E. eugenioides* ('White Stringy bark', #675), *E. resinifera grandiflora* ('Red mahogany', #676), *E. elaeophora* ('Apple [ ]', #679, *E. viminalis* ('Manna Gum or Ribbon Gum', #680), *E. macrorhyncha* ('Brown Stringybark', #681), another *E. tereticornis* ('Broadleafed Red Gum', #682), *E. melliodora* ('Yellow Box', #683), *E. macarthurii* (#684), the silver-leaved *E. cinerea* ('Argyle apple', #685), *E. coriacea* ('Snow Gum', #686), and '*E. Smithii*' ('Gully ash', #687). However, none of these eucalypts were found in the herbarium.

There was the odd specimen of various genera collected from Queensland and bearing the Expedition label, but many had been collected by others in 1923 and apparently sent to Harvard after Wilson's return to the USA. In those first weeks, I could not find plant specimens from south-eastern Australia that had been collected by the man himself.

On trips to the herbarium, I left on an early commuter train from Roslindale Station. From Boston Station I took the metro (the 'T') to Harvard Square and walked to the herbarium. However, as the lack of eastern eucalyptus became apparent, I went back to the archives at the Arboretum at nearby Jamaica Plain to see if I could find a clue to the mystery of the missing specimens. This was a delightful commute on foot through the Arboretum's wooded glades, designed by the famous landscape architect Frederick Law Olmsted. The Arnold Arboretum is at the southern part of Olmsted's 'Emerald Necklace' of green-and-blue parks that make a long chain from the fen areas around Brookline to the Jamaica Plain area, winding its way through part of greater Boston. Olmsted designed the Necklace between 1878 and 1896. It was a pleasure to do the arboretum walk, even in winter, when stops to investigate bark form and colour made the walk very beautiful.

In the warmth of the Arboretum's archives, with lovely rooms and old wooden desks, I looked for some hint to the solution of the growing puzzle of eastern Australia by going through Wilson's letters and American newspaper reports of his visit upon his return to Boston in 1922. Since the 1920s, these reports had been catalogued, put away, and not scrutinised. In doing so, I found the explanation that I needed.

It was a disaster.

Every expedition of discovery and collection of natural objects requires transportation of specimens to somewhere else, often involving long distances. Wilson was well aware of the dangers of transporting any part of a collection. In China in 1908, he had been 'knocked all of a heap' when he heard from Sargent that a collection of precious bulbs that he had sent from Ichang after the 1907 collecting season to the Farquhar Nursery in Boston arrived with more than 95% gone to mush.[8] To replace them, Wilson was content to make a new trip to Ichang and re-collect many. 'Such a trip', he wrote to Sargent, 'will not occupy more than three weeks.'[9] Wilson noted as he prepared for this renewed expedition that he would need to travel down rapids when some were at their lowest and most dangerous. He wrote: 'However, to know and appreciate the dangers is always something towards combating them and I trust the good fortune which has attended us heretofore will not desert us now'.[10]

On the same expedition in China, Wilson noted the need to put his valuable collection in two boats, rather than risk the loss of an entire collection in one boat when the boats were required to run the rapids of the upper Yangtze River. Of this he wrote: 'I of course put the precaution to put half the collections in another boat. But whilst this halves the risks it doubles the anxiety.'[11] Wilson had lost a large quantity of photographic plates in Yangtze River rapids in 1908.[12]

Wilson had written that 'shipments are of the nature of a gamble', and of the 'luck' he usually had on his expeditions in China, Korea, and Japan, despite a serious leg injury that he incurred due to a rockfall in China in 1910. It was in Australia that his luck in the transportation of his collections ran out. Tucked away in the Arboretum archives in the Hunnewell Building was a newspaper comment that Professor Wilson 'experienced one great disappointment' concerning this trip. In eastern Australia, he had gathered 'two large shipments' of plant specimens and glass-plate negatives.[13] They were expected to travel separately, on different ships, to spread the dreadful risk of loss, following the same precautions that Wilson had taken on the Yangtze River years before. However, both consignments ended up on the same steam ship, the SS *Canastota*.

The *Canastota*, a Scottish-owned steamer on charter to the United States and Australia Line, usually sailed between Sydney and Wellington in New

Zealand, then on to San Francisco. From there, Wilson's consignment was to have travelled by rail to Boston.

The *Canastota* had been delayed several times due to problems and debates on the wharf. Tellingly, the dispute between the wharfies[14] and the owners was that the cargo was considered unsafe by the wharfies, who were attempting to unload steel piping. The adjoining cargo in the hold comprised 19,667 cases of Plume benzine, 279 cases of mercury benzine, and 3,864 cases of Plume motor spirit in a hold without bulkheads, which was against British regulations, as was the lack of ventilation for such a cargo. Benzine was known to be a highly unstable cargo.

Wharfies complained that they could smell benzine, which must therefore be leaking into the hold. Two men had been overcome by fumes and taken to hospital. The wharfies refused to continue to load the ship, and they were supported by their trade union. The trade union manager, a Mr Dawson, went onto the ship and was nearly overcome by fumes himself. Although the owners of the cargo, the Vacuum Oil Company, offered higher pay to the workers, they refused to continue to unload, and the ship was delayed in Brisbane due to this industrial action. It eventually sailed, and stopped at Newcastle, further down the coast from Brisbane, where more benzine was loaded. *Smith's Weekly* reported later that this new benzine was housed in cases of bad condition, with many leaking from their thin-walled drums.[15] Wilson wrote of these delays to Sargent; he was anxious that his consignment of plant specimens and photographs get home and be housed safely.

After the delay on the wharf at Brisbane, the *Canastota* finally sailed from Sydney on 13 June 1921, carrying Wilson's entire collection of glass plate images and specimens from eastern Australia. The ship was never heard from again.

The alarm was raised on 20 June, when the *Canastota* failed to arrive at Wellington. The New Zealand Navy put out a search, but no substantial debris of the ship was found at sea, nor had any distress message been sent, suggesting that the ship had been rapidly destroyed. The New Zealand Navy met foul weather, with a major storm in the Tasman Sea. There was some heated exchange between New Zealand and Australia, as the Australian Navy would not take part in a search, possibly due to the weather conditions, and there was adverse comment in shipping circles.[16] In early August, debris matching the *Canastota*'s cargo washed ashore at Lord Howe Island, 700 kilometres (435 miles) north-east of Sydney. Lloyds of London posted the ship as missing in August 1921 and classified the ship as lost in October 1921.[17] Lloyds considered that the volatile cargo of benzine had led to a sudden and complete destruction of the ship and attributed such loss to a presumed explosion at sea. All forty-nine crew members perished; most were under the age of thirty. The *Shipping List* noted that 'The exact circumstances of the loss of the Canastota will probably never be learnt'.[18] The crew of another US ship in Australia at the time, also carrying benzine, believed that 'what probably occurred was that a leakage from some of the cases ran along the bottom of the ship into the engine bilge, where the high temperature caused it to ignite. The wireless operator's cabin was immediately over the hold in which the benzine was stored, and he would probably be killed outright'.[19]

The loss of the *Canastota* caused rage and consternation in Australian newspapers. The weekend paper *Smith's Weekly* had a front-page headline on 21 August 1921 that read: 'CANASTOTA: A FLOATING BOMB. Conditions so bad on ship that wharfies refused to unload her. MEN CARRIED OUT OF HOLD IN BRISBANE IN STATE OF COLLAPSE. TWO DISREGARDED WARNINGS'.

Wilson's consignment from his travels in the eastern half of Australia consisted of images and specimens. What remains are a few annotated lists of plant specimens in his diaries that are a testament to where he collected and when in this period. No physical material from eastern Australia exists except for specimens that seem to have been sent after 1923 by Australian botanical collectors to the Harvard Herbaria, likely requested by Wilson to replace those lost at sea. A few of Wilson's eastern Australian photos from Queensland's rainforests survive, and since some appeared in Australian newspapers, it is likely that those now held at the Arnold Arboretum are also copies later sent. In addition, there are nearly 200 photos of Tasmania sent to Wilson by the well-known Hobart photographer John Watt Beattie;[20] Wilson must have lost his own. Likewise, for Wilson's New Zealand images: a stack of glass-plate negatives from New Zealand are housed in the Arboretum archives, but they are tourist images or forestry activity pictures and do not display Wilson's keen eye for plant species; one shows a display of Dunedin tartan. Other images of New Zealand have a scratched-out text that read 'NZ Forestry Service & Prof Wilson'. Much of the rest of the text is illegible; some show the word Nelson, a city in the North Island.

There is no record of Sargent's response to this huge loss and months of work; perhaps the strong contacts made with forestry and botanical people in Australia sufficed.

Because of the absences of information from eastern Australia, the discussion of Wilson's trip in this book is primarily concerned with Wilson's travel and botanical collection in Western Australia, where he spent longer than he anticipated due to the plant biodiversity that he saw. This region is now called the Southwest Australian Floristic Region, one of the world's great biodiverse regions for flora, and the most floristically diverse area in the megadiverse continent of Australia.

For eastern Australia, I have relied on newspaper articles to frame a picture of where he went and with whom, and what he thought. I have not included information about his trip to New Zealand, where he travelled in January 1921, and I did not look for New Zealand specimens in the Harvard Herbaria. That remains for another researcher familiar with a landscape and flora very different to that of Australia.

NOTES

1. S Dietrich and MR Dietrich, 'Ernest "Chinese" Wilson's re-imagined legacy in Sichuan', *Trans-Asia Photography Review*, 2019, 9(2).
2. Now Hines Hill; I have kept Wilson's names, as they were the ones in use in 1920. 'Centipede bush' appears to be wrongly attributed, as it was the common name for *Templetonia sulcata*.
3. Now named *Grevillea hookeriana* subsp. *apiciloba*.
4. For a discussion of the evidence, see AC Carder, *Forest Giants of the World, Past and Present*, Fitzhenry & Whiteside, Ontario, 1995. Further discussion is found in JE Hickey, P Kostoglou and GJ Sargison, 'Tasmania's tallest trees', *Tasforests*, 2000, 12:105–22.
5. EH Wilson, *Smoke that thunders*, 1985, pp. 141–2.
6. See NJ Caire, 'Notes on the giant trees of Victoria', *Victorian Naturalist*, 1905, 21:122–8.
7. Wilson, diary, 4 January 1921.
8. Letter from Wilson, 27 August 1908, Arnold Arboretum (AA) Archives.
9. Ibid.
10. Letter from Wilson to Sargent, 19 December 1908, AA Archives.
11. Letter from Wilson, 16 January 1909, AA Archives. This is recorded in Spongberg, *A Reunion of Trees. The discovery of exotic plants and their introduction into North American and European landscapes*, Harvard University Press, Boston, 1990, p. 206. The incident occurred in December 1908.
12. Noted in P Chvany, EH Wilson (photographer), *Arnoldia*, 1976, 36(5):190.
13. Recorded in unknown paper, titled 'EH Wilson returns', AA archives, series W XV, box 26, folder 12, titled 'Clippings on lectures by E. H. Wilson on his expedition to the Southern Hemisphere – Gardens of the World, 192-194'.
14. 'Wharfies' is an Australian term for men who work on the docks loading and unloading cargo.
15. *Smith's Weekly*, 20 August 1921, p. 1. This newspaper operated from Sydney from 1919 until 1950.
16. *Daily Commercial News and Shipping List*, 29 June 1921, p. 4.
17. 'The Canastota', *Daily Commercial News and Shipping List*, 4 October 1921.
18. Ibid.
19. 'Canastota, a floating bomb', *Smith's Weekly*, 20 August 1920, p. 1, 3.
20. Beattie, 1859–1930, was of Scottish birth, a Fellow of the Royal Society of Tasmania, and Photographer to the Government of Tasmania from 1896. The note in the AA Archives reads 'Photo by Beattie, sent by E. H. Wilson April, 1922'.

CHAPTER 3

INTO THE OTHER PLANET'S "VERITABLE BOTANIC GARDEN"[1]

After a hot and tedious trip from Ceylon, Wilson arrived in Fremantle, Perth's port, on 14 October 1920, disembarking the *Königin Luise* in the morning. He was immediately struck by the similarity of Perth to southern California, noting the 'abundant sunshine' and that, to 'heighten the Californian delusion', there were many Monterey Pine and cypress planted.[2] Perth has a similar climate to southern California or, as Perth people say, it has the climate Californians think they have. However, beyond the pines and cypress, Wilson would find a totally different flora to any he had seen before. He came to call this the botany of 'the other planet'. These differences arise from the evolutionary history of Australia, of which I include a short discussion later in this chapter.

In Perth, Wilson stayed at the Palace Hotel in St George's Terrace, then a grand Victorian street of sandstone buildings. He wrote that he was 'cordially greeted' by officials.[3] He met Charles Lane-Poole, Conservator of Forests, who 'hails from Dublin' and was trained in France. It was his first meeting of this entire trip, and perhaps his most important.

His arrival in Western Australia was widely reported in major Perth newspapers the *West Australian*[4] and the *Sunday Times*,[5] as well as a suite of small country newspapers from south-western Australia: the *Great Southern Herald* in Katanning,[6] the *Tambellup Times*,[7] the *Albany Despatch*[8] and the *South-Western Times* in Bunbury, south of Perth.[9] Reports of Wilson's travels in the *West Aus-*

tralian and the *Western Mail* and his answers to reporters' questions gave me far more information than Wilson's diaries as to where he travelled, and with whom. These also gave some insights into the man himself, who, it seems, has been a difficult subject for biographers, not least because much of his correspondence was destroyed after his death and his lightly pencilled handwriting in his diaries is atrocious. There are no diary entries for most of his Australian trip.

He stated his intentions very clearly to 'an obtrusive pressman' who was pushing to know why Wilson had come 'all the way from the United States to Western Australia'. The *West Australian* reported that this journalist asked Wilson: 'Which was he representing, his University or his National Government? Was he here as a professor of forestry seeking additional academic knowledge, or as a commercial ambassador with a price to offer for a State sawmill?'[10] The newspaper related that 'the professor smiled'. Wilson, it reported, 'was not particularly interested in the commercial aspect of forestry, except on the theoretical side. He was here as a representative of Harvard, engaged solely in research work'. Wilson stated to them:

> for twenty-one years have we been carrying on the compilation of a world survey of the world's trees. In the past my stamping ground has embraced much of the Chinese and Japanese empires, but on this occasion my journeyings have been diverted to Australasia. Beginning with Western Australia, I purpose visiting every State of the Commonwealth, and going on to New Zealand, calling at India on my return … I am concerned not so much in timber as in trees … And after trees, I am interested in shrubs and plants, in botany generally. Nothing in plant life is too small to engage the attention of those I represent.[11]

Right at the beginning of his travels within Australia the tone of Wilson's trip was set, with a mix of meeting important persons and casual attention. The *Boston Evening Transcript* reported that 'on arriving at Perth, the capital of Western Australia … he was taken by the Conservator of Forests directly to Sir James Mitchell, the Prime Minister, who received him in his room at a hotel'; Mitchell was in fact the Premier of Western Australia, not the Prime Minister. The *Transcript* continued 'the prime minister being in his dressing gown, and seated on the edge of his bed. Mr Wilson was invited to take the only chair in his room, while his introducer leaned against a mantel. In this quite informal manner, which illustrated the democratic simplicity of the country, every possible facility of travel and observation, including a motor car at the public expense, with a guide, was immediately and freely extended to Mr Wilson'.[12] Such, the paper said, was the fame of the Arnold Arboretum.

Wilson's travels have been hard to establish. Table 3.1 and Figure 3.1 give his locations and collecting sites for Western Australia, gleaned from his scant diary, newspaper reports, and dates given on specimens that he collected. This was to be an exhaustive tour or, as a newspaper put it, 'the way America does things'.[13] I have not imposed a strong timeline, and have used the landscape to frame Wilson's trip. Previously, Richard Howard tried to track Wilson's travels in Australia but had only a few dates; this task is best guessed from plant specimens, which are clearest in Western Australia.[14]

TABLE 3.1: Wilson's timetable and collecting sites in Western Australia in 1920

1920	14 October	Arrived Fremantle, Perth's port
	October	Mundaring–Capel–Busselton (the Coastal strip) Toodyay–Northam–Merredin–Westonia (21st)–Merredin–Bruce Rock–Yoting (24th)–Quairading–York (the Wheatbelt)
	November	Albany (6th)–Nornalup–Big Brook–Walpole (southern forests); Coolgardie (22nd)–Kalgoorlie–Coolgardie–Norseman–Widgiemooltha (the Goldfields arid lands) (22nd), departed for Adelaide

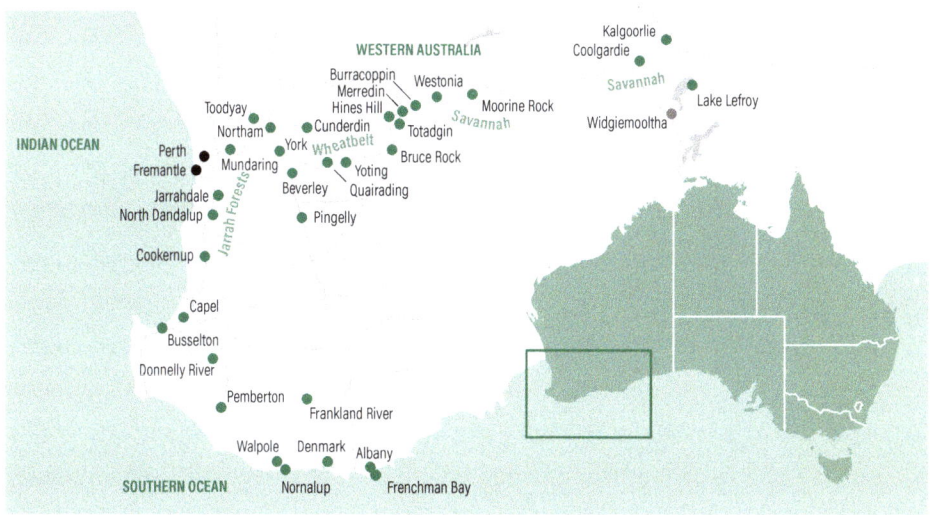

FIGURE 3.1: Locations recorded on Wilson's herbarium specimens and in notebooks revealed extensive collections throughout south-western Australia, from the towering karri *(Eucalyptus diversicolor)* forests, south of Perth, to the semi-desert forests farther east. Contemporary train lines are shown in grey. The Trans-Australian Railway, completed in 1917, connected Kalgoorlie with eastern cities. Image: Arnold Arboretum. Scale to be fixed.

Charles Lane-Poole, the 'introducer' who leaned against the mantel and introduced the Premier, was Wilson's main host in Western Australia and took him to the local forests east of Perth in the Darling Scarp, to the forests of the southern coastal strip along the Indian Ocean, and then to the southern tall forests of the south-west (see Figure 3.2). Although he had taken up his appointment as Conservator of Forests in Western Australia only a few years before in 1916, Lane-Poole had already achieved a great deal by drafting Western Australia's first *Forests Act*, which was proclaimed in 1919. Lane-Poole is now regarded as the father of national forestry in Australia. The Act 'began the reservation of large areas of forest as State Forests for long-term, sustainable timber production and … a new Forests Department'.[15] A 50,000 hectare (120,000 acre) reservation of jarrah forest (*Eucalyptus marginata*) is now named after Lane-Poole.

Lane-Poole had not long come back from the British Empire Forestry Conference in London, which Wilson was disappointed to have missed.[16] At that conference Lane-Poole had spoken very strongly – he usually did – of the need for better and more sustainable forest management. In Western Australia, he was keen to see a publicity campaign 'to form a strong public opinion regarding the proper management and utilisation of the forest heritage of the State. Some foresters who have visited this State have been so disheartened by the condition of affairs they have found that they have said that there will be no forestry in Western Australia until the last tree has been cut down'. Lane-Poole wrote that 'trees have no votes', a comment that Wilson was to repeat several times while in Australia. Lane-Poole believed that 'when the people develop a forest conscienceness the position will be entirely altered'.[17] Here, Lane-Poole was using a French word, *conscience*, likely due to his training as a forester in France. The loquacious Lane-Poole would have spoken to Wilson about his ideas as they toured the south-west together. Major issues for Lane-Poole were the short-sightedness of forest management, misuse of excellent timber for purposes that did not require quality, disregard of agriculture for trees and the presumption that agriculture would provide more wealth than forests, over-exploitation, the moral responsibility of forest management, and a sense that the future was being impoverished. He was concerned that in cultivated forests 'whole forests of giant trees will no longer be seen'.[18] Viewing Wilson's later comments to the press in Australia, it was clear that Wilson had been listening.

FIGURE 3.2: Charles Lane-Poole, Conservator of Forests, Western Australia (1916–22), with *Eucalyptus lane-poolei*, salmon white gum, in November 1920. Lane-Poole lost his left hand in a shooting accident when he was nineteen and he used a hook. Salmon white gum has a distinctive white bark and smaller fruits than salmon gum, *Eucalyptus salmonophloia*. In the bush to left and right there are many specimens of *Kingia australis*, the grasstree. Salmon white gum is now rare and scarcely seen at this size, a problem that Wilson foresaw for many of the world's trees. Image: Wilson Y-331 AA.

Desmond Herbert (1898–1976) also travelled with Wilson and Lane-Poole. Herbert appears in many images from this trip. He was a young man, having been appointed Government Botanist in Western Australia in 1918 at the extraordinary age of twenty. All the images that I can now find of him besides those of Wilson's are of a much older man. Herbert has been described as 'A large man with a swarthy complexion, he had a commanding presence which belied a gentle and kindly nature'.[19] Herbert was an active Freemason, and the masonic star he wore on a chain helped identify him in Wilson's images. He had been trained at the University of Melbourne, remained in Perth only until 1921, the year after touring with Wilson, worked in the Philippines until 1924, and eventually moved to Queensland. A prime interest throughout his life was the impact of climate on the geographical distributions of plants. During World War II, Herbert wrote, with Cyril White – with whom Wilson was to travel in Queensland in 1921 – a book called *Friendly Fruits and Vegetables*, which was 'a survival manual for members of the Royal Australian Air Force'.[20] In his diary, Wilson referred to Herbert as 'my extraordinary [companion]'.[21]

Many other important people in forestry met Wilson during his tour. In Perth, he met the Minister for Forests in Western Australia, Robert T. Robinson, and collected cultivated specimens of *Eucalyptus lehmannii* (#493 HH), *E. macrocarpa* (#496 HH), *E. oleosa* (#490 HH) and others, in 'Robinson's Garden, Perth' in November. This might well be where he took a picture (Figure 3.3) of *E. macrocarpa* flowers in vases; they are large, red-pink, and stunning. Lisa Pearson of the Arboretum Archives said to me that Wilson had often done this sort of arrangement to show off distinctive flowers. Wilson was to see *macrocarpa* in his travels eastward, at Quairading, east of Toodyay.

Wilson described *E. macrocarpa* in a letter to Sargent from the Palace Hotel as 'a truly extraordinary species. It is a bush of straggling habit from 12 to 15 ft tall with [...] leaves glaucous white in colour. The leaves are sessile, axillary & about five inches across & from orange to crimson in color'. (In one sentence Wilson used the British and US spellings of colour/color.) He went on to indicate to Sargent that this was a good specimen for Californian gardens. 'Please tell Mr Huntington to take special care of it for without exaggeration it is in itself worth the thousand dollars he has [ ] toward the expedition'.[22] Later, he wrote in *Smoke that Thunders*: 'To me this Eucalyptus was a plant to marvel at and I sincerely hope it will thrive in Pasadena where I was instrumental in sending seeds together with those of E. torquata and the mallees'[23] *E. tetraptera, E. pyriformis, E. erythrocorys, E. erythronema*, and *E. preissiana* – these last are now garden favourites in Australia.

FIGURE 3.3: (top): *Eucalyptus macrocarpa*, the mottlecah, showing flowers and bud-caps with operculum still intact (left), and fruits (centre and right). Image: Wilson Y-321 AA; (bottom): Flowers in colour, Kings Park, Perth. Wilson noted that 'About E. macrocarpa Hook. I may add here that the flowers are often 7 inches across!'[24] The Noongar word for this species is martilgarrang, which means 'angry-spirited mallee with hand-sized leaves', referencing another aspect of this beautiful plant.[25] Image: M Grose 00260.

While in Perth, Wilson was taken to the Perth Zoological Gardens, where he collected *Eucalyptus megacarpa* (#489 HH), which he was later to see in the field, *E. cornuta* (#484 HH), and *E. lehmanniana* (#486 HH).

It was already warming up. Wilson had come to Australia in time for the heat and would also be in India in the heat of summer in 1921. As a summer visitor to Australia, Wilson would have seen boys and men dressed in white, playing cricket – a sight that would have been familiar to him as an Englishman, even if he did not play himself. Summer is not the best time to see flowers in Australia. Although many eucalypts flower in summer, many other species are winter-flowering, or they flower in early spring. Wilson found that, being south of the equator, it 'was very difficult to get accustomed to conditions the other side of the line, as both the seasons and the vegetation were so entirely different. It was hard to realize that the north, where the sun was to be seen, was not south, and that September and October were the spring months.'[26]

He headed out of Perth.

Most of Wilson's travels in Western Australia were in Noongar Country. The Noongar (also spelt Nyoongar) are the major Indigenous people of the south-west, who use the term Noongar to describe themselves, their lands, and their culture. Noongar *boodjar*, or Country, is all of the south-west of Australia. Many place names in Western Australia kept their Indigenous names after European arrival, with many 'in' endings in the Wheatbelt, such as Kellerberrin, Kulin, Cunderdin, Merredin, Tammin, Dowerin, Bencubbin, Kununoppin, Mukinbudin, Burracoppin, Muntadgin, Yalbarrin, Corrigin, Kondinin, Babakin, Narrogin, Wagin, and many 'up' endings further south, such as Nannup, Nornalup, Gnowangerup, Ongerup, Tambellup, Kojonup, Manjimup, Kulikup, Qualeup, Mummballup, Woggerup, Woogenellup, Porongurup, Narrikup, Chorkerup, Jerramungup. Some are tiny places and some major towns. The name endings reflect the thirteen language groups within Noongar.[27] Although the language is very similar across all thirteen groups, 'each has its own subtle differences in spelling and pronunciation' and each has its 'own identity and history',[28] revealing a rich cultural diversity across this region that is typical of the whole continent.

The Noongar elder Ken Colbung once said that all Western Australians speak some Noongar as we are in Noongar Country. The first thing that came to mind is the names for trees. Wilson very much liked the names of the major trees of Western Australia, for most of them are Noongar names, such as jarrah (*E. marginata*), marri (*Corymbia calophylla*), karri (*E. diversicolor*), wandoo (*E. wandoo*), tuart (*E. gomphocephala*), and moonah (*Melaleuca lanceolata*).[29] The *Western Mail* reported that Wilson thought them very good names indeed, and said that 'they are easily remembered, easily spelled, and are distinctive. They have a great advantage over the technical names, which are often misleading and difficult to remember and pronounce'.[30] The Indigenous names of many Australian plants are used generally across Australia. This has come about because early plant collectors 'often relied physically and intellectually on Aboriginal guides' in the botanical exploration and annotation of Australia's flora.[31]

The Swan River flows through Perth. This river was called Derbarl Yerrigan by the Noongar. The Dutch explorer William de Vlamingh, who sighted it in 1696, named it the Black Swan River for the prolific number of black swans –

a great curiosity to non-Australian minds. One of Vlamingh's officers wrote of trees that were 'full of thistles and thorns' and that 'by day the flies are a terrible torment';[32] Wilson was to comment on both features. The river descends into Perth from the low 'Hills' officially known as the Darling Scarp or Darling Range; these highly eroded granite hills rise only a few hundred metres above the sandy Swan Coastal Plain with its limestone base. While there are similarities in the vegetation between the Swan Coastal Plain and the Hills, they are distinct entities and different suites of species make up these areas. Wilson described the change as moving from coastal lowlands to the Darling Scarp, 'produced by the faulting of the coastal strip and rising to some 1800 feet above sea-level'.[33]

Wilson's first trip was in the late spring – late October – to the east of Perth into the Hills. They went to Mundaring, at that time separated from the suburbs of Perth by bushland. Today, though seen as a suburb, Mundaring is still strongly wooded because the area adjoins Perth's major water catchment forests. Wilson was immediately thrust into the south-western flora and its globally high endemism and rich speciation. He was astounded. From this small area, Wilson collected many pretty plants, due to the rich diversity of understorey species. He commented to the press that in Perth one needed to go no distance at all to see extraordinary plants: 'even the bush close around your fine capital city, when seen for the first time, is distinctly arresting to [a] visitor from distant climes'.[34]

This bushland is a mixed eucalyptus forest, dominated by jarrah and marri. Jarrah, *E. marginata*, is one of the world's great hardwoods, and it quickly blunted the axes of early settlers; it was called Swan River Mahogany by early British settlers because of its beautiful grain and colour. It grows with marri (*E. calophylla*, now *Corymbia calophylla*), with an understorey that includes *Banksia* species.

Jarrah was the first major eucalypt that Wilson recorded on his Australian trip (Figure 3.4). It is an easily identified tree, with long vertical grooves in its bark; its dark leaves, small fruits, and old presence are like an old soul that has seen much and watches patiently. Wilson felt that it was a great mistake that a wonderful timber like jarrah was used for the mediocre purposes that it was – for railway sleepers and street paving. He said:

> it is in a measure very unfortunate that by force of circumstances a market was made for jarrah and karri in the form of sleeper and paving blocks. You

cannot criticise a thing like that, but it is unfortunate that these timbers in the world's market are only viewed from that point. The cabinet timbers of the world are getting scarcer all the time. Oak, which was formerly so much used, is almost impossible to get, and has gone out of fashion. Mahogany is almost worth its weight in gold, and extremely hard to get … Jarrah is a beautiful cabinet wood, if the shrinking and warping can be got over. The pity of it is that you are cutting up the best of the trees for this sleeper business, when there are millions of inferior trees that would cut sleepers. The buyers won't have them, of course, because you have accustomed them to the very best. If you feed a dog on beefsteak and then afterwards give him a crust of dry bread he won't have it; he wants beefsteak.[35]

Wilson did not want to appear critical, noting that his hosts had all been 'very kind' and that he was, in stating his opinion, only trying to 'give something in return'.[36] In his criticism of using the wonderful jarrah for railway sleepers rather than for fine furniture, he was echoing Lane-Poole, who described the use of jarrah for railway sleepers as 'sheer prostitution'.[37] It is still used for sleepers today, and worse, it is often used as firewood.

FIGURE 3.4: Wilson in the jarrah-wandoo forest; jarrah, *E. marginata*, is on the right, and Wilson is standing by *E. wandoo*. Image: Wilson Y-322 AA.

Of the flora he saw in Western Australia he wrote:

32 PLANT COLLECTING IN ANOTHER PLANET

> The astonishing thing is the extraordinary wealth of flowering shrubs. When not in blossom they all look alike; there are only about two common types of leaves, the thin wire like leaf and the long narrow leaf. But when they are in blossom every other one is different, and they vie with one another to see which can produce the most extraordinary shape and the most beautiful colouring. It is a vast botanical garden, and the poorer the soil the greater is the variety.[38]

In *Smoke that Thunders*, Wilson wrote:

> To a visitor from the Northern Hemisphere, no matter how familiar he or she may be with the forest scenery of the North, Western Australia is a new world. Nay, it might well be part of another planet so utterly different is the whole aspect of its vegetation. Intimate knowledge of the plants of the boreal regions only serves to accentuate the variance.[39]

His comments on newness echoed those of a century before by the great botanists Robert Brown and Daniel Solander.

A 2013 paper on the evolutionary history of the Australian flora over the last 65 million years asked the question: 'To what extent is the Australian flora unique?'[40] The authors asked because Australians know the flora is enormously different, and foreigners agree with them; they noted that thickened foliage (sclerophylly) is not unique to Australia, nor are traits that promote water storage (xeromorphy), nor is fire as a selective genetic force unique. But nowhere else do these traits come together to dominate large proportions of an entire continent.

Wilson's thoughts on the Australian flora show his keen eye and observational clarity. Wilson saw very quickly two leaf features as unique to Australian plants—the angle of leaf repose and the colour of the leaves. He wrote in *Smoke that Thunders*: 'In the North our trees in general have spreading umbrageous crowns, dark, often lustrous green leaves which ... cast a heavy shadow'. Wilson saw with great interest that:

> In Western Australia the dominant trees have open, tufted crowns, gray or glaucous green leaves which ... cast little or no shadow. This difference in the color of the tree-foliage and the fact that the leaves are pendant instead of spreading on the branches may seem to the reader trivial matters, but

in reality they completely change the aspect of the forests and profoundly influence the whole landscape.[41]

To have quickly put his finger on the two points of leaf position and colour is tremendously perceptive.

The angle of repose of leaves is especially notable in the eucalypts, which change their leaf angle in relation to the sun. Many eucalypt species have steep, nearly vertical foliage. Steep foliage has long been seen as a functional trait to reduce midday heat loads, as might be expected in dryland regions like the interior of Australia. However, steep eucalyptus leaves are also found in cooler southern regions of Australia, and are thought to be associated with increased light interception at lower latitudes[42] – an important point in relation to the evolution of *Eucalyptus*.

Wilson's comment on the colours of the foliage of Australian trees was of special interest to me because I have a research interest in tree colour. I have examined colour as an aspect of the difference between Australian flora and Northern hemisphere species that are planted in Australia as exotics – those whose natural range is outside of Australia.[43] My research on colour grew out of concerns about the increased number of exotic trees planted in urban areas – many botanists felt that we were losing the colours of Australia in urban areas. This apparent loss has been exacerbated since the mid-twentieth century, with widespread removal of original vegetation in cities and their replacement with exotic species in suburban development. Colour was a visual aspect of conservation that appeared to be completely neglected, thus it is noteworthy that Australian leaf colours were seen as strikingly different by Wilson.

Planting exotics might not mean colour changes everywhere. In American cities, tree introductions are usually from nearby, internal sources rather than distant sources, and if more-distant exotics are planted these are mainly from the 'North', as Wilson termed it – the Northern Hemisphere, the old Laurasia.[44] However, in Australia, introduced trees are overwhelmingly from distant sources – from Europe, North America, and north-eastern Asia – again, Wilson's 'North'. Many of these are deciduous and are placed into the overwhelmingly evergreen continent of Australia.

It is not just the obvious evergreen-versus-deciduous contrast that attracts attention. Wilson noted 'gray or glaucous green leaves' in Australia. 'Glaucous' might need explanation: it is greyish-blue, or the pale-grey, bluish-green appearance of a leaf; there are several glaucous beauties among the eucalypts, including the *Eucalyptus macrocarpa* that was so admired by Wilson. As he had noted, the greens are not the same as in the North. What colour, after all, is green? What colour is your green? If you live in the Northern Hemisphere, it is certain to be different from the green of the eucalypts.

Plant leaf colour is usually recorded in a very general way, perhaps due to a prevailing belief that leaves have little colour variation in comparison to flower colour, and because colour is not an immutable feature of a species.[45] Most botanical descriptions of leaf colour remain qualitative, with descriptions such as 'mid-green', 'dark green' or 'glossy green'.[46]

When Wilson had trained from 1893 at the Birmingham Botanical Gardens, then at the Royal Botanic Gardens, Kew, there were several colour charts for plants, most created by artists and naturalists. The most well-known colour system, used by Darwin in his voyage on the *Beagle*, was by the Scottish artist Patrick Syme, titled *Werners' Nomenclature of Colours*, and published in 1814. This book contains 110 colours based on the colour system of Abraham Werner, a German mineralogist. It contains just sixteen greens. As late as 1957, a writer in *Arnoldia* was still grumbling about the lack of accuracy in descriptions of colour in plants, repeating the words of an article from ten years prior at the beginning of his argument:

> The time has come for American Horticulture to adopt some uniform standard by which color can be accurately measured and uniformly judged and described the country over. Many industries have done this. Horticulture seems to be far behind … Those of us who are constantly studying plants realize better than most, the necessity for having an accurate standard by which we can compare the colors of flowers, foliage and fruit, and afterwards to describe those colors in uniform terms understood by other individuals who have not seen the plants themselves.[47]

The writer might well have been frustrated at the lack of colour charts for plants. Soil colour has long been determined in a very standard manner using the

Munsell Soil Color System, begun by the American art instructor Albert Munsell in 1905 and still widely in use today in soil science.[48]

Early plant colour charts, even some of which evolved from the 1940s, cannot accommodate the greens of Australia; the Australian greens just aren't there. The eye can see them, but the charts cannot accommodate what is seen. Closer colour matches are found using the Natural Colour System of Sweden, which was developed in 1979, because this system includes a far wider range of grey-greens, bluish-greens, and the dull hues of low chroma that are typical of Australian flora.

Australia trees are less chromatic than exotics, and are also darker – they were, as the poets always told us, greyer and more sombre than those in Europe.[49] Wilson noted to reporters in Boston that 'In the matter of its trees Australia is rich, but decidedly strange'.[50] Colour was one of the major observations behind his comment.[51] Wilson also found that the colours of trees and vegetation in New Zealand were totally unlike those in Australia. He noted that 'Gray-green foliage does not characterize New Zealand's trees, neither is gorgeous blossom her floristic feature. Quite the contrary'. He likened New Zealand's colours to those of Oregon, Washington, and British Columbia. 'If we are to visualise New Zealand's plants we must forget entirely Australia's characteristic vegetation',[52] he wrote. New Zealand has a very different botany to that of Australia.

At the beginning of his trip in Western Australia, Wilson might also have been struck by the close botanical associations between southern Africa and Australia, particularly in the Proteaceae. As botanical explorers travelled from Europe to Australia by boat via Cape Town, they were keenly aware of the floral similarities between the two continents, despite the thousands of miles of ocean between. In southern Africa are *Protea*, *Leucadendron*, *Leucospermum*, and other species that have a remarkable similarity to those seen in Australia in *Banksia*, *Dryandra*,[53] and *Isopogon*. The Proteaceae were the second family that I explored in the Harvard Herbaria, searching for specimens from Wilson's expedition.

As I opened the compactors during my research at the herbarium, I could see banksias very easily because their folders were bursting with pillar-shaped inflorescences made up of flowers in their hairy thousands. Some cones – the flower heads or remaining fruiting bodies – measure up to 43 centimetres (17 inches) long, and thus many were stored separately in boxes due to their size. These large inflorescences of banksias attracted a good deal of attention from taxonomists who passed by my workstation in the Harvard Herbarium. It was a pity that they could not see them on the living trees instead of these dead, dry relics. *Banksia grandis* was collected by Wilson in the Darling Scarp with Lane-Poole and Herbert (Figure 3.5). Wilson thought banksias to be 'one of the most strikingly handsome and remarkable groups of woody plants'.[54]

FIGURE 3.5: Lane-Poole with a fine mature *Banksia grandis* in the jarrah forest to the east of Perth, October 1920. Image: Wilson Y-343 AA.

The banksias of the Perth region's forests and woodlands are mid-storey plants. *Banksia grandis* is the tallest, with knobbly grey bark, giant yellow inflorescences, and large leaves that give it the name 'grandis'. *Banksia menziesii* is smaller, gnarled, with beautiful, usually red, inflorescences and named after the botanist Archibald Menzies who was on George Vancouver's expedition along the south coast in 1791; Wilson thought that 'these Banksias ought be cultivated in California'.[55] Figure 3.6 shows why.

FIGURE 3.6: *Banksia petiolaris*, one of the prostrate banksias, in Kings Park, Perth. Image: M Grose.

Though younger than eucalypts, banksias are an ancient genus,[56] with an impressive fossil record.[57] The fossilised banksia cone in Figure 3.7 is 40 million years old but is extraordinarily like a banksia alive today.[58] Like eucalypts, the age of this genera enabled speciation, notably in Western Australia, where banksias had the greatest speciation

 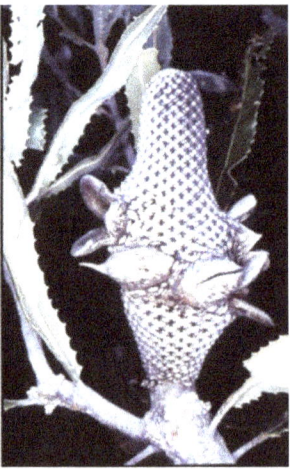

FIGURE 3.7: An ancient banksia (left - Fossil *Banksia archaeocarpa*) and modern banksia (right - *Banksia mensiesii*), showing remarkable similarity in the form of their fruits. The structure was for many years taken as an adaptation to fire,[59] but this is now thought to be unlikely because the climate 40 million years ago was wetter, with drying mainly occurring in the last 23 million years.[60] The current explanation is that the structure is a defence against cockatoos.[61] A form of parrot, cockatoos emerged in the Eocene 52–30 million years ago in Australia, and speciated substantially from 38–18 million years ago.[62] Images: (left) Ken McNamara, with permission, (right) Hesperian, Creative Commons.

Banksias appear resilient to hot conditions. I had an insight into their tenacity during my PhD, which examined stress on banksias. The university had constructed new growth-room chambers, a group of six chambers in one new unit. Each chamber of about 3 metres by 4 metres (about 10 feet by 13 feet)

was kept at a self-contained temperature. Staff in my faculty had argued for an Australian-made air-conditioner, but someone in the university administration over-rode that decision and opted for a chic French system. The faculty members objected strongly, on the grounds that the French company could not guarantee their air-conditioner if the temperature rose above 34°C (93°F), a limit unrealistically low for summer. The higher echelons ignored this, and the French system was installed in winter. I had an experiment in one chamber, with the temperature set at 25°C; other chambers were housing wheat and rice, from 10° to 15°C.

One Sunday morning in late October, I went into the glasshouse area to water; I arrived about 11 am, after going to the beach for an early swim, with a hot easterly wind blowing and the temperature already rising to what we refer to as a 'scorcher' or a 'stinker' (over 40°C; 106°F), the very first hot day for that summer. As I approached the new unit, I could feel heat emanating from the front door. On opening it cautiously, a blast of terribly hot air escaped; the six chambers were reading between 61°C and 75°C (142–167°F); I shut down the entire system immediately. Every plant was dead, except my banksia seedlings, which appeared shaken after some hours of 71°C. With a good watering and relief outside in the shade, every single *Banksia menziesii* recovered and ended up being planted in a primary school revegetation programme. Even though they were babies only inches high, their sclerophylly, and stomata capable of closing tightly, likely enabled them to survive that morning; the grasses had stood no chance at all.

Although Wilson might have lamented coming to Australia in summer's heat, he was thrilled to see the West Australian Christmas tree, *Nuytsia floribunda*, in full flower during his visit (Figure 3.8). He described it thus:

> A remarkable tree of the sand-plains is Nuytsia floribunda which is parasitic on the roots of many plants and is first cousin to our Mistletoes. Imagine a gigantic mass of Mistletoe in the form of a tree 10 feet or more tall and as much in diameter, a trunk 5 feet in girth, and leaves little in evidence on the rounded green shoots, and you have the Nuytsia as it appears for the greater part of the year. In December it bursts into bloom, huge panicles of brilliant orange-colored flowers terminating every branch … It is in blossom one of the gorgeous floral sights of the world … it is worth a journey to Western Australia to see this feast of wonderful color.[63]

FIGURE 3.8: Western Australian Christmas trees, *Nuytsia floribunda*, 'a feast of wonderful color' in the eastern margins of the Swan Coastal Plain. Image: enjosmith/Wikimedia (CC BY-SA 2.0).

Nuytsia floribunda are extremely important trees to the Noongar, who know the tree as moodjar or mungee, with long-held customs specific to this species, which is regarded as a teacher of how to behave in a community.[64] They believe that, on death, the souls of the newly dead are housed in the flowers of the tree.[65] A person's spirit camps in the tree's flowers before they depart for a final resting place, Kurannup, the heavenly home of the dead. No living Noongar 'ever sheltered or rested beneath the shade of the tree of souls; no flower or bud or leaf of the tree was ever touched by child or adult; no game that took shelter beneath it was ever disturbed. But … they loved it, but held it sacred for its spiritual memories'.[66] For this reason, flowers are not cut, nor are they sold in the floricultural trade due to its high spirituality, and because it is forbidden and sacred to the Noongar. This species is protected, both in Indigenous tradition and government legislation.[67]

Daisy Bates (1859?–1951), an Irish-born ethnographer who studied Aboriginal customs in Western Australia, recorded the words of Joobaitch, a Noongar man from the eastern side of the Swan Coastal Plain who died in 1907. He said:

> When I die I shall go through the sea to *Kurannup* where all my *moorurtung* (relations) will be waiting on the shore for me, waiting with meat and drink for me … all who have gone before me. *Kurannup* is the home of my dead people and I must go to them, and my *kaan-ya* must be free to rest on the *kaan-ya* tree *(Nuytsia floribunda)* before it journeys through the sea. Since Nyitting (cold) times (long time ago) all *Bibbulmun kaan-ya* have rested on this tree on their way to *Kurannup*; and I have never broken a branch or flower, or sat under the shade of the tree because it is the *kaan-ya* tree only *winnaitch* (forbidden, sacred).[68]

Nuytsia extracts food from other plants by slicing and strangling the roots of other species to obtain some, but not all, of its nutrients – it is a hemi-parasite, which photosynthesises. *Nuytsia* is different from other members of the southern mistletoes, the Loranthaceae, in that is a tree and not a vine or a shrub attached to its tree host. Instead, *Nuytsia* can attack from afar, and is strong enough to slice though electrical cables in the hope of extraction of food. As a significant tree, *Nuytsia* 'primarily reproduces by cloning, sending out suckers up to 100m from the parent plant to produce identical copies'. The result, as so commonly seen, is patches of Nuytsia 'gathered together in tight-knit populations'.[69]

One hundred years ago none of this was known, although Wilson described 'wonderfully beautiful' *Nuytsia* as a 'remarkable tree' that 'possesses a whole catalogue of peculiarities'.[70] Nor was much known then about the extensive mycorrhizal associations found in Australia. Research in this area expanded after the 1980s, and the distribution of mycorrhizal types and non-mycorrhizal plants with other specialised strategies for nutrient acquisition, and nitrogen-fixing associations, are now major areas of study. These complex root strategies appear to have evolved rapidly, and many of the nonmycorrhizal plants with specialised nutrient acquisition strategies have been traced back to Gondwana. All seem to be associated with increased speciation rates after the separation of Australia from Antarctica. The south-west of Western Australia, where Wilson saw such abundance, has greater riches in these types of plants than anywhere else in the world. It appears that plants in this landscape experimented with roots during evolution more than elsewhere, and they found solutions to the very nutrient-poor soils of these old landscapes. As a result, this little part of Earth has three interesting features. It has 'one-third of all ectomycorrhizal plants (those with a symbiotic relationship between fungi and roots where the mycorrhizal fungus does not penetrate deeply into the roots)'. As well, this region has 'many nonmycorrhizal plants, especially those with cluster roots' (cluster, or proteoid, roots being masses of short rootlets along lateral roots of a plant that aid in nutrient uptake in low nutrient soils).'[71] Finally, it also has 'one fourth of all carnivorous plants in the world'.[72]

Wilson saw his first members of another great Australian genus – *Melaleuca*, another Myrtaceae (Figure 3.9 and Figure 3.10) with Lane-Poole and Herbert. Wilson described them in 'Notes from Australasia': 'In the swamps and along the sides of streams several species of Melaleuca grow and since they have a white thin bark which peels off readily are known as Paperbark trees'.[73] As their bark is peeled off, layer after layer, this exposes the increasingly pink or pale cream inside (Figure 3.11). Strong winds can rip off their bark in strips. Bark was also removed by Aboriginal people for babies' cradles, to wrap and store food, or as a cooking wrap—the latter practices being adopted by early settlers. Their name, melaleuca, a Greek mix of melas (black) and leuca (white), is derived from the likely first sighting of them by French explorers as being a little burnt after a fire. The more usual colours of the unburnt paperbark are creamy whites and pale browns and pink-greys, with many darker stripes. *Melaleuca* are mainly small trees and are often used as street trees in Australia due to their generally deep shade, which is so valuable in summer.

FIGURE 3.9: Charles Lane-Poole with a grand old paperbark tree, *Melaleuca preissiana*, in a *Banksia-Eucalyptus* woodland, near Perth, Western Australia, 1920. The canopy is made up of pillows of dense foliage. Image: Wilson Y-342 AA.

FIGURE 3.10: A typical scene in the Australian bush in south-west Australia. This is the swamp paperbark, *Melaleuca rhaphiophylla*, by a creek. Their soft pinky-cream bark peels off like sheaves of paper and was used by Aboriginal people for babies' cradles and to wrap food. The man is Desmond Herbert, the Western Australian Government Botanist. Debris in the tree indicates previous flood levels. Image: Wilson Y-379 AA.

FIGURE 3.11: Paperbark, showing the sheets that can be peeled off and the lovely colours. Image: unknown author.

Another famous plant in Western Australia is the grasstree, known as balga, which is the Nyoongar name for *Xanthorrhoea preissii*, a monocot. The base is a hollow pseudo-trunk comprised of leaf bases and is usually black. In broken balga, these leaf pieces can be picked apart; they are highly flammable, and when I was a child we used fallen leaf pieces to light a barbeque fire in the bush. The tops of balga consist of long, very slender, geometric, and very brittle leaves that Wilson thought gave a 'mop-like' appearance (Figure 3.12).[74]

FIGURE 3.12: *Xanthorrhoea preissii*, in the Darling Scarp 1920, with Wilson's camera case to give an indication of size. Wilson described them as 'a tree with a frowsy top, somewhat resembling a palm, with descending masses of grass at the top and waist'.[77] Image: Wilson Y-395 AA.

Wilson shared his thoughts with the press on seeing *Xanthorrhoea*, the only genus in the subfamily Xanthorrhoeoideae, which consists of about thirty species. The *Albany Advertiser* reported: 'The grasstree of West Australia, in his opinion, is probably the oldest living thing, and it is astonishing that, side by side with this and the blackboy and zamia palm, one can see that highly developed modern product, the eucalypt, which is the latest thing out in trees'.[75] The idea that eucalypts are young is not correct, but seems so when presented alongside the grasstrees, which belong to two very ancient families that grow on poor soils and are famous for surviving fire. Xanthorrhoea is the first plant to spring out green shoots following a fire.

Kingia australis, the sole member of the genus *Kingia*, were also seen by Wilson for the first time. *Kingia* belongs to Daspogonaceae (Figure 3.13). Though unrelated to *Xanthorrhoea*, they are similar in appearance except for the flowers, which are quite different in shape. In *Xanthorrhoea*, flowers are borne on a very long spike, while in *Kingia* flowers are in 'drumstick-like' clusters.[76] Perhaps the most famous attributes of these two genera are their great age, slow growth, and their rapid regrowth after fire. *Kingia* live for centuries and grow at a rate of about 1.5 centimetres per year; tall *Kingia* are therefore ancient. Tall *Xanthorrhoea* are likewise of great age and are respected multi-centenarians.

FIGURE 3.13: *Kingia australis*, showing their drum-stick inflorescences. This group are very old *Kingia*. Image: Wilson Y-367 AA.

Zamia palm, *Macrozamia riedlei* (Figure 3.14), is a cycad and was mentioned by Wilson as being 'strange … with its short massive trunk, spreading crown of long, dark green, fern-like leaves, huge fruits with bright-colored seeds, poisonous in a raw state to cattle and man, as some of Vlaming's [*sic*] men were first to discover from unpleasant experience'.[78] Wilson was well-informed. Willem de Vlamingh, a Dutch navigator and commander of the ships the *Geelvink* and *Nijptangh*, went ashore in 1697 on the Perth coast. Some of his men saw the fruits of the zamia and ate a few kernels. The upper-surgeon, Mandrop Torst, noted that the kernels appeared in shape like 'the Drioens' – this is a reference to the seed of the tropical fruit, the durian[79] – and that they tasted 'like our Dutch broad beans and those which were less ripe, like a hazelnut. I ate five or six … but after an interval of about three hours I and five more of the others who had also eaten of the said fruit began to vomit so violently that there was hardly any distinction between death and us'.[80] This was the first time that Wilson had been introduced to the many poisonous plants found in Western Australia. He would see and hear of more of these as he travelled further east into the Western Australian Wheatbelt.

It was not only trees and strange, ancient forms that attracted Wilson. Many smaller species caught his attention, such as *Guichenotia* and *Grevillea endlicheriana* (#423 HH), though the visit was not recorded in his diary. At Mundaring, Wilson also visited the site of the beginning of one of the great engineering feats of the nineteenth century, the Goldfields Water Pipeline. The nearby jarrah forest had been quickly identified by surveyors as a water catchment area and had been protected. When gold was discovered in Western Australia in the 1890s, hundreds of miles inland, in arid desert country to the east, water was desperately sought. The solution was the building of the Goldfields Water Pipeline, from Mundaring to Kalgoorlie, and which now has numerous feeder pipelines serving the inland. Lane-Poole knew that Wilson was to travel to the Goldfields collecting a few weeks later, so took Wilson to see Pumping Station Number 1, where water starts its long journey to the Goldfields. Wilson said: 'I was … impressed with the Mundaring scheme, the pumping of water 364 miles to Kalgoorlie. That, I consider, a big thing worthy of a big country. It was a tremendous man that conceived and carried that scheme through'.[81] He was to follow along its path later in his trip. That 'tremendous' man was the Irish engineer Charles Yelverton O'Connor, who conceived and

FIGURE 3.14: *Macrozamia riedlei*. Note axe for size (top image) and fruiting body (bottom). The fruits are poisonous without extensive preparation. Images: (top) Wilson Y-339 AA; (bottom) M Grose 0443.

designed the Goldfields Water Supply Scheme, of which the Goldfields Water Pipeline was a part. O'Connor had endured many criticisms of his scheme due to its high cost, public forecasts of failure, and false allegations of personal corruption and mismanagement made under parliamentary privilege. Ten months before the scheme was completed and only a few weeks before Pumping Station Number 1 became operational, O'Connor rode into the surf south of Fremantle and took his own life with a shot to his head.[82]

From Pumping Station Number 1 at Mundaring, potable water is pushed along by a series of pumping stations lifting the water 400 metres (just over 1,300 feet) to Kalgoorlie, 566 kilometres (over 350 miles) away. In Wilson's time there were eight pumping stations, and today there are twenty; the original pumping stations are beautiful works of architecture. The scheme is listed in Australia as a National Engineering Landmark and by the American Society of Civil Engineers as an International Historic Civil Engineering Landmark. On the train or in a car today, one sees the pipeline disappearing and reappearing, climbing hills and dipping down slopes, running on raised trellises across salt lakes, catching the sun and forever moving forwards alongside your travels east like a faithful friend.[83]

Even in this first brief visit to Australian bushland around Mundaring, with its weir and pumping station, Wilson was struck by the enormous differences of this unfamiliar new Southern Hemisphere flora with that of the Northern Hemisphere of which he was familiar. As one who grew up in the 'veritable botanic garden', northern forests appear to me homogeneous. Or, as a colleague said to me bluntly, 'Northern Hemisphere forests are boring'. In contrast, Australian forests have a heterogenous mix of plants in both the top storey trees and the understorey, with many suites of plant species in close association that shift and change, with subtle changes in soil, aspect, and elevation. It is always a mixed forest and is rarely dominated by a few species.[84] The species richness that Wilson saw immediately in Western Australia was, and remains, particularly striking to botanists from afar.[85]

To make sense of the greater species richness in Australia than in the Northern Hemisphere, one needs to examine the history of the continents leading to the evolution of the flora. Australia's species richness and unique biology is bound up with its geological and climatic history, which is different to that of the Northern Hemisphere, or to other continents that were once linked to Australia.

At the time of Wilson's visit in 1920–1, Wegener's theory of continental drift was still under intense debate among geologists. He had noted, as had many before him, that the coasts of South America and Africa had a jigsaw-like fit, as did their continental shelves. The question of how the continents came to be as they are today and how links between continents, such as mineral deposits, might have occurred had been explained for many years by the idea of the rise and then demise of land bridges between the continents, despite the enormous distances across oceans that this theory required.

Wegener proposed that the continents drifted and were floating on deeper material. He noted that Scandinavia had been steadily rising in response to the loss of the weight of ice from the last ice age, suggesting a buoyancy of the continents – that they were floating on something. He published his idea of continental drift in 1915, titled *Die Entstehung der Kontinente und Ozeane* (*The Origin of Continents and Oceans*). His idea was supported by some but criticised by others, because he could not explain the engine that could drive the drifting continents as they were buoyed along the vast distances that now separated them. During Wilson's last decade, the debate continued.

Struck by the differences between Australia's flora and that of New Zealand, Wilson wrote that New Zealand 'is totally unlike Australia and the differences are much greater than the some 1200 miles of ocean that separates them would appear to warrant. No Eucalyptus, Acacia, Grass-tree, Banksia, Melaleuca, Casuarina, Callitris nor Cycad'. Wilson visited New Zealand late in 1921 and noted that 'the country itself is but a remnant of a vast continent which once linked together South America on the one hand and Tasmania and eastern Australia on the other'. In a lecture to naturalists in Queensland in 1921, Wilson commented on the similarities between the flora of eastern Australia and Chile, and that this 'seemed, with other evidence to indicate land connection between Australia and South America long ago'. The possibility of a link across the Pacific had been alluring, but continental drift was the explanation to the puzzle many had sought, and Wilson was likely no exception.

In 1937, the great South African geologist Alexander Du Toit wrote a book titled *Our Wandering Continents: An Hypothesis on Continental Drifting*.[86] In this, Du Toit proposed that there were two great primordial continents, Laurasia in the north and Gondwana in the south. In a review of this book in the journal *Nature*, Du Toit was 'congratulated on a brave project' – the writing of earth history in terms of continental drifting. The reviewer stated that 'Readers should be warned in advance that they will find du Toit's book difficult to digest, and that they may feel affronted by uses that have been made of fragmentary evidence. On the other hand, it is to be hoped that they will have sufficient vision to be grateful to a courageous pioneer'.[87] By 1944, the great British geologist Oliver Holmes included continental drift in his textbook, *Principles of Physical Geology*, which was to be a standard text for students of geology for decades. The mechanism for drift was only explained in the 1960s by plate tectonics. By 1970, Du Toit's understanding in his early book was hailed in the journal *Nature* as 'a triumph of imaginative synthesis' that new data expanded and supported.[88] The 1920s was the beginning of what is now referred to as a revolution in the earth sciences. Oddly, this story is far less widely known than the story of DNA, yet it is an essential and major intellectual paradigm shift concerning our understanding of our planet. For Wilson, this knowledge was beyond his lifetime.

Gondwana, the large landmass in the Southern Hemisphere, had fragmented into South America, Africa, Madagascar, India, Australia, and New Zealand before and during the Cretaceous period. The drivers of the break-up were volcanism and mantle spreading. Some of these huge fragments bolted north, while others left slowly.[89, 90] South America split off from Antarctica about 138 million years ago, moving north at 63 kilometres (39 miles) per million years. Africa's movement is 'still enigmatic due to missing information', but it seems that it began to draw away from what is now Antarctica about 157–139 million years ago.[91] The Indian subcontinent split off from Antarctica 136–126 million years ago and began to 'unzip' from the micro-continent of Madagascar 94–84 million years ago[92]. It then travelled north at a staggering pace before meeting the landmass of Laurasia and creating the Himalayas with that joining. By the time India unzipped, Laurasia in the north had also broken into its constituents and 'Earth had entered the thalassocratic state of dispersed continents'.[93] All this while, species were moving and evolving on every landmass.

Australia remained attached to Antarctica for millions of years. When Australia was still attached, 50 million years ago, Gondwana was warm and wet, at least near the coast, with parts of the East Antarctic coast warm enough to support palm trees, indicating that winter temperatures were above freezing.[94] Temperate forests existed that were dominated by southern beech (*Nothofagus*), conifers, ferns, and horsetails.[95]

Australia began a long, slow separation about 80 million years ago, finally breaking completely with the eastern Antarctic landmass about 35–38 million years ago, thus taking nearly 50 million years to make that final break. Evidence of this split can be seen today in the towering cliffs along the Great Australian Bight. The departure of the Australian landmass opened a seaway that led to the start-up of the Antarctic Circumpolar Current, which is of profound importance today for cooling the planet because it is the only current in the world that circumnavigates the globe – around and around, from west to east.[96] With the beginning of this current there began a substantial change in the planet's climate; exchanges of water from the Northern Hemisphere to the south ceased and Antarctica began its descent into an icebox.[97] Species that had flourished on Antarctica for millennia disappeared. Warmth-loving plants were replaced by cooler vegetation, including beech, mosses, and ferns. Tundra vegetation survived in places on the Antarctic landmass until about 15 million years ago.[98] The communities of *Nothofagus* that Wilson was to see in Tasmania are remnants of their once extensive forests, which exist today only in cold, damp areas.

The time from the original assembly of masses into the supercontinent of Gondwana to its break-up had been 'a 500Myr odyssey'.[99] These great Gondwanan pieces now adrift as separate continents were not vacant blocks of land, but living rafts full of species, floating over millions of years and either evolving to deal with changing climates and conditions, or failing and becoming extinct, or vastly contracting in extent and reducing in number. As Australia drifted north, the plants it held continually evolved, with a loss of rainforest about 25 million years ago, the onset and increase in sclerophylly, and the increasing dominance of *Eucalyptus*, *Acacia*, and *Casuarina*. Both eucalypts and banksias were on the Gondwanan mass of Antarctica left behind, while some travelled with South America; they eventually became extinct on both landmasses.[100]

Although Australia and New Zealand are often mentioned together today, as if a unit, and were once both part of Gondwana, their geological history and thus biological history diverged long ago.[101] This can be seen in the species present on both land masses. As Wilson had noted in *Smoke that Thunders*, there are no longer any eucalypts, banksias, casuarinas, or acacias endemic to New Zealand. It is a mystery as to why these groups became extinct in New Zealand, where they were once prolific. That eucalypts also became extinct in South America, which has varied landscapes and climate niches that might have aided survival, is also an unsolved puzzle.[102]

Wilson was struck by the links of southern beech, *Nothofagus*, that existed as remnants across South America, eastern Australia, Tasmania, New Guinea, and some Pacific islands, but were not found in Western Australia. All four subgenera of *Nothofagus* had diversified across Gondwana.[103] The genus *Nothofagus* was of great interest to Wilson because *Nothofagus gunnii* is the only deciduous tree native to the whole of southern Australia. He commented that 'there are no Willows, Birches, Poplars, Alders, Elms, Walnuts nor Hickories in Australia and the great family to which belong our Oaks, Hazels, Chestnuts and Beeches, so all-important a feature of our northern forests, is represented solely by Nothofagus'.[104] For Wilson, the sole *Nothofagus* extant in Australia highlighted to him that he was on a very different part of the planet.

Laurasia was also moving and changing towards what we recognise today. The Atlantic was created about 120 million years ago as North America broke away and swung to the west. South America joined up to North America about 9–10 million years ago, leading to the Great American Biotic Interchange (GABI), the pronounced transfer of species between these two landmasses, one landmass that arose from Laurasia and one from Gondwana, giving an unusual and dynamic mix of species that is in complete contrast to the stability and isolation of Australia.

The 'another planet' of Australia of which Wilson wrote has several key differences to the Northern Hemisphere. The landscapes of the Northern Hemisphere have been ground down and reworked by glaciers, creating new soils of higher mineral content, and the coming and going of glaciers and ice over the last 2 million years led to major extinctions, creating a lower level of biodiversity in the north. In the great landmass of Eurasia, most mountains run east–west, and this single feature led to a death zone, as plants, moving south to avoid cooling lands, met mountains with cooler conditions, not warmer; for many species, there was nowhere to go. North America, with its north–south mountain chains, did not fare so badly, but severe ice sheets still led to extinction, contraction, and the narrowing of genetic material. Australia largely escaped glaciation, with less impact of the last ice age on the Southern Hemisphere than on the Northern Hemisphere. It has aged landscapes that have resulted in a land of deeply weathered and largely infertile soils, was isolated from major impacts such as GABI, and has been climatically buffered by surrounding oceans for millions of years.

The contrast between the hemispheres has led to questions from scientists about the '"unifying schemes" served up mainly from Europe and North America' that do not fit the ecologies and biogeography of the Southern Hemisphere.[105] The search for global generalisations seems a peculiar notion in ecology and botany, both fields that describe and exult in the variety of the world. Major differences between the Northern and Southern Hemispheres would indicate a need for a 'broadening of mainstream theory' and new approaches to conservation and management.[106] Dominating in Australia are landscapes that have been described as 'old, climatically buffered, infertile landscapes', which contrast with the younger, often disturbed and more fertile landscapes that are mostly typical of Northern landscapes.[107] These concepts give insight into the heterogeneity of landscapes in Australia that led to conditions ripe for extensive speciation and niches for specialisation and adaptation, giving rise to Wilson's 'veritable botanic garden'.

After the breakup of Gondwana, Australia was the only continent to have no other continental impact on its flora.[108] These major differences between the Northern and Southern Hemispheres and the Australian continent's particular history of isolation and slow climatic changes are at the very heart of Wilson's astounded response to the 'rich, but decidedly strange'[109] Australian flora, with his keen 'Northern' eye and extensive botanical and horticultural knowledge. Wilson wrote, in the preface to *Smoke that Thunders*, that 'to be able to recognize and interpret the things around vastly increases the joy of living'.[110] I am sure that Wilson would be fascinated to know the more-complete story of plant history that we have now, one hundred years later.

PLATE 3.1 *"Banksia grandis* Willd. small tree 15 ft – 20 ft flo yellow forests Darling Mt near Perth" 25 October 1920. Number 197 on plant tag.

PLATE 3.2: *"Anthocercis littorea* Mundaring" 14 October 1920. No. 413.

PLATE 3.3: "*Dampiera linearis* [ ] [ ] Mundaring near Perth" October 1920. No. 452.

PLATE 3.4: *"Eucalyptus pyriformis* collected Zoological Gardens Perth" November 1920. No. 485

PLATE 3.5: "*Eucalyptus rudis* Applecross nr Perth "15 October 1920. No. 415.

PLATE 3.6: *"Verticordia acerosa* Mundaring, near Perth" 14 November 1920. The date of 1921 on the typed label is incorrect.

PLATE 3.7: "*Hypocalymma angustifolium* Mundaring" 14 October 1920. No. 443.

PLATE 3.8: *"Petrophile linearis* Mundaring" 14 October 1920. No. 431.

PLATE 3.9: "*Banksia grandis* The Lakes Darling Range W.A" 27 October 1920. No. 197 on plant tag.

PLATE 3.10: *Hakea ambigua*, collected here as "Acacia Bush 8 ft. Mundaring near Perth" October 1920. No. 479.

PLATE 3.11: "*Grevillea bipinnatifida* Mundaring" 14 October 1920. No. 469.

PLATE 3.12: "*Diplolaema Dampieri* Mundaring" 14 October 1920. No. 425. Now classified as D. drummondii.

NOTES

1. EH Wilson, 'Notes from Australasia. No. 1', *Journal of the Arnold Arboretum*, 1921, 2(3):160–3.
2. Letter to Sargent, 19 October 1920, Palace Hotel, Perth.
3. Letter to Sargent, 19 October 1920, Palace Hotel, Perth, AA Archives, W.XIV, box 2, folder 11.
4. *West Australian*, 15 October 1920, p. 6.
5. 'Perth prattle', *Sunday Times*, 30 October 1920, p. 6.
6. *Great Southern Herald* (Katanning), 20 October 1920, p. 3.
7. *Tambellup Times* (1912–24), 20 October 1920, p. 3.
8. *Albany Despatch* (1919–27), 4 November 1920, p. 1.
9. *South-Western Times* (Bunbury; 1917–29), 30 October 1920, p. 5.
10. 'Western Australian trees, of interest to Harvard', *West Australian* (Perth), 21 October 1920, p. 7.
11. Ibid.
12. *Boston Evening Transcript*, 30 September 1922, magazine section, p. 1, 4.
13. 'Research work', *Daily News* (Perth), 15 October 1920, p. 4. This comment was likely made because the article was lamenting the lack of research funds available in Australia.
14. RA Howard, 'EH Wilson as a botanist', *Arnoldia*, 1980, 40(4):154–93.
15. M Calver, H Bigler-Cole, G Bolton, J Dargavel, A Gaynor, P Horwitz, J Mills and G Wardell-Johnson, 'Why "A forest conscience-ness"?', in Michael Calver (ed.), *Proceedings of the 6th National Conference of the Australian Forest History Society Inc, Augusta 2004*, Millpress, Rotterdam, 2005.
16. Comment about Wilson made in 'Research work', *Daily News* (Perth), 15 October 1920, p. 4.
17. CE Lane-Poole, *Statement prepared for the British Empire forestry conference, London, 1920*, WA Government Printer, Perth, 1920.
18. CE Lane-Poole, *Notes on the forests and forest products and industries of Western Australia*, Forests Department of Western Australia, Perth, 1920.
19. HT Clifford, 'Herbert, Andrew Desmond (1898–1976)', *Australian Dictionary of Biography*, Australian National University, adb.anu.edu.au/biography/herbert-andrew-desmond-10488/text18607, published first in hardcopy 1996, accessed online 3 December 2021.
20. HT Clifford, *Australian Dictionary of Biography*, volume 14, entry for Herbert, 1996. The book noted here was published by Melbourne University Press in 1943.
21. Wilson, diary, 27 October 1920.
22. Letter from Wilson to Sargent, from the Palace Hotel, Perth, 17 November 1920, AA Archives, W.XIV:A, box 21, folder 11. Henry Huntington (1850–1927) was a railways magnate who established a botanic garden and educational collection in his estate in San Marino, out of Los Angeles, in 1919. He would have been very keen to hear of Australian plants suitable for Californian sites. This is the sole reference to Huntington being a funder of the Australasian expedition.
23. *Smoke that thunders*, p. 118. For a definition of mallees, see Chapter 4.
24. EH Wilson, 'Notes from Australasia. No. 1', *Journal of the Arnold Arboretum*, 1921, 2(3):162.
25. CJ Yates, CR Gosper, SD Hopper, DA Keith, SM Prober, and MG Tozer, 'Mallee Woodlands and Shrublands – The mallee, muruk/muert and maalok vegetation of southern Australia', in DA Keith (ed.), *Australian Vegetation*, 3rd edn, Cambridge University Press, 2017, pp. 570–98.
26. 'Wilson warmly welcomed. Given a right royal greeting by the Horticultural Club of Boston', *Horticulture Magazine*, 25 September 1922, AA Archives, W.XV, box 26, folder 12.
27. Wilfred Douglas, *The Aboriginal languages of the south-west of Australia*, Australian Institute of Aboriginal Studies, Canberra, 1976; Clint Bracknell, 'Kooral Dwonk-katitjiny (listening to the past): Aboriginal language, songs and history in south-western Australia', *Aboriginal History*, 2014, 38:1–18.
28. Michelle Johnston and Simon Forrest, *Working two way: Stories of cross-cultural collaboration from Nyoongar Country*, Springer, Singapore, 2020, p. 30.
29. These are commonly known names.
30. 'A study of trees: Professor Ernest H Wilson's visit. Our wonderful flora', *Western Mail*, 25 November 1920, p. 9.
31. PA Clarke, *Aboriginal plant collectors: Botanists and Australian Aboriginal people in the nineteenth century*, Rosenberg Publishing, Kenthurst, 2008.
32. M Torst, 'From the Nijptangh journal', in G Schilder, *Voyage to the Great South Land: Willem de Vlamingh 1696–1697*, Royal Australian Historical Society, 1985, p. 156.
33. Wilson (1927, 1985) *Smoke that thunders*, page 107.
34. *West Australian*, 21 October 1920, p. 7.
35. *Western Mail*, 25 November 1920, p. 9.
36. Ibid.
37. Lane-Poole, *Statement prepared for the British Empire forestry conference, London, 1920*, p. 32.
38. 'A study of trees: Professor Ernest H. Wilson's visit', *Western Mail*, 25 November 1920.
39. Wilson, *Smoke that thunders*, 1985, p. 106.
40. MD Crisp and LG Cook, 'How was the Australian flora assembled over the last 65 million years? A

molecular phylogenetic perspective', *Annual Review of Ecology, Evolution, and Systematics*, 2013, 44:303–24.
41 Wilson, *Smoke that thunders*, 1927, p. 107.
42 D King, 'The functional significance of leaf angle in Eucalyptus', *Australian Journal of Botany*, 1997, 45(4):619–39.
43 MJ Grose, '"Green above, paler below": Descriptions in the literature of the biodiversity of the colours of trees from southwest Australia', *Journal of the Royal Society of Western Australia*, 2007, 90:179–94; MJ Grose, 'Plant colour as a visual aspect of biological conservation', *Biological Conservation*, 2012, 153:159–63; MJ Grose, 'A finer-scaled reading of the colours of plants', *Journal of Landscape Architecture*, 2014, 1:42–7; MJ Grose, 'Green leaf colours in a suburban Australian hotspot: Colour differences exist between exotic trees from far afield compared with local species', *Landscape and Urban Planning*, 2016, 146:20–8.
44 ML McKinney, 'Species introduced from nearby sources have a more homogenizing effect than species from distant sources: Evidence from plants and fishes in the USA', *Diversity and Distributions*, 2005, 11(5):367–74.
45 BJ Glover, 'The diversity of flower colour: How and why?', *International Journal of Design & Nature and Ecodynamics*, 2009, 4(3):211–18
46 MJ Grose, 'Green above, paler below', 2007.
47 D Wyman, 'The new horticultural color chart', *Bulletin of Popular Information of the Arnold Arboretum, Harvard University*, 1957, 17(10). David Wyman discussed the problems with the Royal Horticultural Colour Chart, which was expensive and largely unavailable in the USA, and the new Nickerson Color Fan just available through the American Horticultural Council's Office at the Arnold Arboretum, which was also much cheaper.
48 AH Munsell, *Soil Color Charts*, Munsell Color Company, 1905.
49 MJ Grose, 'Green leaf colours in a suburban Australian hotspot: Colour differences exist between exotic trees from far afield compared with local species', *Landscape and Urban Planning* 2016, 146:20–8.
50 *Boston Evening Transcript*, 30 September 1922.
51 See MJ Grose, 'Green above, paler below: descriptions in the literature of the colour in trees from southwest Australia', *Journal of the Royal Society of Western Australia*, 2007, 90:179.
52 These quotes from Wilson, *Smoke that thunders*. Waterstone, London, 1927 (1985 edition), Chapter XXVVII, pp. 216–17.
53 *Dryandra* is now classified in the genus *Banksia*.
54 Wilson, *Smoke that thunders*, p. 159.
55 Wilson, 'Notes from Australasia. No. 1', p. 163.
56 Eucalypts are thought to be at least 110 million years old.
57 RJ Carpenter and LA Milne, 'New species of xeromorphic *Banksia* (Proteaceae) foliage and *Banksia*-like pollen from the late Eocene of Western Australia', *Australian Journal of Botany*, 2020, 68:165–78.
58 D Peyrot, G Playford, DJ Mantle, J Backhouse, LA Milne, RJ Carpenter, C Foster and A Mory, 'The greening of Western Australian landscapes: The Phanerozoic plant record', *Journal of the Royal Society of Western Australia*, 2019, 102:52–82. This paper has images of ancient and extant banksias.
59 BB Lamont, and T He, 'Fire-adapted Gondwanan angiosperm floras evolved in the Cretaceous', *BMC Evolutionary Biology*, 2012, 12:1–11.
60 RJ Carpenter and LA Milne, *Australian Journal of Botany*, 2020, 68:165–78.
61 SD Bradshaw, KW Dixon, SD Hopper, H Lambers and SR Turner, 'Little evidence for fire-adapted plant traits in Mediterranean climate regions', *Trends in Plant Science*, 2011, 16:69–76. They considered that 'exaption' is a better term for the forces that led to the traits observed, rather than adaptation.
62 NE White, MJ Phillips, MTP Gilbert, A Alfaro-Núñez, E Willerslev, PR Mawson, PBS Spencer and M Bunce, 'The evolutionary history of cockatoos (Aves: Psittaciformes: Cacatuidae)', *Molecular Phylogenetics and Evolution*, 2011, 59(3):615–22.
63 Wilson, *Plant hunting* (1985 edition) p. 122.
64 See: A Lullfitz, L Knapp, S Cummings, J Woods, and SD Hopper, 'Talking Mungee – a teacher, provider, connector, exemplar: What's not to celebrate about the world's largest mistletoe, *Nuytsia floribunda*', *Plant and Soil*, 2023, https://doi.org/10.1007/s11104-023-06057-9
65 Kaanya means 'recently dead'. The legend of the Christmas Bush is given in D Bates, *Aboriginal Perth and Bibbulmun Biographies and Legends*, PJ Bridge (ed.), Hesperian Press, Carlisle, Western Australia, 1992.
66 Ibid, p. 153.
67 For example, the 1935 *Native Flora Protection Act No. 37*, an Act to provide for the protection of the native flora of Western Australia, assented to 7 January 1936, lists *Nuytsia floribunda*, along with others that include kangaroo paws (*Anigozanthos* spp.), Black Kangaroo Paw (*Macropidia fuliginosa*), all orchids, all *Kennedya*, all *Boronia*, all *Hovea*, all *Crowea*, all waxplants (*Chamaelaucium* spp.), all *Darwinia*, *Verticordia*

grandis, the Sturt Pea (*Clianthus speciosus*), all flannel or Blanket plants or Lambs' tails (*Lachnostachys* spp., *Physopsis* spp., *Newcastelia* spp., and *Hemiphora Elderi*), and all *Lechenaultia* spp. This list was extant in 1924 in a Flora Protection Act; West Australian Wednesday 3 September 1924, p. 8.
68 Bates,). *Aboriginal Perth*, 1992, p. 14. Fascinating to see the mention of cold times long ago.
69 A Lullfitz, J Woods, L Knapp, S Cummings, and SD Hopper, 'WA's Christmas tree': What mungee, the world's largest mistletoe, can teach us about treading lightly', *The Conversation*, 2023, https://theconversation.com/was-christmas-tree-what-mungee-the-worlds-largest-mistletoe-can-teach-us-about-treading-lightly-205568
70 Wilson, 'Notes from Australasia. No. 1', p. 163.
71 B Dinkelaker, C Hengeler and HJBA Marschner, 'Distribution and function of proteoid roots and other root clusters', *Botanica Acta*, 1995, 108(3):183–200.
72 M Brundrett, 'Distribution and evolution of mycorrhizal types and other specialised roots in Australia', in L Tedersoo (ed.), *Biogeography of Mycorrhizal Symbiosis. Ecological Studies (Analysis and Synthesis)*, Springer, 2017, p. 230.
73 Wilson, 'Notes from Australasia No. 1', p. 163.
74 *Smoke that thunders*, p. 112.
75 *Albany Advertiser*, 5 March 1921, p. 3.
76 *Smoke that thunders*, p. 112.
77 Wilson in the *Boston Evening Transcript*, 30 September 1922.
78 Wilson, *Smoke that thunders*, p. 112.
79 The Noongar have a long process of soaking and cooking to destroy the toxins in zamia. The reference to the Drions was a mystery to the translator of the Dutch, solved with a check on Old Dutch. MJ Grose and K Panegyres, 'An unidentified word in the account of the zamia poisoning on de Vlamingh's expedition of 1697', *Journal of the Royal Society of Western Australia*, 2021, 104:1–3.
80 Willem de Vlamingh, logbook entry 6 January 1697. In,) *Voyage to the Great South Land. Willem de Vlaming 1696-1697*, G Schilder (ed.), C de Heer (trans.), Royal Australian Historical Society, 1985, p. 155.
81 *West Australian* 21 October 1920, p. 7.
82 Richard G Hartley, *River of Steel: A History of the Western Australian goldfields and Agricultural Water Supply*, Access Press, Bassendean, Western Australia, 2008.
83 The pipeline was originally buried but was redone in the 1930s and much laid above-ground. More recent renovations have replaced the pipeline underground as it had originally been laid.
84 One region where only a few species dominate is the Snowy Mountains region of south-eastern Australia, where glaciation occurred in the last Ice Age.
85 SD Hopper, 'South-western Australia, Cinderella of the world's temperate floristic regions 1', *Curtis's Botanical Magazine*, 2003, 20(2):101–26; SD Hopper, 'South-western Australia, Cinderella of the world's temperate floristic regions 2', *Curtis's Botanical Magazine*, 2004, 21(2):132–80; SD Hopper SD and P Gioia, 'The Southwest Australian Floristic Region: Evolution and conservation of a global hot spot of biodiversity', *Annual Review of Ecology, Evolution, and Systematics*, 2004, 35(1):623–50.
86 AL Du Toit, *Our Wandering Continents: An hypothesis of continental drifting*, Oliver & Boyd, Edinburgh, 1937.
87 EB Bailey, Review, *Nature*, 28 January 1939, 3613.
88 AG Smith and A Hallam, 'The fit of the southern continents', *Nature*, 1970, 225:139–44.
89 See W Jokat, T Boebel, M König and Uwe Meyer, 'Timing and geometry of early Gondwana breakup', *Journal of Geophysical Research*, 2003, 108. Jokat et al., based at the Alfred Wegener Institute for Polar Research in German, asked the simple question: 'How and when did Gondwanaland begin to separate?'
90 See JJ Veevers, 'Gondwanaland from 650–500 Ma assembly through 320 Ma merger in Pangea to 185–100 Ma breakup: Supercontinental tectonics via stratigraphy and radiometric dating', *Earth-Science Reviews*, 2004, 68:1–132. John Veevers published,) *Billion-year earth history of Australia and neighbours in Gondwana*, 2000.
91 CO Mueller and W Jokat, 'The initial Gondwana break-up: A synthesis based on new potential field data of the Africa–Antarctic Corridor', *Tectonophysics*, 2019, 750:301–28.
92 AD Gibbons, JM Whittaker and RD Müller, 'The breakup of East Gondwana: Assimilating constraints from Cretaceous ocean basins around India into a best-fit tectonic model', *Journal of Geophysical Research: Solid Earth*, 2013, 118:808–22.
93 Ibid, Introduction.
94 DE Sugden and SSR Jamieson, 'The preglacial landscape of Antarctica', *Scottish Geographical Journal*, 2018, 134(3–4):203–23.
95 J Pross, L Contreras, PK Bijl, DR Greenwood, SM Bohaty, S Schouten, JA Bendle and U Röhl, 'Persistent near-tropical warmth on the Antarctic continent during the early Eocene epoch', *Nature*, 2012, 488:73–7.
96 For a recent review discussion

on these changes see: DK Hutchinson, HK Coxall, DJ Lunt, M Steinthorsdottir, AM de Boer, M Baatsen, A von der Heydt, M Huber, AT Kennedy-Asser, L Kunzmann, J-B Ladant, CH Lear, K Moraweck, PN Pearson, E Piga, MJ Pound, U Salzmann, HD Scher, WP Sijp, KK Śliwińska, PA Wilson and Z Zhang, 'The Eocene–Oligocene transition: A review of marine and terrestrial proxy data, models and model–data comparisons', *Climate of the Past*, 2021, 17:269–315.

97 It has been hypothesised, based on geological data and sea level changes, that ephemeral and localised icesheets existed during these times of warm, high-latitude climates and thus pre-dated this final break-up and thermal isolation of Antarctica. See KG Miller, JD Wright and JV Browning, 'Visions of icesheets in a greenhouse world', *Marine Geology*, 2005, 217(3–4):215–31.

98 Sugden and Jamieson,), 'The preglacial landscape of Antarctica', 2018.

99 RC Blakey, 'Gondwana paleogeography from assembly to break-up: A 500 m.y. odyssey', in CR Fielding, TD Frank and JL Isbell (eds), *Resolving the Late Paleozoic Ice Age in time and space*, Geological Society of America Special Paper, 2008, 441:1–28.

100 For a recent review of the break-up of South America from Gondwana and the complex dual biogeographic unit of that continent, see MA Reguero and FJ Goin, 'Paleogeography and biogeography of the Gondwanan final breakup and its terrestrial vertebrates: New insights from southern South America and the 'double Noah's Ark' Antarctic Peninsula', *Journal of South American Earth Sciences*, 2021, 108(4):103358.

101 MG Laird and JD Bradshaw, 'The break-up of a long-term relationship: The Cretaceous separation of New Zealand from Gondwana', *Gondwana Research*, 2004, 7:273–86.

102 For a review of where eucalypts once grew, see RS Hill, YK Beer, KE Hill, E Maciunas, MA Tarran and DD Wainman, 'Evolution of the eucalypts—an interpretation from the macrofossil record', *Australian Journal of Botany*, 2016, 64:600–8.

103 U Swenson, RS Hill and S McLoughlin, 'Biogeography of *Nothofagus* supports the sequence of Gondwana break-up', *Taxon*, 2001, 50:1025–41.

104 Wilson, *Plant hunting*, 1985 edition, p. 147.

105 C Lusk and P Bellingham, 'Austral challenges to northern hemisphere orthodoxy. IV Southern connections conference: Towards a southern perspective, Cape Town, South Africa, January 2004', *New Phytologist*, 2004, 162:248–51.

106 S Hopper, 'Out of the OCBILs: New hypotheses for the evolution, ecology and conservation of the eucalypts', *Biological Journal of the Linnean Society*, 2021, 133:342–72.

107 SD Hopper, 'OCBIL theory: Towards an integrated understanding of the evolution, ecology and conservation of biodiversity on old, climatically-buffered, infertile landscapes', *Plant and Soil*, 2009, 322:49–86.

108 Bar Madagascar, which is regarded as a continental fragment, and is now also a hyper-diverse region.

109 *Boston Evening Standard*, 30 September 1922.

110 Wilson, *Smoke that thunders*, 1927, preface.

CHAPTER 4

'THE PRICE ... EXACTED' IN WESTERN AUSTRALIA'S WHEATBELT

On his first trip out of Perth, Wilson travelled east. This trip was taken before a journey to the cooler southern forests, likely to avoid as much heat as possible inland as summer approached. He traversed a transect of changing vegetation, all dominated by eucalypts. The transect runs north–south, beginning with the sandy strip of the coast, then the strips of the Swan Coastal Plain's soil and vegetation types, the forest strip of the Darling Scarp – 'the Hills' – and then east into the region known as the Wheatbelt. Along this line, tuart (*E. gomphocephala*) and flooded gum (*E. rudis*) change to jarrah (*E. marginata*), and forests with marri (*E. calophylla,* now *Corymbia calophylla*) give way to wandoo (*E. wandoo*) in the more lightly wooded eastern slopes of the Darling Scarp. This is followed by powderbark wandoo (*E. accedens*) and, to the east, York gum (*E. loxophleba* subsp. *loxophleba*); then in the Wheatbelt, wandoo gives way to salmon gum (*E. salmonophloia*) and wheatbelt wandoo (*E. capillosa*). However, such a list is a simplification, for this is not a country dominated by single species. A host of tree species are found farther to the east – the whole of the south-west and into the interior of the south-west of Australia is typified by complexity in plant species, with often fuzzy edges of distribution. Wilson would come across suites of species that included the genera *Acacia*, *Banksia*, *Casuarina*, *Allocasuarina* (Figure 4.1), *Grevillea*, and a score of others.

FIGURE 4.1: Desmond Herbert with *Allocasuarina huegelliana*, Sighing sheoak, in the Wheatbelt, 1920. The tree's common name is well-suited as it has a distinctive and beautiful sound in the wind. Image: Wilson, Y-389, AA.

Although Wilson would have seen eucalypts in California, the home of *Eucalyptus* was a new world for him. He would have been immediately struck by three new terms for eucalypts that are confined to Australia – mallee, mallet, and maalok. All are Indigenous terms to define major morphological characteristics of the trees. Mallee – a Wemba Wemba word from south-eastern Australia,[1] is a well-known term in Australia that describes the form of a tree that is multi-stemmed from ground level and usually less than 10 metres (around 33 feet) tall.[2] Even small trees with this feature are called mallee and not shrubs, the latter being a term that is rarely used for eucalypts. It is not common globally to use a tree term for a multi-stemmed plant. In the Northern Hemisphere, in the Arnold Arboretum, a tree is defined as a plant that develops 'a single, erect woody trunk', while a shrub is described as 'a woody plant that, when undisturbed, branches spontaneously at or below ground level to produce multiple stems'.[3] This is not so in Australia; it is a different planet. Mallee would have been a firm part of Wilson's lexicon by the time he departed Australia.

A mallet is a less well-known term, though still widely used. A mallet is a type of eucalypt with a slender trunk and steeply angled branches, much like an umbrella; it does not have a lignotuber. Wilson collected the mallet *Eucalyptus astringens* in the Wheatbelt, where he also saw the salt-tolerant mallet *E. sargentii*.

The term maalok (or marlock; sometimes marloch) is also less well known today. A maalok is a single-stemmed small tree, short in stature, with dense and leafy branches that are often almost to the ground such that they form dense and impenetrable thickets. Like mallets, they lack a lignotuber. Maalok were the bane of early European explorers in Western Australia in the nineteenth century. They were cursed due to their denseness and the difficulties that this presented to both men and horses with packsaddles, let alone drays. Early explorers described the maalok as 'detestable scrub' that had to be cut through with a tomahawk.[4] An idea of the type of tree is gained from Richard Belches' evocative description of maalok thickets in 1866:

> this scrub appears ... of various sizes and so close together that they seemed like a number of stakes driven into the ground for the purpose of opposing our progress ... We cut a road through for the horses, and it presented the appearance of a track with a wall of sticks on each side.[5]

There are few species classified as maalok, but they are extensive in their range, and Wilson would have seen thickets of maalok in his travels. Some maalok are well known in cultivation today. *Eucalyptus platypus* is a maalok that is widely planted as a street tree,[6] but it grows taller in domestication than in the wild and does not present the problem of the thickets of sticks.[7] The terms maalok and mallet now appear to be somewhat interchangeable.[8]

Wheatbelt woodlands evolved on the Yilgarn Craton, a huge and geologically stable granite–gneiss craton or shield that consists of some of the oldest rocks in the world, formed 2.6 billion years ago as upthrusts of magma. The Yilgarn Craton has been subjected to weathering by rain and wind for the last 250 million years, and the soils are therefore washed of most fertility. Great age and geological stability have produced tremendous plant diversity, with plants having evolved many strategies to deal with the very nutrient-poor and shallow soils. The Wheatbelt has wide, gently undulating ancient valleys that are commonly 5–8 kilometres (3–5 miles) across and are filled with sediments from millions of years of erosion and drainage. Between the shallow valleys are often caps of laterite, and granite outcrops of extraordinary visual beauty and plant biodiversity.

Wilson travelled through the Wheatbelt with Desmond Herbert, Government Botanist in Western Australia. The Wheatbelt is a large agricultural area, and at this time was dominated by both wheat and sheep. In Wilson's time, sheep for wool were a mainstay of Australia's economy.

Their journey began on 21 October 1920 with a trip to Toodyay, 85 kilometres (54 miles) east of Perth on a 'good road' that 'bends through the Darling range [sic]'.[9] There are various theories as to the origin of Toodyay's name (pronounced 'two-jay'), all linked to the local Nyoongar language. The original version was likely 'Toojie'.[10] Wilson's Wheatbelt travel over a few weeks took him from Toodyay and Northam to Merredin and Westonia, back to Merredin, to Bruce Rock, and then via the hamlet of Quairading to York. In doing so, he passed the small towns or hamlets of Cunderdin, Tammin, Kellerberrin, Doodlakine, Hines Hill, and Burracoppin. Wilson was immediately concerned at what he saw in the Wheatbelt. He wrote:

> There are no good trees left [ ] [ ] them even where many of size [ ] but man in his effort to 'improve the country' has destroyed even where these

were. Agriculture is necessary but its initiation is expensive – extravagantly so, & I fear that someday the price will be exacted.[11]

This was a prescient comment so early in his Australian expedition, and would come to be a dominant impression of his experience of Australia and the warning he gave of future problems.

The vegetation was dominated by salmon gum and gimlet woodland (Figure 4.2). Wilson wrote 'we met with Salmon Gum & Gimlet gums 25 miles east of Northam & they continue with us' as he travelled east. Salmon gum (*Eucalyptus salmonophloia*) was described by Wilson:

> The Salmon Gum is a handsome tree with smooth white to pinkish trunk dividing 20 ft up into several ascending [...] spreading stems which [ ] form a table like [...]. The twigs are reddish, the [...] yellowish & the leaves lustrous dark green & the effect with sunlight [shining] through is striking.[12]

FIGURE 4.2: The iconic salmon gum (*Eucalyptus salmonophloia*), with Herbert standing. The image also shows the extensive clearing for fields, lack of tall trees remaining in paddocks, and scant trees remaining along roads, having not survived clearing. Yet Wilson noted that salmon gum grow to a 'lofty size in regions where one would suppose no tree could possibly exist'.[13] The root system of the salmon gum is shallow, an extraordinary thing in a large dryland tree. Yet no respect seemed to be shown to these magnificent trees, and little has changed to support an increase in their numbers. Image: Wilson, Y-401, AA.

The species is known for its trunk turning a beautiful bronze gold in autumn.

Sir David Hutchins, who had worked for decades in India and southern Africa as a conservator of forests, had visited Western Australia only a few years earlier than Wilson at the behest of the government to advise them on forestry. He noted that 'Salmon gums, Gimlet, and smaller gums are wonders of tree growth in such a climate'.[14, 15] Wilson was asked by the press about his impression of salmon gums and gimlets. He noted that it was a surprise to see larger trees like the wandoo (*Eucalyptus wandoo*) in low rainfall areas, but said that it was 'in keeping. Of course there is a wonderful adaptation of the character of the tree to the conditions, but I have seen similar things in other lands.'[16] Despite this seemingly dismissive comment, he was astounded that salmon gum 'is surface-rooting!' He noted that

> A degree or two of frost is not unknown in the region where it grows and I am told that it is flourishing in parts of South Africa. I cannot help thinking that this tree would be a good subject to plant for forestry purposes in the hot, arid parts of Lower California, Texas, New Mexico and of Arizona. Of course the Salmon Gum like all other plants of this Hemisphere are useless in the Arnold Arboretum but they would be of immense value in California.[17]

Moving east, Wilson and his party travelled along the route of the Goldfields Water Pipeline, whose origin he had seen in Mundaring. Wilson stopped at Doodlakine,[18] between Kellerberrin and Merredin, and collected a bounty of plants. During the gold rush, Doodlakine had been a major stopping place to Coolgardie and Kalgoorlie. Cobb & Co., the major transport and gold escort company of its day, changed horses at Doodlakine. The company, famous in Australia, was founded in 1853 by Freeman Cobb, a Massachusetts man, and three American associates who had arrived in the 1850s in search of gold.[19, 20]

Travel for Wilson was hot. To modern eyes, travel in this region in an open vehicle in late spring or summer has little, if any, appeal. But for many in 1920, travel in a Ford car as Wilson did, would have been unusually good. Only twenty years before, travel on Cobb & Co. had been described as: 'under a broiling sun, veiled only by the dense clouds of choking and blinding dust which, like the pillar of cloud of biblical story, hung over the unhappy wayfarer until they crossed the red sea of grit … But with all its miseries a journey by mail coach was a picnic compared with the former days of travel behind heavily laden waggons, which took a week or 10 days to span the distance [between Southern Cross and Coolgardie].'[21]

The origin and importance of towns like Doodlakine was due to the existence of one of Hunt's wells.[22] Charles Cooke Hunt led four expeditions east of Perth between 1864 and 1866, and established wells at sites of natural waterholes to which he had often been guided by friendly local Aboriginal people. He travelled with an Aboriginal guide, six pensioner soldiers, and several convicts, and these men dug and formalised the wells, which are deep and often perfectly circular granite structures. Hunt himself felt that he had achieved little. He lamented that 'my work will be read with but little interest being but a record of disappointment and drought'.[23] Hunt died of exhaustion and heart failure at the age of thirty-five, not knowing that later explorers, sandalwood cutters, pastoralists in search of winter grazing, and thousands of prospectors travelling to the Coolgardie and Kalgoorlie Goldfields would thank him for opening up much of this arid country by mapping the designated wells along the entire route. The transcontinental rail line, the telegraph, and the Goldfields Water Supply Scheme in 1902 followed his wells. Like Wilson, modern travellers also move along Hunt's route, which defines the Golden Pipeline Heritage Trail between Mundaring and Kalgoorlie.[24]

Between Doodlakine and Merredin, Wilson passed by a wide expanse of salt lake, Baandee Lakes – another wonderful placename full of vowels. Baandee Lakes are part of a wide, curling ancient paleo drainage of salty lake and river systems; at the time of Wilson's visit there was a hamlet at Baandee. Passing through the lake and the twisting series of pans and salt flats runs the ancient Yilgarn River, which is 525 kilometres (326 miles) long, with only the slightest gradient. This is not a river with riparian edges, blue water, and ducks, nor one that rises in hills or mountains, makes its way over plains, and ends in the sea. Few rivers in Australia follow that pattern. Australian rivers are often called backward rivers, or reversed rivers, because they start in flat plains in areas of ancient drainage, as in the Yilgarn River, move through hills, and only sometimes make it to the sea. Many agricultural areas in other parts of the world as large as the Wheatbelt can boast of a long list of rivers. However, in the Wheatbelt, today an area the size of

FIGURE 4.3: Image shows cropping for wheat right up to and between salt lakes, east of Kellerberrin, near Doodlakine, in the Wheatbelt. Image: Google Maps 2025.

England and Wales combined, there are few rivers, and no major ones. From the air, it's possible to see the enormity of the ancient saltway, the Great Salt River, to the north and south of Wilson's immediate route.

Near Baandee, on 21–22 October, Wilson collected '*Salicornia australis*, 6' to 1' 'Samphire' Hyne's Hill from banks of salt lake' (#34, Wilson collection HH).[25] Unlike most samphire in other parts of the world, Australian samphires are nearly all perennials.[26] At the same time, he collected '*Daviesia euphorbioides*, 4' 'Centipede Bush" (#32), and '*Callitris robusta*? 'Pine' 18" (#33).[27]

The salt lake system is extensive, with scores of salt lakes and saline sumps, mainly consisting of the minerals sodium and chloride.[28] Flying over this country, few stare out the plane window in wide-eyed wonder at a fascinating land with myriad coloured circles of salt lakes – some pink, some green, most white or shades of pale white (Figure 4.3), where megafauna of 3-metre (10-foot) kangaroos, hippo-sized wombats and 15-metre (50-foot) pythons roamed when it was a more fluvial land. The majority of travellers are now plugged in watching a movie that has nothing to do with the great browned continent below and its sweeping plains, which support a miraculous diversity of life.

The Yilgarn River was active in the early Tertiary after the extinction of the dinosaurs, when its base was 70 metres (230 feet) deeper than now in one part, and flowed through a deep ravine, with hills 70 metres higher than at present; all has been eroded and deposited over that immense time. Sometimes in its youth, long ago, this river had been 60 metres (nearly 200 feet) across and carried huge flood loads as it ran along. Now it is like a frail old man tapping with his stick, stopping to catch his breath before starting off again in tentative little moves.

This ancient saline system has become hypersaline,[29] due to the extensive clearing that Wilson noted in 1920. Now, only about 7% of pre-European vegetation remains across the Wheatbelt generally.[30] Land clearing was applauded at the time it began, and for long after. For example, the settlement of Doodlakine was described in 1909 as follows:

> Up to some three years back the country was wilderness, with the exception of 'Mooranoppin' and Mr Ripper's fine place, 'Whollandra', and one or two other less prominent farms; now the district is so far from being merely bush, with an occasional oasis-like clearing in it, that it is rapidly becoming one big oasis with a fringe of bush.[31]

Such fringes of bush are now conserved and needed more than ever. A large influx of unemployed men was sent by the government in 1908 to cut down the gimlet, *E. salubris*. The newspapers reported that 'The district has suddenly awoke to find itself famous owing to the arrival of scores of unemployed, provided with tents, water-bags, billy-cans, and sharp axes, ready and willing, apparently, to chop down every tree in our vast salmon-gum and gimlet-wood forests.'[32] In the same newspaper in 1909 was an image of the salt lake at Doodlakine, showing extensive woodland with tall trees (Figure 4.4).

The 'vast' salmon gum and gimlet forest is now gone. The ensuing environmental changes due to rising salinity have been devastating to surviving vegetation. I wondered if the lake stored memory in its salt, and remembered the mallets reflected in its still water and the cry of red-tailed black cockatoos in their thousands. The Noongar people carry this memory very strongly. The pain of this mindless assault must have been unbearable, and it must remain unfathomable. It was a radical erasure.

FIGURE 4.4: 'SALT LAKE AT DOODLAKINE' 1909. This photograph is from page 24 of the *Western Mail*, 21 August 1909. The image shows extensive woodland, with tall trees around the Baandee Lakes before clearing. The taller canopy trees are salmon gum. The land was cleared right to the lake's edge, showing little care about the dangers of salt, even though the danger was known to local Aboriginal people and early explorers. We can repair and remediate this lake and its riparian vegetation, if we are determined. Image: Yeates, CSS (Charles Samuel Sparks) Salt Lake at Doodlakine 1909.

Baandee Lakes had blue water in 2021 and 2022, which were wet years. Yet in 2021, the water was far from the bank, and the bank eroded with random use by walkers and vehicles and lack of care or revegetation, the place beautiful but forlorn. The potential to take better care of this country and allow organised spaces for campers and others is enormous. Places like Baandee are the very sites that many city dwellers long to visit, away from the denser tourist areas of other regions. Here, within the lakes, salt, rocks, and paddocks, are small gems of silence and bush that many in the city ache to find. Higher ground provides a magical view of Noongar country. Seeing our distant forms as we searched for their misty, bounding figures in light rain, kangaroos were alert and curious and moved fluidly to the south.

There is now an artisanal salt farm at Baandee in the middle of the paleo chain of lakes that contains part of the Yilgarn River. The salinity is five times that of the ocean – it is 17% saline; people float with ease, just as in the Dead Sea. A major aim of its owner, John Mizzi, is to replant the area with salt-tolerant native plants and trees. Some wild existing plants are being used in salt infusions and in skin-care products – eucalyptus species, the Gumbi Gumbi tree (*Pittosporum angustifolium*), saltbush (*Atriplex* spp.), bluebush (*Maireana* spp.), tea tree (*Melaleuca* spp.), samphire (*Tecticornia* spp.), and iceplants (*Mesembryanthemum* spp.).

Wilson also collected *Pittosporum* here. Gumbi Gumbi is also called native apricot, due to its small and very pretty, apricot-coloured fruits. Traditionally, these are not a food source; rather, the leaves are used in bush medicine by Noongar as a treatment for chest infections and eczema, and to assist lactation.[33] Used as a tea, with the leaves steeped in hot water, it is believed to reduce inflammation.[34]

Wilson wrote about the soils, sand, wheat, and the ever-present salt lakes, which were indicators of trouble to come:

> The sand [patches] & the [Pindan] are not [ ] [ ] made [salt]. We found one small saltlake & a few shallow ponds but no streams. Water is all that is needed to make this land teem with farms. The Jam country is good wheat land but supports a very meagre flora.[35]

Wilson noted that Jam wattle, *Acacia acuminata*, known as Mangart in Noongar, was not attacked by white ants (termites) and was therefore used for fence posts. 'Jam' as a word in Wilson's diary and 'jam country' would readily confuse anyone not familiar with Western Australian plants and did confuse a transcriber of his notes. Jam is an odd common name, especially as it is just referred to as 'jam' and not jam tree. It gained the name in the early days of settlement because the broken timber of *A. acuminata* smells like raspberry jam, something early settlers from Europe likely missed. It has a beautiful red timber. It is a tree often mentioned by Wilson in his diaries, and he photographed it in his way, with his camera case for scale (Figure 4.5).

Wilson noted in his diary on 22 October 1920 that 'The country was largely rolling plains & much of it under crop of which wheat in many areas was good. It was ripe & ripening.' The wheat in the Wheatbelt is winter wheat, but not the winter wheat of the Northern Hemisphere that, after sowing in September or November, emerges in spring to be harvested in summer or early autumn of the following year. Winter wheat here is sown in autumn with the 'break of season', a term for the arrival of the winter rains in this classic Mediterranean climate of mild wet winters and hot dry summers. It then grows through the winter and is harvested in late spring or early summer. Wilson saw wheat nearly ready for harvest in late October. Today, Australia ranks sometimes sixth and sometimes tenth in the world for wheat production, dependent on the quality of the growing season in this drought-prone continent, after China, India, Russia, and the USA.

FIGURE 4.5: 'Jam', *Acacia acuminata*, with Wilson's camera case. This is a common small tree found across a large area of Wilson's travels in the Wheatbelt. Image: Wilson, Y-405, AA.

Wheat (*Triticum* spp.) in Australia expanded after breeding experiments conducted by William Farrer near present-day Canberra in the late nineteenth century and early 1900s. Farrer, another Englishman, was the son of a tenant farmer but achieved a degree from the University of Cambridge. Tuberculosis led him to abandon medical studies and seek out the warmer, drier climate of New South Wales. There, he set up the first cross-breeding trials for wheat in Australia, largely to combat wheat rust caused by the fungus *Puccinia graminis*, but unlike researchers of wheat rot in the USA, Farrer also aimed to improve the baking qualities of wheat. He wanted to provide wheat varieties that were suited to the environmental conditions of Australia, particularly those of soil nutrient poverty, water stress, weeds, and lodging. In the same vein as Wilson's 'other part of the planet', Farrer wrote:

> In a new country like this, where the climate is so widely different and in many respects the very opposite of that of the country which, only a short time ago, was the home of the vast majority of our farmers, and if not of them, of their fathers … the traditional practices of the old country have to be unlearnt, in addition to entirely new ones learnt.[36]

Farrer's comment about unlearning is one that is ongoing for our understanding of this different continent. Geoffrey Bolton, a noted Australian historian with a strong geographical focus, believed that 'Australians have done best when they discarded ideas and models brought from Britain and North America' and that the worst mistakes were when local understanding was disregarded.[37]

Farrer, later praised by the great Russian agronomist Nikolai Vavilov,[38] worked with Russian, American, Canadian, Hungarian, British, Indian, and Kenyan wheats and others, with names like Fife-Indian, Indian G, Yandilla, Purple Straw, Etawah, Gluyas Early, Jondhala, Yalta, White Odessa, Hussar, No. 2775, and Buivola Red Winter Wheat.[39] He developed twenty-two varieties that became commonly used across the continent, and these varieties ensured the success of Australia's wheat production. His most famous cultivar was Federation, named for the 1901 Federation of Australia as a Commonwealth of States.[40] Varieties developed by Farrer included Cedar, Bayah, Comeback, Florence, Genoa, Bobs – an attempt to cross barley on a wheat – and some with the idiosyncratic Australian names of Bunyip (a mythological creature of billabongs) and Jumbuck (an old Australian slang term for a sheep). Federation wheat was used extensively for decades and spread to every cereal country in the world, including wheat regions in the Pacific

Northwest of the USA and South America. Farrer's success in breeding wheats suitable for drier conditions was a double-edged sword, because it enabled the extension of the wheat-growing areas into drier country than otherwise thought possible, and the further destruction of Australia's woodlands.[41]

It was not just the woodland trees of the Wheatbelt that impressed Wilson, but the

> astonishing [and] extraordinary wealth of flowering shrubs. When not in blossom they all look alike; there are only about two common types of leaves, the thin wire-like leaf and the long narrow leaf. But when they are in blossom every other one is different, and they vie with one another to see which can produce the most extraordinary shape and the most beautiful colouring. It is a vast botanical garden, and the poorer the soil the greater is the variety.[42]

Near Merredin, in the central Wheatbelt, Wilson collected dozens of species from the sandplains, so many that he simply wrote 'N.B. s.p = sand plain' at the bottom of the page (Figure 4.6).

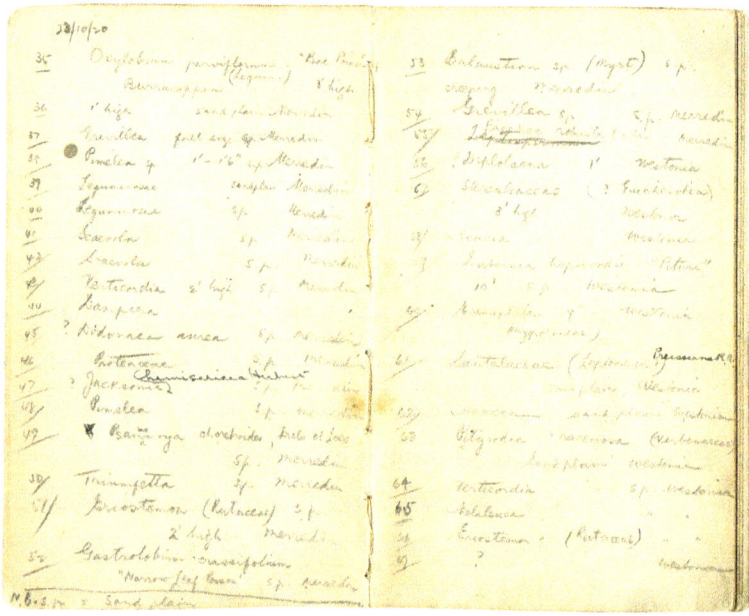

FIGURE 4.6: Diary notes, showing lists of plants collected near Merredin and Westonia, 1920. They included the poisons *Oxylobium parviflorum*, the box poison, and *Gastrolobium crassifolium*, the thick-leaved poison. The note about sand plains is at the bottom left.

Wilson called this area a 'savannah' and collected on the 'sandplains'. But this region is not the flat, grassed lands of prairies or the savanna of few, scattered trees over grassland, but a savanna of a high percentage of tree cover, with a scattered bushy understorey of great species diversity. Wilson was intrigued by the sandplains. They collected everything that caught Wilson's eye. 'On sandy flat gathered a new Xanthorrhoea in a [ ] [leaves?] & a short inflorescence of white flo.'[43] He marvelled at the number and variety of species he saw, writing:

> The extraordinary thing about the sand plains is the quantity of [supportative] plants: usually topped in [? Nature] [ ] 9 – 18 inch high & marvellously floriferous. They are of all colours & [ ] an indigo blue Dampiera was a wondrous sight'——so too was an 18 inch high Datonea with clear primrose yellow flo. Curious plants are the Proteaceous Isopofoma [sic] with cane like fruits, much divided spring foliage & [ ] [ ] white & pink flo. Among the taller [ ] Grevilleas Hakea & [] & Melaleucas abound & Leptospermum are [ ]. Green Parrots are common & so too is the Pied Magpie. Indeed we saw many birds & the old nest of a Mallee Hen. Rabbits unfortunately are not rare here we have seen but one dead snake & the common & racehorse Goanna (Iguana) are the only reptiles. The country is slightly undulating & from the crest of a slight rise the view of sand & [farmed] patches is very expansive.[44]

Wilson likely saw, but does not mention, the magnificent wedge-tailed eagle (*Aquila audax*), the largest bird of prey in Australia, which can often be seen soaring for endless minutes, riding thermals without wingbeat on their near 3-metre (over 9-foot) wingspan. Wedge-tailed eagles are one of the largest eagles in the world, with a long, razor-sharp hooked beak and great steel-like talons that rip up prey larger than themselves and far larger than most of the world's eagles are capable of despatching.

He collected many *Grevillea*, a member of the Proteaceae family, across the Wheatbelt, where the family are very well represented. Many are large shrubs that are key supporters of pollen and sugars beloved of, and a core source of food for, Australian birds, who are major pollinators of Australia's flora.[45] Through the Wheatbelt in October, Wilson collected *Grevillea apiciloba* (#70 and #82,

Westonia),[46] *G. teretifolia* (#78, '4ft', Westonia), and *G. pterosperma* (#101, on the 'Hakea sandplains', Westonia, 21 October). At Merredin he collected *G. eryngioides* (#37, Merredin and #150, Quairading), *G. hakeoides* (#46, Merredin), and *G. paradoxa* near Merredin (#54) shows the very sharp and prickly 'leaves'. South of Kellerberrin, Wilson and Herbert collected *G. uncinulata* (#176, 'sandplains', Yoting) and moving west collected *G. paniculata* (#26, 'bush 4ft', Cunderdin), *G. excelsior* (#27, '10ft fl orange Sandplains', Tammin), *G. pritzelii* (#30, Tammin), and *G. integrifolia* (#29, Tammin), among others. Later, returning to Perth, he collected near York *G. tridentifera* (#216, '4ft flo white', York) and '*G. Wilsonii*' (#199, The Lakes), which was named for Thomas Braidwood Wilson (1792–1843), a Scottish medical practitioner and naval surgeon, who discovered the species in the 1830s.[47]

Other Proteaceae collected included *Banksia*, *Hakea*, and a host of other genera, including the woody pear, *Xylomelum angustifolium* (Figure 4.7). Hakeas are brilliant flowering shrubs, usually highly bird-attracting. Included in Wilson's collection are *Hakea incrassata* (#184, '1ft young fruit glutinous', Quairading), '*H. Meissneriana*' (#71 and #84, Westonia); and *H. commutata* (#208, York).

Wilson wrote in his diary on 24 October 1920 that after breakfast in Merredin they travelled the 70 miles to Bruce Rock:

> From Merredin we journeyed south-west through flat scrub-clad country with here & there agricultural clearings. At Bruce Rock where is much farms & sown fields were [ ] [intended] to yield 18 bushels to the acre. The rainfall is from 12-14 inches so such crops are [remarkably] good. With water everything is possible here.

At Bruce Rock, Wilson made another comment about his accommodation, noting the 'very excellent hotel' and tariff. They stayed the night with comfort enough for Wilson to write quite a lot in his diary. There was a near full moon that night. Did the party look at the moon over the white salt lakes, rising in the magnificent way it does in this landscape, all pink, lilac, and thick cream?

FIGURE 4.7: Collected at Yoting was the woody pear, *Xylomelum angustifolium*, a Proteaceae known as the Sandplain woody pear, which is often a small and twisted tree. Wilson wrote of the woody pear in *Plant Hunting* (page 121): 'An ugly tree but with handsomely figured wood is Xylomelum occidentale with an inedible pear-like fruit from which the name Wooden Pear [*sic*] is derived'. The seeds from the fruits were roasted and eaten by Noongar.[48] Wilson notes here on the label: 'Bush 12 ft', 25 October 1920.

To follow Wilson's site visits across this region, my friends Pauline and James Scott from Kellerberrin took me to Totadgin Rock, south of Merredin (Figure 4.8). This granite outcrop is properly called an inselberg. Totadgin is today a square nature reserve surrounded by land stripped of most native vegetation,

The rock and its immediate surrounds are wonderful woodland and heathland. Extraordinarily, its top has deep canyons and mounds, and is not a smooth surface, with many gnamma holes (often quite shallow water storage areas in the rock) and deep crevasses. At Totadgin, Wilson collected the very pretty mallee *E. erythronema* (3 sheets, no numbers given, 23 October 1920), noting 'Bush 6ft flo red "Mallee" common around Totadgin' (Figure 4.9), and also collected *Acacia, Melaleuca, Daviesia,* and *Kunzea*.

FIGURE 4.8: A gnamma hole at Totadgin Rock and the view beyond. Totadgin Rock, 11 kilometres (7 miles) south of Merredin in Totadgin Conservation Park, is a granite inselberg that rises above the surrounding plain and was visited by Wilson. Most inselbergs have unique granite-loving vegetation. In Western Australia, many rocks are undercut by chemical action millions of years ago and appear with wave-like forms, as at Totadgin. One of Hunt's wells is at the base. When Hunt's men built it in 1866, they also created a garden (now lost) for use against scurvy, a problem of long periods of exploration in the bush by European explorers at that time, just as if on an ocean voyage.[49] Image: M Grose 1324.

FIGURE 4.9: The 'Mallee' *Eucalyptus erythronema* collected at Totadgin Rock (no number given, Wilson, 23 October 1920). Wilson notes: 'bush 6 ft. flo red'

Thickets of *Hakea* trees abound about the base of Totadgin Rock. Common is *H. recurva*, a bush or small tree known as jarnockmert. Many of the hakeas bear names that refer to needles and pins, making it clear what sort of plants these are – the sharp points are on the leaves, and they prick you. *Hakea* is a major genus in Australia and in the family Proteaceae. Hakea has been identified as the world's most sclerophyllous genus,[50] with sclerophylly thought to be related to the impoverished soils of this entire region.[51] *H. recurva* is armed with tough, prickly leaves, and has dry, open seed capsules that often take years to open. They are small, a little over 2 centimetres, and when open they are shaped like mini avocados; the expelled seed has left a little shape, just as in a halved avocado with its seed removed. Aboriginal people used the fruits for decoration; there is a wonderful form to them.

This hakea has a lignotuber, from which it resprouts, and the roots can be a source of water in this dry environment. Aboriginal people would dig up the roots and put one end in a slow fire, and water would be forced out the other end. An early description of water trees (which may be one of several species) was given in 1894 by Mr Brooks, a settler:[52]

> Of the four kinds of water found in the river-less interior I need not describe 'bring gabby' (soakages), or 'karo gabby' (rockholes), but pass on to 'womar gabby'. Let those in need of water, while travelling through a forest, keep a lookout for trees with three or more large branches springing from one butt; should there be a dead stick protruding from the fork, pull it out, and nine times out of ten water will be found within; to reach it, roll a thin piece of bark into a tube and suck the water through it. The aborigines are adepts in the art, and in their natural state carry a hollow bone stuck through the nose for the purpose. After drinking replace the stick, not only to keep away dogs, emus, etc., but to prevent impurities from accumulating in the receptacle. I have found numbers of trees containing from ten to fifteen gallons, though more frequently only a few pints. Many of the trees have never been opened; by tapping them with the finger a native knows at once if they are good or not; if it gives a hollow sound cut a small hole in the fork, having attention to drainage, and generally water will be found. Every forked tree has not a cavity; every cavity will not hold water; trees which have held water perfectly for years will suddenly cease to do so. When the rock holes and 'womar' water are exhausted as they always are after a long spell of dry weather, the natives fall back on 'cooran gabby,' the best of all, obtained from the roots of the tree.[53]

Wilson would have no doubt been interested in their dagger-like foliage and beauty. He wrote that 'the variety of flowers was astonishing. Superficially they all look alike but in blossom the variety is manifest. They are mostly prickly in character & many of them especially so.'[54]

It is a fascinating thing to ponder why a plant leaf might need to be so sharp. Many Australian shrubs exhibit this feature, and they tend to be harsher than their cousins in parallel vegetation communities in southern Africa (see Figure 4.10).[55] It has long been asked if these daggers, prickles, needles, pins, spines, and thorns are a response to previous and now-extinct megafauna,[56] such as *Diprotodon optatum*, a massive wombat-like marsupial megaherbivore that inhabited Australia until about 60,000 years ago.[57] Diprotodon, the first of the megafauna (over 44 kilograms) of Australia to be identified, by Richard Owens FRS in 1838,[58] was a browser of leaves, whereas most marsupials and most herbivores in Australia today are principally grazers.[59] Thus ancient leaves needed more protection than today and the prickly character persists. Like large modern mammals such as those still found in southern Africa, Diprotodon migrated seasonally – as did mammoths in the Northern Hemisphere[60] – moving from Australia into what was the Sahul (Pleistocene Australia–New Guinea) and back when sea levels were low.[61] Many of the megafauna existed until as recently as 40–45,000 years ago.[62] Though their extinction is more likely due to environmental changes, as humans were not present at most sites,[63] some were recorded by Aboriginal artists in large, naturalistic paintings.[64]

Wilson thought that many of the plants he found in Western Australia were so strange and different that they suited the world of dinosaurs – monsters of 'the remote Lizard Age'.[65] Defences against megafauna and likely dinosaurs seem to have developed with many Australian shrubs, such as needlewood, that did not grow tall enough to avoid the attention of these big herbivores. Dinosaurs inhabited Gondwana when it was still in its entirety, and as it was breaking up. Many endured low, polar latitudes as 'dinosaurs of darkness' 120–100 million years ago.[66] It is well to recall that plants evolved alongside such animals when Australia was still far south and attached to Antarctica.

At Totadgin and other preserved bushland granite areas, the mind can move freely back in time and be in any age. It is easy to imagine Hunt's party digging a well just over there, dusty and hot, or finding one of the Old People, the pre-colonial Aboriginal population, collecting bush tucker, or sighting great megafauna, such as the giant 2.5 metre (8 foot) kangaroo that became extinct about 10,000 years ago. All this time, the trees were watching over these changes. But on the agricultural fields, it is only the present that one senses easily, and it is harder to conjure up the old landscape and the soil that is now missing the life it once knew. When one does imagine the old landscape, it is trees that are seen as ghosts of sorrow on the land.

Moving east, Wilson saw the Number 1 Rabbit Proof Fence – 'wire net … 4 ft' – near the little town of Burracoppin, where they had lunch. Burracoppin had been the depot for the boundary riders who maintained the fence. Here, Wilson collected the distinctive pear-fruited mallee (or Dowerin rose) *E. pyri-*

FIGURE 4.10: A grevillea collected by Wilson at Merredin, 23 October 1920, showing the very prickly nature of many of these plants. This specimen was classified by Moore in 1961 as *Grevillea paradoxa*. Image: #54, Wilson collection, HH.

formis, *Persoonia*, *Adenanthos*, sandalwood (*Santalum spicatum*) and gimlet, as well as a 90 centimetre (3 foot) box poison, *Oxylobium parviflorum*, now known as *Gastrolobium parviflorum*.

Further north and east near Westonia, Wilson collected a large suite of species under Herbert's guidance. At the time of Wilson's visit, Westonia was noted as a mining town, where gold was extracted at the Edna May Mine. The many species collected included a ten-foot (over 3-metre) tall *Duboisia hopwoodii* 'Pituri', which has narcotic properties, *Acacia*, *Eremophila*, *Diplolaena*, *Verticordia*, *Melaleuca*, and *Pityrodia*. He wrote in his diary that evening:

> Westonia is a gold-mining camp—a scattered neat village of corrugated iron houses & shops including two Hotels. We stay in the Edna-May which is named for the goldmine. Things are no longer brisk here & the mine seems to be [panning] out. Saturday night in a mining camp is not a quiet place. Singing & chatter extended far into the night. Hotel people very civil & obliging.[67]

Westonia today is a charming little town with a historic streetscape and a modern 'donga' mining settlement at the entrance.[68] The town has had an inventive vision that celebrates its history and landscape and the beauty of the woodland all about; it is really a showcase of how to preserve history while modernising. Few little towns have been so successful, and good design is evident.

North of Westonia is Sandford Rock, another massive inselberg (Figure 4.11). This is an extraordinary place, with a sense of stepping back in time thousands of years and reminding us of our short span on this earth. This 800 hectares (1977 acres) of rock, with its wind and striking hollows, would have been very remote in Wilson's day. Part of the Yilgarn Craton, the rock is 2,600 million years old. The glorious grey-silver-leaved tree *Eucalyptus crucis* subsp. *crucis*, the Southern Cross silver mallee,[69] grows scattered at the base of the rock, no doubt benefiting from run-off water. *Callitris* also abounds.

One hopes that Wilson had time to feel and simply sit among these ancient outcrops. At Sandford, we had become very quiet and moved about solo, as if the Rock instils silence through its own spirit. While my friend Pauline climbed, with the shadows of cloud changing the colours of the rock, I wandered, and James sat, perhaps pondering the landscape that he had lived in all his life. The silence was eroded by the wind in sheoaks. Three emus were startled by seeing a human being, and their indignant toss of their heads seemed to ask, 'What are you doing here?' Caves suggested old occupation, old songs.

FIGURE 4.11: Sandford Rock, north of Westonia. The figure top centre shows the size of just part of this large inselberg. Image: M Grose 1327.

FIGURE 4.12: Silver-leaved *E. crucis*, Southern Cross silver mallee (centre back), at Sandford Rock, north of Westonia, showing the complex vegetation assemblage, with other species of eucalypts in the distance. Image: M Grose 285.

As we made our way back to Kellerberrin from Sandford Rock Nature Reserve in 2022, cutting north, west, north, and eternally west, there were vast acreages of farms of 15–20,000 hectares (37,000 to around 50,000 acres). This landscape has been nearly totally cleared, with rolling plains where trees are found only along the edges of enormous paddocks and as thin strips along road verges, odd creek lines, or as single and childless sentinels in the middle of a paddock, where they must endure isolation and emptiness. No recruitment of young trees in the paddocks was seen. Large old trees left in paddocks have a desolate fate. Without offspring, surrounded by crops, endangered by spraying drift and the force of heavy machinery about them, they will not survive. Their fate is sealed. Yet many were saved by the original farmer because they were beautiful, or the oldest, or a grand specimen. Where are such old trees to be found in future centuries, when these old survivors are gone and there are no youth next to them to replace them?

We call woodland a community of species for a reason. Trees left in paddocks appear to be dying of loneliness, with their roots never finding another woody friend or their microbial associates in that complex three-dimensional world beneath the surface. In the endless sea of cereals or lupins that come and go with the passage of the year, there are no stayers to grow old with them or young trees for old trees to oversee, as we now know that many old mother trees do. Wilson had been appalled by what he saw one hundred years ago, yet vastly more clearing has been done since his journey.

On my visit, some gravel roads had been closed due to rain. One very wooded road was closed permanently; this would provide a wonderful opportunity for a revegetation corridor but might instead be taken up by one of the adjoining farms for yet more cropping and fewer trees. However, one can hope that different attitudes will prevail, and such spots will be tended in future for native vegetation, to assist in drought-proofing Country, and for biodiversity and wildlife corridors. Perhaps, as Wilson noted before he left Australia, many do not realise what they have in terms of beauty and rarity. Hidden gems abound in this landscape. At Baladjie Rock, another wonderful inselberg of the undulating plains, the setting sun turned the entire rock a glowing orange and a nearby salt lake turned pink in the gathering dusk. In the silence, faint voices of a few campers could be heard. That evening Pauline served a delicious acacia ice-cream made from *Acacia micro-*

botrya, manna wattle, which she and James grow on their farm to harvest seed for bush tucker cuisine. That evening too, I found *Eucalyptus crucis*, the Southern Cross silver mallee that had been at Totadgin, in the book *Eucalypts of Western Australia's Wheatbelt*, by Malcolm French. He described *E. crucis* as a large and striking tree that stands out in the landscape, with grey-leaved foliage and strong branches up to 9 metres (30 feet) tall, 'often contrasting spectacularly against the brown background of its granite rock habitat or against the glossy, bright green leaves of its common associate, *E. loxophleba*',[70] the smooth-barked York gum, and other plants. This glaucous foliage has led to it being widely planted as an ornamental and used in floral arrangements.[71] This species would be suitable for California.

Malcolm, the world authority on Western Australian Wheatbelt eucalypts, lives near Kellerberrin. The next morning, a cold, wet winter's day on the shortest day of the year, 21 June, Malcolm gave me a tour of his farm at Mount Caroline, south of Tammin and Kellerberrin and close to the 600 hectare (around 1500 acre) Charles Gardner Reserve, which is named after one of Wilson's travelling companions in the Goldfields later in his tour. Malcolm was astounded that someone from Harvard had come through this region in 1920, looking at trees and other flora. Originally from Victoria, he had purchased his farm because of a wonderful subspecies of *E. caesia* found at Mount Caroline. Wilson would have loved to see it because it would be an excellent species for Californian gardens and streets. 'Caesia' is a popular garden tree now, but the subspecies at Mount Caroline outshines the common one (Figure 4.13).

Malcolm also pointed out the salt – seeping salt, creeping salt, rising salt, moving and extending salt. Wilson was not aware of salt in 1920 when he visited, but had a feeling of foreboding. He could see that something was already amiss. He was struck by the extensive clearing of the woodland for wheat and sheep. It was while in Bruce Rock, south of Kellerberrin, that Wilson wrote his significant comment 'There are no good trees left', and 'I fear that someday the price will be exacted.' His handwriting is exceptionally poor here, even for Wilson, and I did wonder if his hand quivered with anger at what he saw as he travelled through this region.

The inveterate explorer and government surveyor John Septimus Roe had been the first European to see much of this region in his many expeditions that criss-crossed the south-west of Australia between 1829 and 1849. He had noted

FIGURE 4.13: *Eucalyptus caesia* subsp. *caesia*, type locality Mt Caroline at Mount Caroline, near Kellerberrin, in heavy rain. This form survived the extensive local extinction from extreme clearing for agriculture in this district. Images: M Grose.

the extensive naturally saline conditions of this land, the suites of salt lakes and saline claypans, brackish water, and the proximity to good soils 'of so much salt', salt lakes, and saline water.[72] Such nuances were lost on subsequent settlers, who flocked towards 'good soils' not long after Roe's comments.

When European farming began in this landscape, the idea was to clear to 'improve' the land. Warnings had been given that this was a different type of land than previously met by European settlers, and that with the loss of native vegetation, the water table would rise. In some countries, this would not be a calamity and might even be welcomed; you might be rewarded with fresh water. However, Western Australia has an ancient hydrological system, and salt has been locked away deep in the soil profile.

There are important differences between native vegetation and a crop such as wheat or barley, for which the trees of this land were felled. Native vegetation is physiologically active all summer, and most species are deep-rooted, extracting water well, with a strong requirement in summer when rain is scarce and temperatures in the Wheatbelt region are in the high 30s (degrees Celsius) to low-to-mid 40s. Before clearing, native vegetation used most of the rainfall intercepted in the soil, including using in summer some of the water reserves held in groundwater. Native vegetation is long-lived and draws up water from underground all year. Indeed, eucalypts extract groundwater vigorously, and are known to steal water from wells. After clearing, run-off increased and the groundwater recharged, rising and bringing ancient saline water to the surface and into creeks and rivers.

During the 1910s, scientists had already been concerned at the hydrological changes wrought by agriculture and widespread land clearing in Western Australia. Though there was evidence of the very different physical and chemical features of this ancient landscape as compared to anywhere else one hundred years ago, and some voiced this, these earlier warnings about excessive land clearing went unheeded. In 1917, just a few years before Wilson's visit, Robert Bleazby wrote in the *Proceedings of the Institute of Civil Engineers, London*, that 'settlers in many districts found that when the land was cleared and cultivated, tanks which they had excavated while the land was still bush became too salt for stock to drink. At this stage it became evident that salt in the soil needed careful investigation', and he noted the way that clearing had completely changed the hydrology of the land and that there was a need to 'reserve the catchment to protect it from becoming salt'.[73] By the 1920s, the warnings were there and salt's ancient curse, though yet to manifest, was largely foreseen by contemporary agricultural scientists.[74]

Debate and caution about land clearing ran against the desire to expand agricultural land in a region that gave so much promise of being a major global wheat and sheep area, upon which so much of the Australian economy came increasingly to depend through the twentieth century. In the years after World War I and Wilson's visit, Australia exported 92% of its wool, with about 50% going to the UK's textile industries. In 1924, wool accounted for 78% of Australia's pastoral wealth.[75] By 1928, Australia had nearly 20% of the world's sheep population and produced 25% of the wool supply. Millions of sheep earlier in the twentieth century contributed to the ecological damage done to fragile Country. Unlike other lands, Australia has no native hoofed animals, and soils were broken by the impact of hard hoofs that were far different to the soft pads of kangaroos and other native grazers.

Wilson might have been informed by Lane-Poole of the concerns of scientists about the extent of clearing in the Wheatbelt. Lane-Poole had himself cautioned about land clearing, noting publicly and to the government the 'wasteful and destructive cutting'[76] of forests and woodlands. Lane-Poole was at that time in a titanic struggle with the Government of Western Australia to protect and conserve woodlands. As Conservator of Forests, he took the word conserve to be a major part of his job, while the government appeared to see his role as increasing the production of timber for the wealth of the State.[77] Lane-Poole saw forest and woodland destruction being enacted before policy could be established to protect trees. In October 1921, not long after Wilson's visit, Lane-Poole resigned and took up a post in Papua New Guinea, where he was based through 1922–4; tellingly, during his time in Papua New Guinea, his focus was forest ecology, not timber production.[78]

Despite warnings from knowledgeable voices, clearing of the woodlands in Western Australia continued until the early 1980s, supported by government policies and widespread poor cultural attitudes to the environment.[79] The policy of clearing was pushed by the Western Australian Government; it was a condition of purchase that the purchase-owner was compelled under legislation to ringbark the timber and use the land exposed for cultivation.[80] Clearing of the land in this region was supported until the early 1980s by the State Government of Western Australia, with severe penalties for not clearing most of a block.[81] In addition, the feeling of the day, supported by the majority of the public and the press, was that great work was being done creating a new nation, stabilising food supplies, laying down lines of settlement, and creating great wealth for the country through the expansion of wheat and wool. They were operating under the ideas that they knew. Ignoring the worries and warnings of many farmers, soil scientists, and foresters, clearing went on and was typical across Australia.[82]

Despite warnings, salinisation caught most European settlers by surprise, because they did not understand this landscape, and cleared trees without mercy.[83] As Wilson had predicted, nature would exact a penalty; 8% of the Wheatbelt is now saline and unable to support either cereal cropping or sheep,[84] or even the prior rich native vegetation, and is expected to become much more extensive.[85] It is an environmental catastrophe on the scale of the Dust Bowl of the United States.

By the 1980s, a 'salinity crisis' was declared by the government, the press, and scientists. In the Wheatbelt, 'an area the size of a small European nation is being poisoned by salt'.[86] Sixty years earlier, in 1920, Wilson had been disturbed by what he saw, and had said so – in his diary, to the newspaper men, and to his hosts. Wilson said plainly to the press:

> There are not laws to prevent wholesale destruction of the trees; the laws are as bad as in some parts of America – they could not be worse. The man who destroys every tree on his place is unwise, purely from the point of view of his own interests; he is doing himself an injury. How long is this sort of thing going to go on?[87]

Of importance is that Wilson clearly saw that the problem was the government's policy of development at all costs for the ideas of advancement and expansion, and an indifference to caution or the findings and warnings from scientists about the rise of salt in the profiles of these low-fertility areas. This devaluing of science, critical to the spread of salt in the Wheatbelt, must have irked a man such as Wilson, who admired the knowledge and toil of botanical exploration and scientific inquiry.

Newspaper men noted to Wilson that the Lands Department was responsible for insisting on the removal of trees, with land cut of all timber going from a pre-cut value of 2 shillings and sixpence per acre to 5 shillings an acre – it was thus 'improved'.

What Wilson saw was just the beginning of a radical erasure of the woodlands of this now-major agricultural region. He made prescient comments about land clearing, and was clearly dismayed at the foolishness of these actions and attitudes. He witnessed the misunderstanding of the real, deeper landscape, a landscape seemingly tough and harsh, but in reality, fragile, ancient, finely balanced, complex, and susceptible to damage. If Wilson were to return and see it now, to find that clearing went on for another sixty years after his visit in 'one of the heaviest assaults on virgin bush in Australia', what would he think?[88] It is easy to guess his response.

Wilson later commented:

> Western Australia has largely cleared out her big timber areas, and the result is that water which was sweet has become brackish. Australia has subscribed large sums to relieve the farmers and stricken people in China, but Australians do not realise that much of the famine is due to the way in which timbered areas were cleared out. Water sheds were denuded, and when there was excessive rainfall away went the land. I don't know if the Chinese are to be blamed as much as we are. The Chinese took thousands of years to clear their trees. America has done it in about a century, and in Australia they had done it in less than that time.[89]

Wilson related the story told to him in Western Australia by an alarmed but 'witty' farmer that 'the object' of clearing is 'of making one blade of grass grow where there used to be two trees'.[90] Wilson was so struck by this comment that nearly ten years later he repeated the theme in his book *Aristocrats of the Trees*, where he wrote about the waste of trees carried out by the colonising nations – 'wherever he has gone he has laid waste the tree wealth of the lands in an effort, often in vain, to make a blade of corn grow where two trees grew before.'[91] Wilson might or might not have been aware that the farmer had been paraphrasing Jonathan Swift, in his 1726 satire *Gulliver's Travels*.[92]

The zoologist Barbara York Main (1929–2019), who grew up in the Wheatbelt near the town of Tammin, later wrote powerfully about erosion and loss of good soils in her book on the natural history of the region, *Between Wodjil and Tor*.[93] She wrote poignantly, in 1967, that:

> Little more than a generation ago this creek passed the whole of its route through virgin bushland and it flood course was impeded by trees and litter and its wide flow, rather than scouring the ground, had an aggrading effect, depositing silt and debris as it flowed slowly past. But now for the most part its flood spills out over bare paddocks or stock-grazed timber country, where it gathers surface soil into its stream, thereby eroding and guttering.[94]

Wilson again felt a foreboding of bad times to come. Even though he was writing before the infamy of the Dust Bowl of the Colorado–Kansas–Oklahoma Great Plains area of the United States in the early 1930s, when overcultivation and poor land management of sensitive soils led to vast amounts of topsoil loss, Wilson saw the clearing as worryingly excessive, with little vegetation left. He noted that when degradation occurs, it often occurs as a creeping loss, a slow departure of soil health. His prime concern was loss of topsoil, an effect York Main later lamented. He chatted to the agricultural editor for the *Western Mail*, who reported his comments as 'things of no small interest to readers'. When asked by the agricultural editor what he thought about the timbers of the Wheatbelt trees, Wilson went on: 'My special interest in the trees has been to observe the timber trees of wide renown. The [ ] gum, gimlet etc are not known commercially and are only small in comparison with the other trees you have got – you have such giants here'. He added: 'wandoo and York gum are splendid woods'. Wilson photographed York gum, *Eucalyptus loxophlobia*, which has a dark trunk (Figure 4.14).

Wilson asked:

> In regard to the Wheatbelt area, do you think that the method of clearing is wise, where the farmers are destroying the trees over miles of country? First and foremost, they are letting in the wind, which carries off the moisture, which they want to conserve [ ] also carries off some of the soil, which they want to conserve. It is very unwise to clear large areas absolutely.[95]

The agricultural editor of the *Mail* wrote in November 1920: 'It is to be hoped that settlers will take heed of his warning in regard to ruthless destruction of trees.'[96] But rich vegetation like that seen by Wilson at Totadgin is found today as scattered remnants, and is a sad indication of loss and the rampant denaturation of the wider region, which is now a land of patches of trees, with long distances between treed patches for species to move across agricultural paddocks.

Efforts are now being made to ameliorate the damage due to tree loss, largely through replanting native vegetation. Salt, and evidence of creeping salt, needs to be dealt with early. Some farmers who cleared land decades ago are now realising that they got it wrong and were advised badly, and are revegetating and lightly stocking, with greatly reduced sheep numbers, if any. These farmers are looking forward, and are thinking of how to ensure a positive future by using different practices and a total-ecosystem approach to farming. In many ways, it is remarkable that after bringing tried-and-true farming practices from Europe to Australia, in scarcely three generations lessons have been learnt that these ideas and presumptions were wrong and do not work in this part of the planet. New approaches are being worked on to deal with a very different environment, where

FIGURE 4.14: York gum, *Eucalyptus loxophleba*, with Desmond Herbert in the Wheatbelt, 1920. Though the land is much cleared, one can see in this image many trees on the horizon. This view is now largely gone due to clearing.

trees can assist the farmer in keeping salt at bay, preserving soil, and subduing wind erosion of fine-particled soils.

One of the prices to be exacted is replanting with native vegetation. A key to ameliorating salt today is to plant to maximise water extraction by trees, to lower the groundwater that brings salt to the surface. Australia might be in a good position to return 'much of its native ecosystems to more functionally resilient states given its relative prosperity and the momentum of change that has gripped the country in the last few decades'.[97] One of the concerns of new planting is to prepare for a predicted drying climate. Initiatives are being carried out at various scales.

As an example of the small scale of the individual repair on a farm, a former ram paddock on the way to Yoting is now planted with trees. The very size of modern machinery is, perhaps curiously and ironically, assisting in the replanting of some Wheatbelt areas, though their very size has done great damage by removing large trees. Modern machines, now much larger and heavier than the old horse-drawn ploughs and seeders, have wide turning circles, meaning that many machines cannot get into some areas in the paddock that were previously accessible to early clearing. Where the big machinery cannot get, farmers are abandoning cropping or grazing. This means that odd corners, areas between granite mounds, granite sheets, and granite grouped like standing stones that are scattered in paddocks, could provide small opportunities for revegetation, increasing, if the opportunity is taken, the patchiness and connectivity in landscapes farm by farm. In addition, farmers tend to deep-rip the land today, and to prevent damage to their machinery, they are abandoning cropping in those areas where granite is close to the surface. These points show the mix between pragmatism and the re-creation of habitat. It is by needs a balance between ecology and production.

Wise farmers have been replanting with native vegetation. Expert in Wheatbelt eucalypts, Malcolm French, pointed out the land of a neighbouring elderly farmer who had originally cleared most of his land because it was the thing to do and was expected of a farmer. He is now replanting extensively and planting against salt. While we might now criticise the way advice was not heeded, or the short-term thinking involved, no sensible farmer wants to injure their land or disable it for their own children or grandchildren, though this is what has occurred.

However, some are not changing their old ways. Many of the bigger farmers are making money and have a good lifestyle that they do not wish to alter, or see no reason to alter. But as Wilson noted one hundred years ago, they are mining their soil, removing its wealth of moisture and biota that took thousands of years to develop.

Wilson would find today that his words of one century ago are still pertinent. Carbon farming is, perhaps surprisingly, an example of a new initiative that sounds green but echoes past statements and beliefs that Wilson criticised. Carbon farming aims to increase the amount of carbon stored in the soil and in vegetation. One programme that I came across focuses on reforestation by environmental or mallee plantings. Yet for this initiative to be adopted, the land must have been cleared of all native vegetation for at least five years prior before it can be replanted to 'improve soil health' and 'to improve the productivity and sustainability of their land'.[98] This is still the old claim for improving the country via its destruction.[99] Native bushland saved by a farmer cannot take part in this scheme – thus the farmer who did not clear land is penalised. Farrer's unlearning – that the traditional practices of the old country have to be unlearnt – is still required. Some good people are trying to change the ideas and rules so that unlearning's wisdom can prevail.

Now, many who live in the Wheatbelt are replanting, filling in, putting in lines of trees, revegetating creek lines, assessing the large old trees left behind without children, working with salt-tolerant, native trees and shrubs, translocating native animals and removing feral predators who hunt them.[100] With it come the near-impossible tasks – all requiring time – to remake and kickstart some complexity: hollows, old trees, layered vegetation, assemblages, and their inhabitants. Barbara York Main made the point that reclamation of cleared, ploughed land is slow and that there 'are no remnant stocks of mallee, melaleucas, *Petrophile* in the ground ready to sprout new growth. There are not tubers of orchids to send up seasonal blooms. There are no fibrous butts … or tuber of yams or bases of flax lilies … no seeds of spores sheltering in the soil'.[101] There has been no care of the small things, those that York Main drew attention to in *Between Wodjil and Tor* – spiders, bugs, insects, twigs, beetles, centipedes, night moths, grasshoppers – all living in tiny spaces. One of her studies included Number 16, a trapdoor spider, and the oldest known spider in the world. Number 16 was born in 1974 and was killed by a wasp in 2016; she occupied the same self-made 'round, flat, litter' door, sealed with silk, all those years in the Wheatbelt.[102]

This landscape has inspired such poetry and literary prose, but these often speak of loss and salt, and what is more usually heard about the Wheatbelt today is negative. That the Wheatbelt is ruined Country. It is destroyed. Drive through. Or fly over and don't look down on your way to Sydney, the Great Barrier Reef, or shopping in Melbourne. However, those that do this are missing out on the wide skies, the stars, the flocks of colourful birds, and the sense of distance and the depth of time. There are signs that many areas can be recovered using the trees to regenerate land, bit by bit, in determined steps, patch by patch, especially if salted land is repaired as it newly appears, as it comes creeping along. The creep can be seen, like a soil fog, a slight stunting, mocking wheat row by row.

Malcolm French believes that in another one hundred years the Wheatbelt will be a better place, with restored ecosystems and thus with a more sustainable future. It is, he feels, a case where 'we will get better as we learn'. His motto is that 'trees are our promise to the future'. Just as Wilson had noted the 'vast botanical garden', so Malcolm called this landscape 'a big garden'. Like all gardens, nurturing is needed to bring both produce and beauty. Steps are slowly being taken today too for 'working two ways' to protect Country and culture in many parts of Australia, working with Indigenous traditional owners who have detailed ancestral knowledge, and with scientific ideas from local farmers and academic specialisations.

While the focus of revegetation is always on the cleared agricultural land today, there is also a deal to be done in the towns, which are often forgotten in discussions of revegetating rural areas. Although greening regional hot areas is now a priority,[103] many Wheatbelt local governments appear to remain cautious and restrained, perhaps by budgets, or even irresolute in ambitions for greening their towns. There are many hot, tree-poor Wheatbelt towns, most like one another, with hundreds of lost opportunities for trees in streets, edges, corners, parks, and businesses. Just as earlier farmers 'improved' their land by destroying trees, there is an absence of big trees, often lost with 'improved' town roads, where existing mature trees have been torn out and with them decades of time and growth lost to the community. Many sons and daughters of this region, attracted to quite literally greener pastures, have left. Will they and others be attracted back if they can see a better, greener environment? As Wilson noted, people appeared to need education to see trees as valuable assets that increase land values, or to even see trees at all. It does not seem that we have learnt the lessons Wilson warned of one hundred years ago, when there are so many tree-poor towns today, still making rough and hot grey roads where thirty trees grew before. Echoing Wilson, we can ask: *How long is this sort of thing going to go on?*

With thoughts and concerns about excessive removal of trees in the Wheatbelt, Wilson turned back west towards the coast, and visited the old town of York, 100 kilometres (about 60 miles) east of Perth. He noted in his diary: 'York is as old as Perth & famously was the startixng point for the trek to the goldfields. It is a nice little town on the banks of the Avon which is really the Upper Swan.'[104] The York region was settled by Europeans in 1830. Wilson had a good hotel bed and an excellent stay in York at the Palace Hotel, noting in his diary that 'The Hotel a [palace] kept by T C [Snow] is a model of what such place should be. Clean well-kept with good bed & sanitary arrangements & with courteous & obliging help it is all a man can want for.'[105]

In York, Wilson was introduced to the local chemist and druggist, another Mr Sargent, who collected plants and was an authority on the botany of the York area. Wilson described Oswald Sargent in his diary:

> He is a little bearded man & evidently well acquainted with the local flora. We spent an hour with him in the evening. He is at work on a handbook of the flora of Western Australia & has been for the last 5 years. He has a private herbarium & has promised to duplicate specimens. Pretty plants.[106]

Oswald Sargent was born in Birmingham in 1880, and had emigrated to Australia with his parents in 1886.[107] His father had been a chemist and druggist in Birmingham, and started a chemist shop in York. When his father died, Oswald continued the family business. While at the University of Western Australia, he was influenced by the chemist and botanist Alexander Purdie (1859–1905), who had a deep interest in native orchids.[108] Oswald's major passion was the flora of the York region, and he was an assiduous collector who carried out extensive fieldwork. In 1928, only a few years after Wilson's visit, Oswald donated his entire collection to the Western Australian Herbarium.[109] Three orchid species are named after him, *Pterostylis sargentii*, *Thelymitra sargentii* and *Prasophyllum sargentii*.[110, 111] *Eucalyptus sargentii*, the Salt River mallett,[112] which is found almost entirely in the Wheatbelt along salty swamps, is also named after him. Like many eucalypts, it has a variable growth in the wild, being between 3 and 12 metres tall at maturity – this feature of wild seed makes them a wild card in any estimation of their final size for domestic gardens or street-tree planting.

Within only a few years of the settlement of Perth in 1829, settlement moved into this region. Pastoralists and other settlers began to find that their sheep and cattle were exhibiting signs of apparent poisoning. While exploring east of York in October 1835, John Septimus Roe, the first Surveyor-General of Western Australia,[113] came across dead cattle from a party who had gone ahead of his own. Horrified, as an explorer depending on animal transport would be, Roe wrote: 'we … had our worst fear realized … by coming on 7 fine bullocks lying dead by the road side, & 2 carts in consequence left behind.'[114]

York Road poison, *Gastrolobium calycinum* Benth., was the first plant to be understood as a poison to stock in the 1830s, and was classified by the English botanist John Lindley FRS in 1839.[115] In 1842, the English botanist James Drummond noted that a very large proportion of the plants responsible for the poisoning belonged to members of the pea family, Papilionaceae, and the genera *Oxylobium*, *Gastrolobium*, *Isotropis*, *Gompholobium*, and others,[116] with the first two genera causing most of the problems.[117] The unusual feature of the poison plants is that there are dozens of them, and they are widely found in the south-west of Australia.[118] By the end of the nineteenth century, botanists realised that Western Australia had 'the unenviable distinction of possessing more than an ordinary share of poisonous plants' and that they 'assume an importance not usually attained by noxious plants of other countries'.[119, 120]

Wilson was intrigued by these poisonous plants and collected *Gastrolobium* with Desmond Herbert in the Wheatbelt.[121] These are pretty flowering plants, with wonderful hues of oranges and reds (Figure 4.15). Dr Alexander Morrison, a medical doctor and botanist to the Department of Agriculture, noted in 1898 that 'though noxious to animal life', the poison plant *Oxylobium retusum*, the bloom poison now known as *Gastrolobium retusum*, was 'a very handsome little shrub and worthy of cultivation for ornamental purposes'.[122] Wilson was keen to find pretty plants that could be grown in the climate of southern California, where they might have been well adapted, and we can only speculate what he might have thought of these very toxic plants and the dangers that would have ensued if they had escaped from any horticultural environment overseas, or the dangers to both children and adults. These plants will make a person ill even if parts of the plant are simply used in food preparation, as discovered by several explorers in the nineteenth century. In 1866, four escaped convicts caught a fish and smoked it using sticks of a poison bush. They all became violently ill, and before night 'were all lying at death's door'. They were still 'helplessly ill' the following day, when they were relieved to be recaptured.[123]

FIGURE 4.15: *Gastrolobium melanocarpum*, box poison, a shrub that grows up to 1.5 metres (around 5 feet). The image shows the ornamental value of this group of plants, if they were not poisonous; it flowers a brilliant orange-red, and profusely. Image: With permission William Archer, esperancewildflowers.blogspot.com/2012/02/gastrolobium-melanocarpum-box-poison.html

The main poison plants[124] of the York area killed stock due to their high levels of the toxin sodium monofluoroacetate. Wilson would have seen specimens of *Gastrolobium* and *Oxylobium* with Oswald Sargent in York, because they were important and needed to be easily identified by farmers and travellers in the Avon Valley. These poison plants can be eaten by native animals such as possums, woylies, boodies, kangaroo rats, kangaroos, wallabies, and others, without impact. However, when feral or domestic dogs or cats eat a native animal that has died of other events but has fed on the poisonous plants – either leaf or seed – they are killed by the poison in the carcass, particularly if they eat the crop of birds or intestines of marsupials. The problem of the transfer of poison was identified as early as 1834, with a settler reporting the death of five of his neighbours' dogs who had fed on the carcass of a sheep that had been poisoned.[125]

The transference of poison is interesting for two reasons. First, native animals have evolved to deal with the fluoroacetate in the poison bushes,[126] and second, supporting the growth of new stands of poison plants in revegetation programmes has been suggested as a method to assist in the control of feral predators such as cats, dogs, and foxes.[127] Because of the poisonous nature of these pretty plants, and the death of stock, farmers have over the decades eradicated these species from many areas. An ambition now is to bring back the poisonous plants to assist in the control of feral cats, dogs, and foxes, with the idea that if native marsupials have fed on the poison plants, they will, if eaten, poison their feral killer, who will not live to kill native fauna again.

The entire story of these poisons shows the complexity of this old region, with native animals perfectly adapted to these poisons, and plants completely adapted to poor soils, harsh summers with long years of drought, and isolation. The complexity of these Australian landscapes growing on poor soils in many regions would continue to intrigue Wilson as he travelled eastwards.

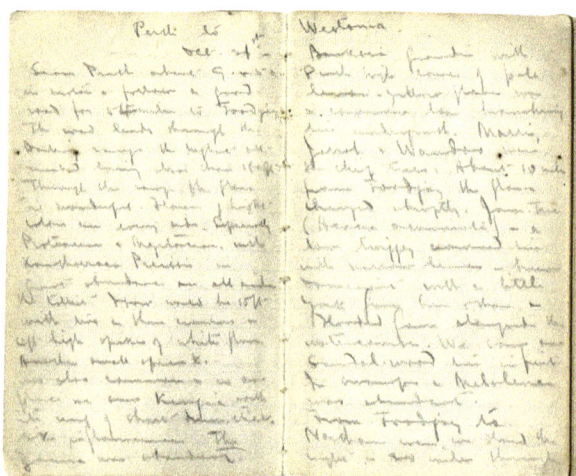

FIGURE: 4.16: 'Perth to Westonia' is the title of this section of Wilson's diary notes. These two pages are quite easy to decipher; many are not easy or are nearly impossible.

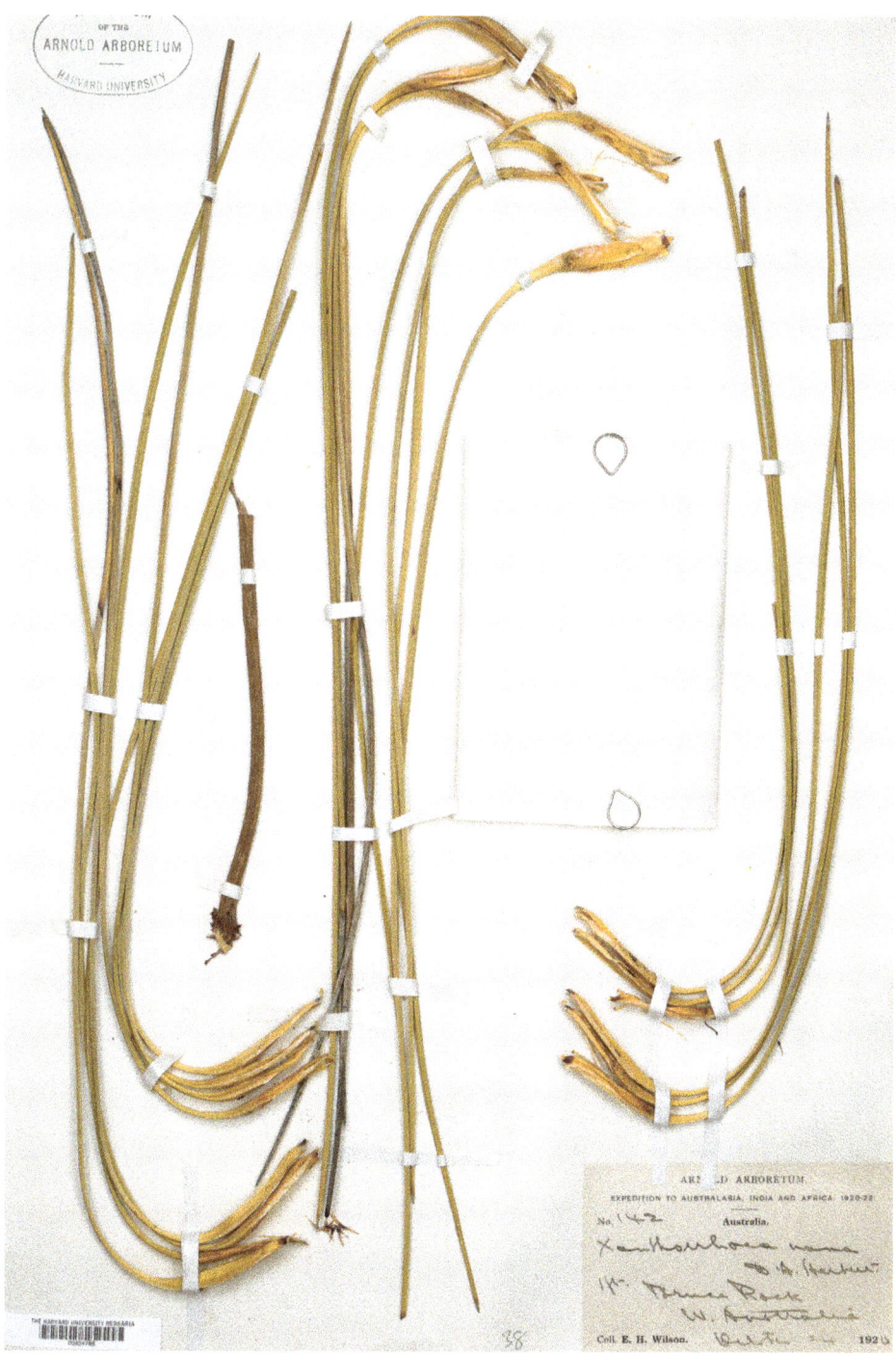

PLATE 4.1: "*Xanthorrhoea nana* D.A. Herbert 18 ft Bruce Rock W. Australia" 24 October 1920. No. 142.

PLATE 4.2: *"Banksia grandis* Tree 20 ft Yoting W. Australia" 25 October 1920. No. 197

PLATE 4.3: "Balaustion sandplains Merredin" 23 October 1920. No. 53. Classified as *Balaustion pulcherrimum* Hook. The common name is Native pomegranate.

PLATE 4.4: "Calythrix 2 ft Sand plains Yoting". 25 October 1920. No. 175 Classified as *Calytrix stringosa*.

PLATE 4.5: "Callistemon 4 ft Bruce Rock" 25 October 1920. No. 113. Classified as *Calothamnus homalophyllus*.

PLATE 4.6: "Melaleuca 5 ft Bruce Rock" 25 October 1920. No. 153. Classified as *Melaleuca glaberrima*.

PLATE 4.7: "Isopogon 11/2 ft Bruce Rock W.A." 24 October 1920. No. 147. Classified as *Petrophile circinata*.

PLATE 4.8: "Persoonia sand plains Westonia" 23 October 1920. No. 81. Classified as *Persoonia saundersiana* var. *diadena*.

PLATE 4.9: "Isopogon 1 1/2 ft flo reddish Bruce Rock W.A." 24 October 1920. No. 115. Classified as *Isopogon divergens*.

PLATE 4.10: "*Hakea platysperma* Hook 3 ft Sandplains Yoting W. Australia" 25 October 1920. No. 192. Known as Sandplain woody pear.

PLATE 4.11: "Grevillea 10 ft flo orange Sand plains Tammin W. Australia" 23 October 1920. No. 27. Classified as *Grevillea excelsior*.

PLATE 4.12: "Grevillea 4 ft Westonia" 23 October 1920. No. 78. Classified as *Grevilea teretifolia*.

PLATE 4.13: "Hakea 6 ft sand plains Westonia" 21 October 1920. No. 101. Now classified as *Grevillea pterosperma*.

PLATE 4.14: "Grevillea Merredin" 23 October 1920. No. 54. Classified as *Grevillea paradoxa*.

PLATE 4.15: "Hakea 1 ft young fruit glutinous Quairading" 25 October 1920. No. 184. Classified as *Hakea incrassata*.

PLATE 4.16: "Hakea Westonia" 23 October 1920. No. 71. Classified as *Hakea Meissneriana* [sic] from H. Moore.

PLATE 4.17: "Dryandra Jarrah forests near the Lakes east of Perth W. Australia. 27 October 1920. No. 203. Given as *Dryandra bipinnatifida* by Moore 1967; now classified as *Banksia bipinnatifida*.

PLATE 4.18: "Eriostemon Sand plains Westonia". 25 October 1920. No. 66. Classified by Moore as *Eriostemon thrytomenoides*.

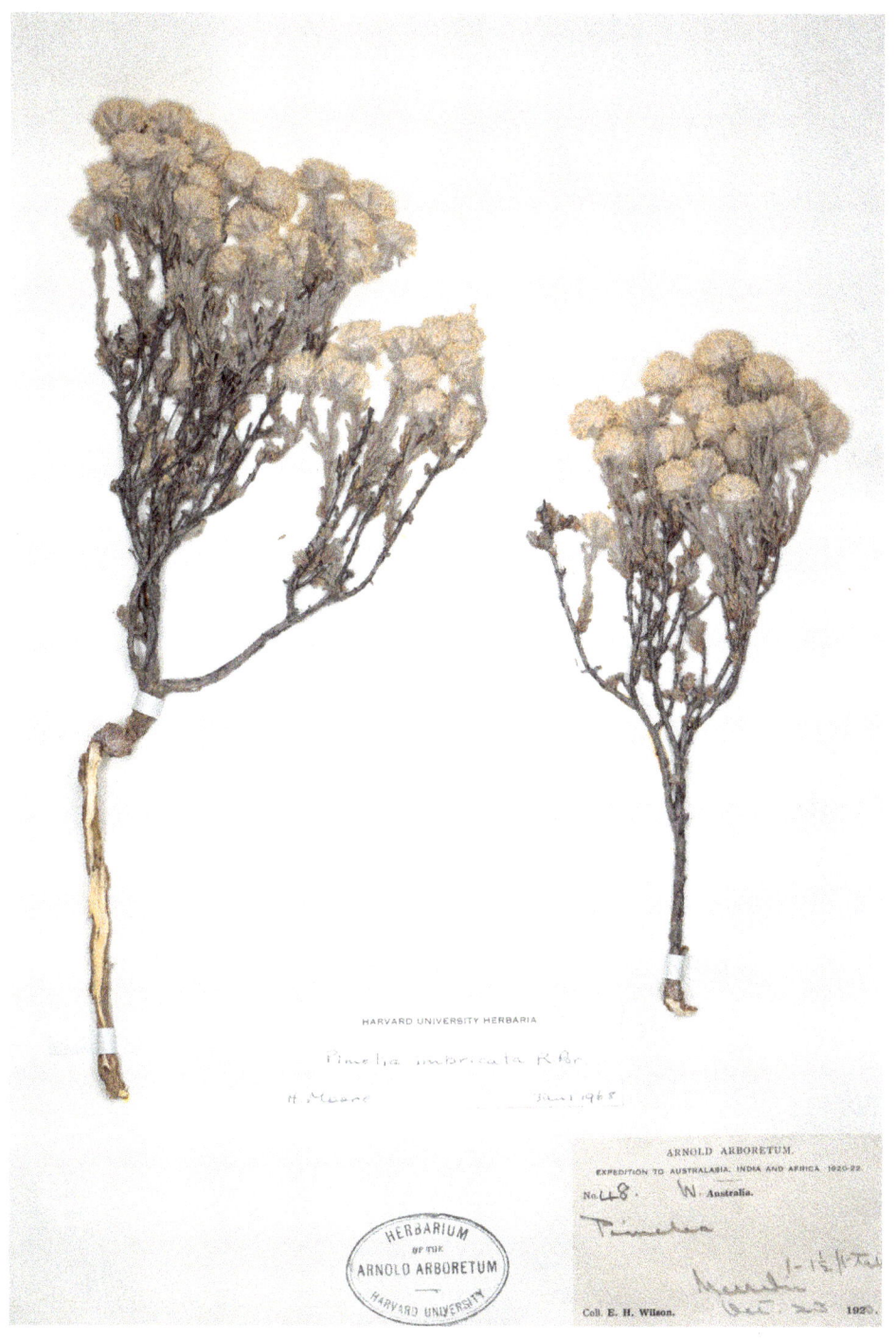

PLATE 4.19: "Pimelea 1-1½ ft tall Merredin" 23 October 1920. No. 48. Classified as *Pimelia imbricata*

PLATE 4.20: "Duboisia Hopwoodii "Pitubi" 15 ft Westonia" 23 October 1920. No.59.

PLATE 4.21: "Duboisia? 2 ft Merredin 22 October 1920". No. 122. Classified as *Anthocercis anisantha*.

NOTES

1. The history of the origin and use of the term mallee is given in R Broome, C Fahey, A Gaynor, and K Holmes, *Mallee country: Land, people, history*, Monash University Publishing, Clayton, 2020.
2. A discussion of the terminology and historical origins is also given in CJ Yates, CR Gosper, SD Hopper, DA Keith, SM Prober, and MG Tozer, 'Mallee woodlands and shrublands: The mallee, Muruk/Muert and maalok vegetation of southern Australia', pp. 570–97, in DA Keith (ed.), *Australian Vegetation*, 3rd edn, Cambridge University Press, 2017.
3. P del Tredici, 'Sprouting in temperate trees: A morphological and ecological review', *The Botanical View*, 2017, 67:121–40. Of note is that it is for temperate regions.
4. See p. 405, the Belches and Taylor expedition of 1866, and various entries for marlock, in PJ Bridge and K Epton, *Exploration Eastwards 1860-1869*, Hesperian Press, Carlisle, Western Australia, 1866.
5. Ibid., p. 403.
6. Charles Austin Gardner recommended it as a street tree in 1963: p. 371 in 'Trees of Western Australia ... 104. The Moort (*Eucalyptus platypus* Hook)' *Journal of the Department of Agriculture, Western Australia, Series 4*, 1963, 4(6).
7. Dean Nicolle, of the Currency Creek Arboretum Eucalypt Research in South Australia, has noted the problems of these definitions due to the alteration of their growth habitat when given different opportunities. See Nicolle, 'A classification and census of regenerative strategies in the eucalypts (*Angophora*, *Corymbia* and *Eucalyptus*—Myrtaceae), with special reference to the obligate seeders', *Australian Journal of Botany*, 2006, 54:391–407. However, I keep the distinctions here, as the terms are still used and would have been in Wilson's time.
8. Yates *et al*, 2017.
9. Wilson, diary, 21 October 1920.
10. Daisy Bates, *Aboriginal Perth and Bibbulmun Biographies and Legends*, (PJ Bridge, ed.), Hesperian Press 1992 (first published 1902), p. 1.
11. Wilson, diary, 24 October 1920.
12. Wilson, diary, 22 October 1920.
13. Wilson, *Notes from Australasia*, no. 1, p. 162.
14. Cited in M French, *Eucalypts of Western Australia's Wheatbelt*, 2012, p. 186. Hutchins's visit was in 1914.
15. J Dargavel, *The Zealous Conservator: A Life of Charles Lane Poole*, 2008, p. 40. Hutchins had studied at the same institution in Nantes, France as Lane-Poole did decades later, had mentored Lane-Poole, and recommended him for the position of Conservator of Forests that he took up in 1916.
16. *Western Mail*, 25 November 1920, p. 9.
17. Wilson, *Notes from Australasia*, No. 1, p. 162.
18. Doodlakine was jokingly suggested as a potential site for the national capital after the hunt for a capital site began with Australian Federation in 1901. At first, Melbourne became the temporary capital of Australia. Western Australia's first federal parliamentarians, like most, appear to have been almost unanimously against either Melbourne or Sydney becoming the permanent capital. Doodlakine (population approximately 25) was not considered, but the possibility is still celebrated in the town today, with car number plates that read 'Doodlakine: The Forgotten Capital': see MJ Grose, 'A forgotten capital', *Australian Planner*, 45(4):42–4.
19. They had gone to the Victorian Goldfields but saw greater financial opportunities in transport. Freeman Cobb's biography notes that he was twenty-three years old when he started the company. His three friends came from New Hampshire, New York, and Kansas and their average age was twenty-two. Cobb died insolvent in South Africa in 1878, long after he had sold the company in Australia. KA Austin, 'Cobb, Freeman (1830–1878)', Australian Dictionary of Biography, National Centre of Biography, Australian National University, adb.anu.edu.au/biography/cobb-freeman-3237/text4883, published first in hardcopy 1969, accessed 2 August 2021.
20. On the Western Australian Goldfields in 1897, Cobb & Co. travelled nearly 322,000 miles. 'A wonderful coaching business', *Western Australian Goldfields Courier*, 1897, 20 November 1897, p. 11.
21. Ibid.
22. The original Doodlakine was a little north of the present town.
23. Plaque recording Hunt's Well, Doodlakine, Western Australia.
24. Grose, 'A forgotten capital', 2008.
25. Now given as *S. quinqueflora*, synonym *Sarcocornia quinqueflora*, this halophyte is a member of the Chenopodiaceae, tribe Salicornieae. This species is also found in New Zealand. However, most Australian samphires are endemic.
26. B Datson, *Samphires in Western Australia: A field guide to Chenopodiaceae Tribe Salicornieae*. Department of Conservation and Land Management, Government of Western Australia, 2002.
27. 'Centipede bush' here incorrectly attributed. See Appendix 2, specimen #32.
28. Western Australia, Department of

Water; Avon Catchment Council; Western Australia, Dept. of Water, *Riparian condition of the Salt River: Waterway assessment in the zone of ancient drainage*, Department of Water, Perth, 2008.

29 A study done in 2009, 100 years after the comment in the newspaper below, gives a reading for total dissolved salts (mg/L) as 23,571: *Waterway Assessment of the Yilgarn River: Hines Hill to Lake Campion*, Report Water Resource Management series no. 56, Department of Water, Perth, 2009, p. 9..

30 See CJA Bradshaw, 'Little left to lose: Deforestation and forest degradation in Australia since European colonization', *Journal of Plant Ecology*, 2012, 5(1):109–20. Bradshaw notes that Australia has lost nearly 40% of its forests, with much that is left fragmented. Much of the rapid decline has occurred since 1950.

31 Gifford Hall, 'Settlement around Doodlakine', *Western Mail*, 21 August 1909.

32 'Doodlakine Notes', *Western Mail*, 1 February 1908, p. 19.

33 tuckerbush.com.au/gumbi-gumbi-pittosporum-angustifolium/

34 There is great interest in 'bush tucker' (foods) and bush medicines. Examples of recent publications about bush tucker are V Hansen V and J Horsfall, *Noongar Bush Medicine: Medicinal Plants of the South-West of Western Australia*, University of Western Australia Publishing, 2016; C Williams, *Bush remedies*, Rosenberg Press, 2020; D Coulthard and R Sullivan, *Warndu Mai Good Food: Introducing native Australian ingredients to your kitchen,* 2019, Hachette Press.

35 Wilson, diary, 22 October 1920.

36 W Farrer, *Agricultural Gazette, Department of Agriculture NSW*, 1900, 111:142–5. Farrer was pointing out that old counties had accumulated experience, but Australia needed to build up knowledge from experimental farms. For a discussion, see Archer Russell, *William James Farrer: A biography*, 1949, pp. 150–5.

37 Cited in A Gaynor and T Griffiths, 'Tearing down and building up. How Geoffrey Bolton's environmental history made a difference', https://insidestory.org.au/tearing-down-and-building-up/, 2017.

38 Noted in LT Evans, 'Response to challenge: William Farrer and the making of wheats. Farrer Memorial Oration 1979', *Journal of the Australian Institute of Agricultural Science*, 1980, pp. 3–13.

39 Russell, 1949, pp. 172–84.

40 The Commonwealth is made up of New South Wales, Queensland, Victoria, South Australia, Western Australia (which is the second-largest country political subdivision in the world after Yakutia in Russia), Tasmania, and the Northern Territory, with the (then) newly formed Australian Capital Territory (ACT) the seat of national government. Regarding the largest political subdivisions: Greenland is the fourth, Nunavut in Canada fifth, Queensland sixth, and Alaska the seventh; Texas is the twenty-sixth.

41 There is a story for all of us in Farrer's life. Although he received an annuity from a wealthy relative in England in the early days of his research, it was later demanded of him that he return to England or be disinherited. He chose to stay in Australia and continue his work, rejecting what might have been the comfortable life of an English gentleman for a precarious but meaningful life that resulted in him being for many decades the face on Australia's five dollar note: noted in Russell, 1949 pp. 35 and 71.

42 Wilson, diary, 22 October 1920.

43 Wilson, diary, 24 October 1920; 'flo' is flowers.

44 Wilson, diary, 24 October 1920. The green parrots are the 'Twenty-Eight-parrots', endemic to the south-west of Western Australia, given that name because their call is said to sound like the words 'twenty-eight'. They are *Barnardius zonarius semitorquatus*. The handwriting is very poor and very difficult to ascertain in many places. Where good estimates have been made of the text, I have added the word but in brackets; where the pencilled word is too difficult, I have made an empty bracket. This applies to all diary quotes.

45 Note, as in G Vaughton, 'Pollination disruption by European honeybees in the Australian bird-pollinate shrub *Grevillea barklyana* (Proteaceae)', *Plant Systematics and Evolution*, 1996, 200:89–100. The evolution of bird pollination in Australia has occurred in the absence of social bees and is a great point of difference to Wilson's 'north'.

46 Now *Grevillea hoookeriana* subsp. *apiciloba*.

47 Thomas Wilson (1792–1843) had a fascinating life. As a very young man, he had overseen the health of convicts on the long voyages to Australia and succeeded in achieving good survival of his charges by insisting on care and good hygiene, lime juice to ward off scurvy, and good behaviour. He described the soil of Fremantle as 'so sandy that it could be run through an hour-glass'. Gwendoline Wilson, 'Wilson, Thomas Braidwood (1792–1843)', Australian Dictionary of Biography, National Centre of Biography, Australian National University, adb.anu.edu.au/biography/wilson-thomas-braidwood-2806/text4007, published first in hardcopy 1967, accessed

48. Hansen and Horsfall, *Noongar Bush Medicine*, 2016, p. 216.
49. Bridge and Epton, *Exploration Eastwards 1860–1869*, 2018, p. 433.
50. BB Lamont, T He, and SL Lim Hakea, 'The world's most sclerophyllous genus, arose in southwestern Australian heathland and diversified throughout Australia over the past 12 million years', *Australian Journal of Botany*, 2016, 64(1):77–88.
51. D Perot et al., 'The greening of Western Australian landscapes: the Phanerozoic record', *Journal of the Royal Society of Western Australia* 2019, 102:52–82.
52. JP Brooks of Balbinia Station, 'Water Trees', in the Shire of Esperance Letter to the Editor, *West Australian*, 30 April 1894. The letter is dated 4 April. Many of the species mentioned by their Indigenous names in this article are eucalypts. The word 'gabby' is widely used as a word for water in south-west Western Australia by both Noongar and Yuat people to the north of Perth.
53. A recent study of the many sources of water utilised by Aboriginal people in desert regions (Bayly 1999) describes water trees in similar ways to that of Brooks in his letter to the editor a century earlier. IAE Bayly, 'Review of how indigenous people managed for water in desert regions of Australia', *Journal of the Royal Society of Western Australia*, 1999, 82:17–25.
54. Wilson, diary, 22 October 1920.
55. PJ Grubb, 'A positive distrust in simplicity—lessons from plant defences and from competition among plants and among animals', *Journal of Ecology*, 1992, 80:585–610.
56. KC Burns, 'Are there general patterns in plant defence against megaherbivores?', *Biological Journal of the Linnean Society*, 2014, 111:38–48.
57. GJ Price, KE Fitzsimmons, AD Nguyen, JX Zhao,YX Feng, IH Sobbe, H Godthelp, M Archer, and SJ Hand, 'New ages of the world's largest-ever marsupial: Diprotodon optatum from Pleistocene Australia', *Quaternary International*, 2021, 603:64–73.
58. R Owen, 'On the fossil mammals of Australia—Part III. *Diprotodon australis*', *Proceedings of the Royal Society, London*, 1870, pp. 519–78, with additional plates.
59. WR Barker, RM Barker, and L Haegi, 'Introduction to Hakea', *Flora of Australia Volume 17B Proteaceae 3. Hakea to Dryandra*, A Wilson (ed.), ABRS/CSIRO, Melbourne, 1999, p. 23.
60. MJ Wooller, C Bataille, P Druckenmiller, GM Erickson, P Groves, T Howe, J Irrgeher, D Mann, K Moon, and BA Potter, 'Lifetime mobility of an Arctic woolly mammoth', *Science*, 2021, 373(6556):806–8.
61. GJ Price, KJ Ferguson, GE Webb, Y-x Feng, P Higgins, AD Nguyen, J-x Zhao, R Joannes-Boyau, and J Louys, 'Seasonal migration of marsupial megafauna in Pleistocene Sahul (Australia–New Guinea)', *Proc. R. Soc. B*, 2017, 284:20170785.
62. B David et al., 'Late survival of megafauna refuted for Cloggs Cave, SE Australia: Implications for the Australian Late Pleistocene Megafauna extinction debate', *Quaternary Science Reviews*, 2021, p. 253.
63. SA Hocknull, R Lewis, LJ Arnold et al., 'Extinction of eastern Sahul megafauna coincides with sustained environmental deterioration', *Nature Communications*, 2020, 11:2250.
64. PSC Tacon and S Webb, 'Art and megafauna in the Top End of the Northern Territory, Australia: Illusion or reality?', in B David, P Taçon, J-J Delannoy, and J-M Geneste (eds), *The archaeology of rock art in Western Arnhem Land, Australia*, ANU Press, 2017, pp. 145–61.
65. E Wilson, *Plant hunting* volume 1, 1927, p. 113.
66. See TH Rich and P Vickers-Rich, *Dinosaurs of Darkness: In Search of the Lost Polar World*, Indiana University Press, Bloomington, 2nd edn, 2020.
67. Wilson, diary, 22 October 1920.
68. A donga is a temporary, usually transportable, dwelling and they are used extensively in mining in Australia.
69. It is named the Latin 'crucis' (cross) after the nearby town of Southern Cross, itself named after a star cluster that is a marker in the Southern Hemisphere, and noted on the flags of Australia, New Zealand, and Chile: M French, *Eucalypts of Western Australia's Wheatbelt*, 2012, p. 240.
70. French, *Eucalypts of Western Australia's Wheatbelt*, 2012, p. 240.
71. Ibid.
72. See *The Western Australian Explorations of John Septimus Roe, 1829–1849*, (edited and with an introduction by Marion Hercock), Hesperian Press, Carlisle, Perth. See, for example, p. 106.
73. R Bleazby, 'Railway water supplies in Western Australia – difficulties caused by salt in soil', *Institute of Civil Engineers London, Proceedings,* 1917, 203:394–400. Of note is his closing comment: 'Water-supplies, however, particularly in a dry country, should be designed to last for all time, and the safest method of dealing with the matter would seem to be, in the first place, to secure the catchment with the forest on it and construct the reservoir. Later, if the water remains good, and an increase of salt would not be detrimental to boilers, a small part of the catchment might be carefully cleared, and

the clearing gradually extended as far as the results of analysis of the water would permit. In this way possibly, in the course of time, the salt would be washed away without doing harm, and the supply of water would be gradually increased to fulfil the requirements of the growing population'.

74 Q Beresford, 'Developmentalism and its environmental legacy: The Western Australia Wheatbelt, 1900–1990s', *Australian Journal of Politics and History,* 2001, 47(3):403–14.

75 Australian Government Statistics Yearbook, 2000, accessed 22 March 2022. https://www.abs.gov.au/Ausstats/abs@.nsf/90a12181d877a6a6ca2568b-5007b861c/3852d05cd2263db-5ca2569de0026c588 OpenDocument.

76 Western Australian State Record Office: Acc 934/3116/1920/0704, Charles Lane Poole to Minister for Forests, 18 November 1920. Cited in Dargavel, 2008, p. 69.

77 Detailed examination of the struggle of Lane-Poole against the government at that time is given in J Dargavel *The Zealous Conservator: A Life of Charles Lane Poole,* University of Western Australia Press, 2008.

78 J Dargavel, 'From exploitation to science: Lane Poole's forest surveys of Papua and New Guinea, 1922–1924', *Historical Records of Australian Science,* 2005, 17(1):71–90.

79 Beresford, 'Developmentalism and its environmental legacy', 2001.

80 C Piesse, *WA Planning* 46, 782, dated 27 August 1913. Biologist Ian Abbott notes that this would also have 'removed habitat, den sties, and food (foliage)' for many animals. See Abbott, 'Historical perspectives of the ecology of some conspicuous vertebrate species in south-west Western Australia', *Conservation Science W. Aust.,* 2008, 6(3): 1–214, accessed August 2021, https://www.dpaw.wa.gov.au/images/documents/about/science/cswa/articles/25.pdf.

81 A summary of land clearing is given in J Squelch, 'Land clearing laws in Western Australia', *Legal Issues in Business,* 2012, 9:72–86.

82 For example, the New South Wales Government legislation of 1913 notes that improvements 'included necessary ringbarking suckering scrubbing clearing': *NSW Crown Lands Consolidation Act,* No. 7, Part VIII, Division 4, p. 149.

83 D Bradshaw, 'Conservation and land mis-management in the south-west of WA: A cautionary tale', *Australian Biologist,* 2002, 15(1):13–25.

84 PA Caccetta, J Simons, S Furby, N Wright, and R George, 'Mapping salt affected land in the South-West of Western Australia using satellite remote sensing', CSIRO Report Number EP2022-0724, CSIRO, Australia, 2022.

85 For more detail see, for example, DJ McFarlane and DR Williamson, 'An overview of water logging and salinity in southwestern Australia as related to the 'Ucarro' experimental catchment', *Agricultural Water Management,* 2002, 53(1–3):5–29.

86 Beresford Developmentalism and its environmental legacy. Beresford notes even 'open hostility' from governments about the 'detrimental effect scientists were having on their development plans', 2001.

87 *Western Mail,* 25 November 1920, p. 9, accessed 27 November 2020.

88 Shire of Jerramungup Municipal Inventory, undated. www.jerramungup.wa.gov.au/documents/166/shire-of-jerra-mungup-municipal-inventory. This records a common method of destruction of the bush: 'A 250ft length of anchor chain weighing approximately five tons was attached to two tractors which dragged it between them, pulling down the scrub in its path. Quiet a lot of the mallee was torn out by the roots.' This 1950s 'onslaught' was referred to as 'chaining'.

89 'Our timber reserves. Australia's great asset. American expert's views', *Ballarat Star,* 21 April 1921, p. 2.

90 These stories and quotes from the *Creswick Advertiser,* April 1921, and the *Ballarat Star* of the same date, p. 2.

91 Wilson, *Aristocrats of the Trees,* 1930, p. 1.

92 'Whoever could make two ears of corn, or two blades of grass grow upon a spot of ground where only one grew before, would deserve better of mankind, and do more essential service to his country, than the whole race of politicians put together.' Jonathan Swift, *Gulliver's Travels,* Part 2, Chapter 7.

93 Barbara York Main, *Between Wodjil and Tor,* Jacaranda Press, 1967. Wodjil is the Noongar word for thickets dominated by *Acacia* species, with a mixture of other small trees and shrubs that includes hakeas, grevilleas, and *Casuarina* species. York Main noted that, to the early settlers, wodjil indicated poor country; application of trace elements changed this. An old professor of mine, who also grew up in the Wheatbelt, met Barbara at a university hall of residence after World War II. Taken with her prettiness, he was pleased to be invited up to her room one evening. However, Barbara's passion was spiders, and he found her room was crammed with scores of bottles of dead and living arachnids, with some living poisonous ones in open

shoe boxes within inches of her pillow; he exited quickly. Barbara became a world authority on Australian spiders.
94 Ibid., p. 66.
95 'A study of trees: Professor Ernest H. Wilson's visit. Our wonderful flora', *Western Mail* (Perth), 25 November 1920, p. 9.
96 Ibid.
97 Corey J. A. Bradshaw, 'Little left to lose: Deforestation and forest degradation in Australia since European colonization', *Journal of Plant Ecology*, 2021, 5(1):109–20, https://doi.org/10.1093/jpe/rtr038
98 Carbon farming meeting, Kellerberrin, June 2022, cited from documents given.
99 The point about reforming carbon without harm was made by Buesseler et al., 'Removing carbon dioxide: First, do no harm', *Nature*, 28 June 2022.
100 Unfortunately, most small Australian marsupials are nocturnal, while feral predators such as cats and foxes are also nocturnal. This is a terrible conjunction.
101 York Main, *Between Wodjil and Tor*, 1967, p. 110.
102 A *Gaius villosus* trapdoor spider. Description from *Between Wodjil and Tor*, p. 31.
103 Western Australian Local Government Authority, Urban Forest Conference, Curtin University, 2023.
104 Wilson, diary, 26 October 1920.
105 Wilson, diary, 26 October 1920. The Palace Hotel, York, was well known and is still operating as the York Palace Hotel and considered one of Australia's finest period hotels. It was opened on New Year's Day 1909; the owner was Matthew Ryan, a local farmer. Its accommodation at the time, not changed when Wilson visited, was described as follows: 'The hotel has been planned and constructed on most up-to-date lines ... The accommodation throughout is admirably arranged, and sufficiently extensive to cope with the demands of the public for some time to come. The bathrooms are lofty and well ventilated, while the private sitting rooms are replete with every comfort and convenience. The well-finished bath-rooms are equipped with hot and cold water services, while the sanitary conveniences are modern and complete, the septic tank system being in vogue. ... An acetylene gas plant supplies the lighting throughout the premises.' *Eastern District Chronicle*, 30 January 1909, p. 2. Just a year after Wilson stayed, in 1921, the licensee threatened to blow up the hotel because the mortgagee had sold the freehold, and he was found by police hiding on the premises with gelignite, fuse, and detonators—all readily available in a town close to mining interests.
106 Wilson, diary, 26 October 1920. Oswald never completed his handbook on the flora of the region.
107 NG Marchant Sargent, 'Entry for Oswald Hewlett (1880–1952)', *Australian Dictionary of Biography*, vol. 11, Melbourne University Press, 1988.
108 Australian National Herbarium, 'Biographical Notes for Purdie, Alexander'. The article notes that 'Purdie gave a lecture on 'Our Native Orchids' before the Mueller Botanic Society, 24th September 1900, illustrated with lantern slides. This lecture is given as part. 8 of the *Proc. Mueller Bot. Soc. W.A.*, with blocks from his lantern slides'.
109 Ibid.
110 Noted in Karrakatta Historical Walk Trail Two, Karrakatta Cemetery, Perth, accessed April 2020. http://www.mcb.wa.gov.au/our-cemeteries/karrakatta-cemetery/historical-walk-trails/karrakatta-historical-walk-trail-two Plaque found at crematorium rose garden 7B.
111 Reported in *Journal of Botany* 59:175. Oswald later named the local orchid *Caladenia doutchiae* O.H.Sarg., collected at Datatine near Katanning, 288 kilometres south of York, by Miss L Doutch. In 1914, Oswald had been recognised for his work on orchids by the British Association for the Advancement of Science.
112 CA Gardner, 'Trees of Western Australia. 101. The Salt River Mallett (Eucalyptus sargentii Maiden). 102. The Two-Winged Gimley [sic] (Eucalyptus diptera C.). 103. Eucalyptus Burdettiana Blakely et Steedman. 104. The Moort (Eucalyptus platypus Hook)', *Journal of the Department of Agriculture, Western Australia*, 1963, series 4, 4(6), article 6. Available at https://researchlibrary.agric.wa.gov.au/journal_agriculture4/vol4/iss6/6. Note that 'gimley' seen in this title is now given as 'gimlet'.
113 John Septimus Roe (1797–1878) is a famous name in Western Australia due to his great influence through his many pivotal explorations; he was Surveyor-General from 1829–1870, an extraordinary 41 years. He was born in Berkshire, England and from the age of ten to fourteen attended the Royal Navy's mathematical and navigation school at Christ's Hospital in London. This was known as the Blue Coat school because it prepared young men as navigators for the Royal Navy, which he entered as a midshipman in 1813. During his time in the Navy, he travelled in European waters, Newfoundland, Burma, Arabia, and China, and had spent time in New South Wales and the eastern seaboard of Australia. In 1828, he took leave from the Admiralty and moved to the Swan River colony to take up the civilian position of Sur-

veyor-General. One of his many legacies was the reservation of Perth's Mount Eliza for public open space and recreation, which was gazetted as a public park in 1872, after Roe's death. He remained in Western Australia until his death. Details and further references are given in *The Western Australian explorations of John Septimus Roe 1829–1849* (edited and with an introduction by M Hercock).

114 Page 80 in *The Western Australian Explorations of John Septimus Roe*. Roe wondered about 'similar cases which had occurred in the same kind of country, where the dry scrubby nature of the animals food would not admit of its passing into the second stomach, & death has in consequence.'

115 J Lindley, 'A sketch of the vegetation of the Swan River Colony', appendix to the first twenty-three volumes of *Edwards's Botanical Register*, James Ridgeway, London, 1839.

116 'The poison plants of Western Australia', *Western Mail*, 2 December 1898, p. 6.

117 TE Aplin, *Poison plants of Western Australia: The toxic species of the genus Gastrolobium and Oxylobium*, Department of Agriculture and Food, Western Australia, Perth. 1973, bulletin 3772.

118 GT Chandler, MD Crisp, LW Cayzer, and RJ Bayer, 'Monograph of Gastrolobium (Fabaceae: Mirbelieae)', *Australian Systematic Botany*, 2002, 15:619–739.

119 'The poison plants of Western Australia', *Western Mail*, 2 December 1898, p. 6. Substance of an address delivered to the Mueller Botanic Society by Dr Morrison, botanist to the Department of Agriculture. Dr Alexander Morrison was a medical doctor and plant collector who had a large private herbarium of Australian plants, which on his death in 1913 was bequeathed to the University of Edinburgh.

120 The definitive book on the poisonous plants, *The Toxic Plants of Western Australia* (West Australian Newspapers, Periodicals Division), was written in 1956 by Charles Gardner, one of Wilson's companions in 1920, and HW Bennetts, a Doctor of Veterinary Science. There is an earlier book written on the subject: Edgar Dell and CA Gardner, *Poison Plants of South-Western Australia*, Perth, West Australia. *The Toxic Plants of Western Australia* is a treasure. The book gives details of the extensive number of poisonous plants in south-western Australia, and contains fifty-two plates of watercolours of the major poisonous plants, many done by Gardner, who was a talented artist with a great love of his subject matter. When Charles Gardner died in 1970, his original watercolours were bequeathed to the Benedictine Community, New Norcia Monastery. New Norcia is 132 kilometres (82 miles) north-east of Perth and is Australia's only monastic town. It is a gem of continuing monastic life in the Wheatbelt. Gardner had a long association with the Benedictine Community of New Norcia.

121 Seven specimens in the Western Australia Herbarium, Perth, collected in November 1920.

122 *Western Mail*, 2 December 1898, p. 6.

123 See Bridge and Epton, *Exploration Eastwards 1860–1869*, 2018, p. 495.

124 Gardner and Bennetts, *The Toxic Plants of Western Australia*, 1956.

125 DE Peacock, PE Christensen, and BD Williams, 'Historical accounts of toxicity to introduced carnivores consuming bronzewing pigeons (*Phaps chalcoptera* and *P. elegans*) and other vertebrate fauna in south-west Western Australia', *Australian Zoologist*, 2011, 35(3):826–42. This article gives lists of anecdotes and reports of poisoning of non-native animals.

126 DR King, AJ Oliver, and RJ Mead, 'The adaptation of some Western Australian mammals to food plants containing fluoroacetate', *Australian Journal of Zoology*, 1978, 26:699–712.

127 Peacock et al., 'Historical accounts of toxicity to introduced carnivores consuming bronzewing pigeons (*Phaps chalcoptera* and *P. elegans*) and other vertebrate fauna in south-west Western Australia', 2011.

CHAPTER 5

INTO SOUTHERN FORESTS TO SEE 'THE ADONIS AND HERCULES OF THE TREE WORLD'

On his return from the Wheatbelt, Wilson went south along the Indian Ocean coastal strip and into the forests of the south-west corner of Western Australia with Charles Lane-Poole and Desmond Herbert. Wilson later wrote: 'the Conservator himself was my guide through all the important forest areas.'[1] No diary entries exist for this part of his trip, which was taken in two journeys: the first along the coast and the second by train to Albany and thence to the Frankland River, 58 kilometres (36 miles) west of Denmark. Of this second trip Wilson said that 'the scenery is just about as beautiful as you could have anywhere in the world. It so impressed me that I did not want to go anywhere else in West Australia'.[2] In this chapter I have focused on the major eucalypts of the south-west that make up this landscape, and I discuss the evolution of the eucalypts because their history says so much about the differences between the Northern and Southern Hemispheres, and the 'other planet' seen by and spoken of by Wilson.

Travelling by motor near the coast along the Bussell Highway, then only a minor road, Lane-Poole took Wilson to Ludlow (Figure 5.1) to see the only pure stand of tuarts (*Eucalyptus gomphocephala*) in the world. This area is a place of extraordinary large tuarts set amid a grass-like parkland as understorey. Tuart is a Noongar word. These eucalypts are limestone lovers with grey, tessellated bark and a certain majesty. A feature of this tree is that they often branch very close to the ground, and this makes them excellent climbing trees for children. In the Ludlow Forest near Busselton, which Wilson visited, they tend to be taller and have a straight trunk before branching. The largest tuarts at Ludlow reach heights of over 33 metres (108 feet), with girths of over 10 metres (33 feet). Lane-Poole was fond of this area and championed its preservation as a reserve. He had proposed a forest school for apprentices at Ludlow and this was built in 1921, the year after Wilson's visit, but closed in 1927.[3] Despite its short life, graduate apprentices of the school became strong advocates of forest conservation, having had training in forest management and ecology, as Lane-Poole supported.

Tuart timber is a wonderful hardwood, strong and durable, yet good to work. Ludlow's large, open forest began to be logged in the 1830s, soon after the settlement of Perth. The tuart forest once stretched all along the Swan Coastal Plain, but has been largely destroyed by extensive clearing for suburban development in the twentieth and twenty-first centuries. Larger houses on smaller suburban house lots in the last few decades have also led to less room for the retention of large trees, and in-fill housing in older suburbs has eliminated many trees that survived the initial bursts of suburban development in earlier decades. Because tuart has been the tree most impacted by suburbanisation, it is now vastly reduced from its former range,[4,5] with a great loss in tree health due to changing ground-water levels from water extraction, reduced fire from changes to its understorey, insect attack, and soil microbial changes.[6] Due to Lane-Poole's foresight, the Ludlow Forest is today a treasure and is now the Tuart Forest National Park. He shared with Wilson a strong concern about the over-exploitation of forests and their loss worldwide.[7]

Genetic work on these great trees of the coastal plain has shown that tuarts shifted westward during the last Ice Age onto land now covered by the Indian Ocean, and that their centre of major genetic diversity is now lost beneath the waves.[8] Noongar stories carefully passed down for hundreds of generations tell of a time when the coastal plain extended far out into today's ocean, and the flat plain was 'heavily treed',[9] being full of banksias and eucalypts; the major tree was tuart.

FIGURE 5.1: Charles Lane-Poole, the Conservator of Forests for Western Australia 1916–22, with the great eucalypt of the coastal plain of the western coast – the limestone-lover *Eucalyptus gomphocephala* in the Ludlow Tuart Forest, near Busselton, south-west of Perth, Western Australia, when he travelled with Wilson. This woodland is noted for its open understorey. Image: Wilson Y-344 AA.

Wilson said little if anything about tuart, but it is a magnificent tree, although it might appear a little odd to non-Australian eyes. This is because it carries old dead limbs for decades.[10] These old limbs are key habitat for birds such as cockatoos, who nest in the empty cavities of the trunk. In *Smoke that Thunders*, Wilson commented that the peppermint tree, *Agonis flexuosa*, which often grow as part of the rich understorey of tuart, would be suitable for California.[11]

Wilson saw the three largest banksias of the coastal plain – *Banksia attenuata*, *B. grandis*, and *B. menziesii*, which are found with *Eucalyptus gomphocephala* in the eucalyptus–banksia woodlands along this coast. A feature of many banksias is the great colour contrast between the upper (adaxial) and lower (abaxial) surfaces. In the evolution of leaves, the physiological purpose of these two different surfaces had been established by 420 million years ago. Banksias have many different types of hard, sclerophyllous leaves, with striking inflorescences. All banksias are found in mixed woodland with eucalypts and many other species. Figure 5.2 shows a typical forest mix, as photographed by Wilson.

Besides these major trees, Wilson collected more of the other planet's botanic garden as he travelled south. Near Busselton, his collection included *Grevillea quercifolia*, '4-5ft flowers yellowish with its very banksia-like leaves' (#291 HH, Sabina River) and inland of Bunbury he found *Hakea ceratophylla*, 'a low spreading shrub up to 3ft sandplains' (#236, Waterloo) – both members of the Proteaceae family – as well as *Boronia* and *Pimelea*, both possessing very pretty flowers in season.

Western Australia's tallest forests are in the south-west corner of Australia, and the eucalypts there dominated Wilson's attention. Because the eucalypts represent the continent botanically, and are widely known, Wilson thought that 'a casual observer might remark that there are no others'.[12] They dominate all forests and woodlands in Australia and thus provide key habitat within many differing ecosystems. 'Eucalypts' is a casual nametag. Once the eucalypts were all considered part of the genus *Eucalyptus* (Myrtaceae), exclusive of *Angophora*. However, phylogenetic studies now give seven genera in the eucalypt complex – five are very minor and together make up a little more than a dozen species. The genera *Eucalyptus* and *Corymbia* dominate and make up more than 900 species.[13]

Astoundingly, eucalypts comprise up to 75% of the total plant biomass of Australia.[14] They are iconic Australians and are found almost exclusively on the continent, with only a handful of species in the eucalyptus complex ranging north into the islands of New Guinea, Celebes in Indonesia, and Mindanao in the Philippines.[15] Outside of that natural range, many species are used all over the world for shade and in plantations, where they have considerable economic importance. They are almost without exception the tallest plants in any vegetation community in Australia.[16] Wilson was to criticise Australians' lack of knowledge of eucalypts,

FIGURE 5.2: The mixed and complex understorey typical of Australia's forests. Here can be seen *Banksia grandis* (far left, with large, serrated leaves), eucalypts, and an *Allocasuarina* on the right, with a rich and mixed understorey. Image: Wilson Y-313 AA.

saying that 'it is a remarkable thing that you Australians don't know anything about your own eucalypts.'[17] He would surely be pleased if he could come back today and learn so much about these remarkable trees.

Stephen Kessell, who became Conservator of Forests in Western Australia after Lane-Poole, in 1922, gave a description of the eucalypts that is quite perfect. He wrote:

> The Eucalyptus derives its name from two Greek words which may be translated as 'well-covered' – a name applied to the little cap which protects the unopened flower, and one which aptly describes what is perhaps the leading feature of the genus. The most noticeable feature about a Eucalyptus flower is the absence of both sepals and petals, and the presence of the operculum or bud-cap which protects the stamens in the bud stage. The operculum usually falls off entirely as the flower expands, but sometimes remains hinged on to the calyx after the flower opens. These characteris-

tics, together with the presence of the inferior ovary and the conspicuous stamens of indefinite number, serve to distinguish the genus Eucalyptus from all other flowering plants.[18]

Wilson photographed Herbert with wandoo, *E. wandoo* subsp. *wandoo*, on the way south (Figure 5.3). Desmond Herbert would have pointed out to Wilson the scar marks on this tree, made by Noongar and likely over two hundred years old at the time. Scar trees like this are now protected. Wilson had likely also seen powderbark wandoo (*E. accedens*), which has a powdery outer bark that rubs off when touched, in the forest east of Perth.

As Wilson went further south into the loftiest forests of Western Australia, he was entering a world with some of the tallest trees on Earth. All the forest giants in Australia are eucalypts.

Americans all know that they possess the tallest trees in the world, with giant sequoias reaching above 100 metres (328 feet) in height. All the large North American species from the west coast are conifers, which are gymnosperms (flowerless plants with cones and seeds) – *Sequoia sempervirens* (coast redwood), *Pseudotsuga menziesii* (Douglas fir),[19] *Picea sitchensis* (Sitka spruce), *Sequoiadendron giganteum* (giant redwood) and *Abies procera* (noble fir). Less well known is that some angiosperms (flowering plants) attain comparable heights. In Australia these giant angiosperms are *Eucalyptus regnans* (mountain ash), the (then) tallest known flowering plant in the world, *E. globulus* (southern bluegum), and *E. viminalis* (manna gum) in south-eastern Australia.[20] Less well known are the karri (*E. diversicolor*) and red tingle (*E. jacksonii*) of south-western Australia.[21]

Eucalypts tend to speed to the sky, stay there, and senesce after 400–500 years; they do not attain the enormous ages of the slow-growing giant conifers of North America. There are exceptions to this; some snow gums (*E. pauciflora*) are thought to be 1500–1700 years old.[22] Snow gums form the treeline in the higher country of Australia's south-east in the coldest and windiest places, and are famously trunk-twisted and striped with wonderful colours, all made more vibrant when seen with snow (see Figure 7.3 on page 181).

The train service to rural Australia was more extensive in 1920 than it is now, and Wilson was able to travel to the south coast on the train, to Albany and then Denmark. Today, one must drive to the south coast. With Lane-Poole,

FIGURE 5.3: Desmond Herbert with wandoo, *E. wandoo* subsp. *wandoo*, in the drier forest of the southwest. The cutting out on this tree means that it is of cultural significance to Noongar people. The purpose might have been for a utensil or as a spiritual scar, and the cutting would have taken at least a day. This image is also used in Kessell and Gardner's 1924 book (page 96), where the species is given as *E. redunca* var. *elata*. Image Wilson Y-365 AA.

he found a very rich collection of not just trees, but species of considerable horticultural beauty. While liking big trees, Wilson also collected many pretty flowering bushes at Nornalup and in nearby Albany, as seen in his notes and in the herbarium. Wilson collected a large suite of species, but only the names are included in his diary (Figure 5.4), which consists solely of species lists and the heights of the plants for this part of his trip; there are no travel notes. Species include: *Leucopogon*, *Conospermum*, the purple-flowered *Hovea chorizemifolia* (holly-leaved hovea), *Stylidium scandens* (a trigger plant), *Callistemon*, *Epacris*, *Melaleuca*, *Banksia*, *Casuarina fraseriana*, '*Symphea?*', *Xanthosia*, *Pimelia*, *Isopogon occidentalis*, *Banksia coccinea* (now a well-loved floriculture flower with its crimson flowerheads), *Eucalyptus marginata* (jarrah), *Andersonia*, *Jacksonia*, several *Leptospermum*, *Xanthosia rotundifolia* ('Southern Cross'), *Acacia cyclops*, *Dryandra*, 'Isopogon or Petrophile', 'Pultenaea?', and *Santalum*.[23]

Near Denmark and the Frankland River area, Demark being a small town on the south coast, Wilson collected the major trees *E. megacarpa* ('80ft 30 miles W Denmark', #395), *E. jacksonii* ('Red Tingle 180ft'), *E. diversicolor* ('Karri', #402), *E. guilfoylei* ('Yellow Tingle up to 100ft Frankland River', #388), *E. ficifolia* ('40ft', #389), and *Banksia littoralis* ('50ft', #407) and *Agonis* ('12ft'), as well as a host of large shrubs such as *Boronia* ([ ], '8ft').

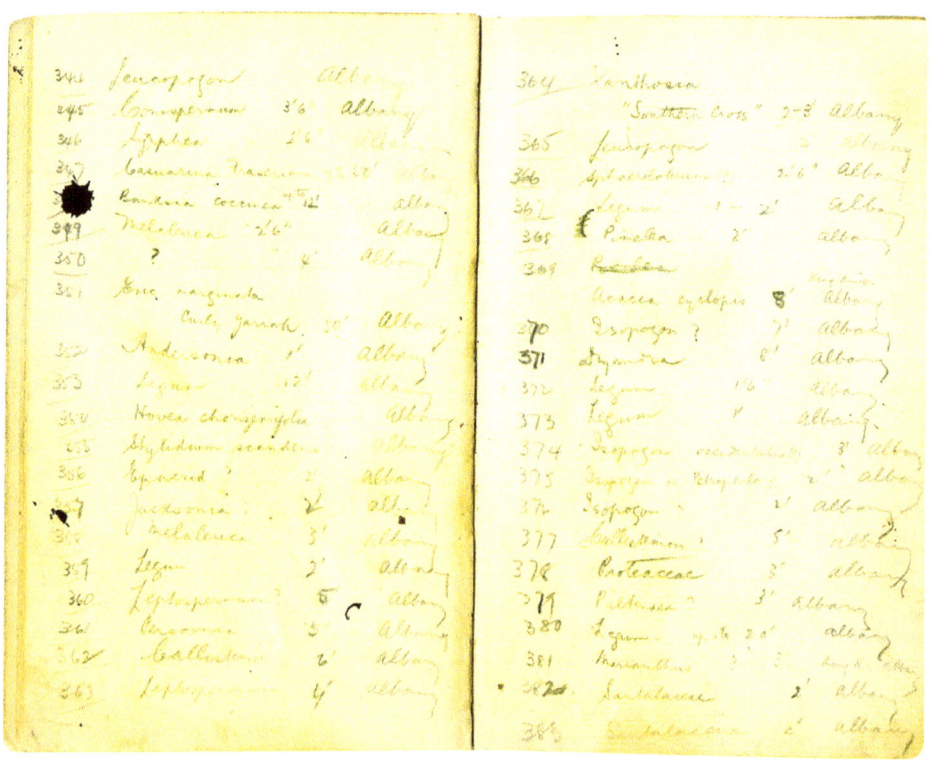

FIGURE 5.4: Wilson's diary notes from collecting in the Albany area, which was first settled by Europeans in 1826. In 1791, on his way to America, George Vancouver, famous for the city that bears his name and his exploration of the north-western coast of North America, named the magnificent harbour King George the Third's Sound. It is now more simply King George's Sound and is the site of Albany. The botanist Archibald Menzies travelled with him and collected here; *Banksia menziesii* is named after him.

Wilson loved the southern forests. Nornalup 'calls up for Professor Wilson visions of West Australia's ideal watering place of the future'. He was quoted as saying: 'You have got everything there, wonderful forest scenery, mountains, landscapes, sea-scapes, boating, fishing. It is one of the most beautiful sights I ever saw in my life. It is a fascinating place. I don't know of any other place that appealed to me in the same way. It was so un-expected.'[24]

FIGURE 5.5: Nornalup, of which Wilson spoke fondly. Image: Wilson Y-319 AA.

Of Nornalup Inlet (Figure 5.5), among giant trees of karri (*E. diversicolor*) and red tingle (*E. jacksonii*), Wilson said again to the press:

> Nornalup is with me the culmination. I compare that with other scenes in the world, and only a few are in the same category. I like a big tree. I have been thousands of miles to see a tree. A karri tree is the Adonis and Hercules of the tree world in one, it is a symbol of beauty and also of strength.[25]

Wilson was taken into an area now called the Valley of the Giants, in the Walpole region of the south-west and where tingle (Figure 5.6) and karri (Figure 5.7) are found.

FIGURE 5.6: Wilson with red tingle, *E. jacksonii*, as identified by Gardner.[26] This species has stringy, red-grey bark from its base to high in its canopy. Image: Wilson Y-330 AA. A copy of this image is held in the Forests Department of Western Australia.

FIGURE 5.7: Lane-Poole with a close-up of the varied-coloured trunk of the karri, hence its name *E. diversicolor*. Wilson was struck by the beauty of the karri, which he called the Adonis and Hercules of the tree world. Image: Wilson Y-373 AA.

FIGURE 5.8 (opposite): The tiny figure of Lane-Poole in the karri forest (can you find him?) is seen at the bottom of the image just right of centre. Wilson described being in the karri forest as 'standing in great columned rows, underneath which may be seen a limitless carpet of brackens.'[27] Image: Wilson Y-375 AA.

Karri is one of the tallest trees in the world, as can be seen from the very small form of Lane-Poole in Figure 5.8. Wilson wrote that karri were the 'tallest and most beautiful of the Western Australian species', with trees 'little short of 300 feet tall with a trunk clean of branches for fully 150 feet common'.[28] The *Albany Advertiser* reported that Wilson 'compared a walk through the karri forests with passing among the pillars of a vast and beautiful cathedral'.[29] Wilson said that the karri forest was 'wonderful'. At the time of his visit, the newspaper reported:

> The karri giants he found to be the finest broad leaved trees he had ever seen, and very impressive. He did not see why the national park there should not become as famous as some in America if properly handled. The man of science took second place to the keen appreciator of beauty, as the professor commented meditatively, 'These karri forests are something to dream about as long as one lives.'[30]

FIGURE 5.9: Wilson and an unidentified man with horse inside the cavity of a forest giant, likely *E. jacksonii*, the red tingle, 1920. Image: Wilson Y-327 AA.

The *Albany Advertiser* reported that 'The professor loved the forests of West Australia, and enjoyed every minute of his journeys through them.' As Wilson had predicted, this is now a tourist area, though surprisingly less well known than it should be, and less well known than either Tasmanian or Queensland forests. The Valley of the Giants (Figure 5.9) is still a feature today.

In mid-November, Wilson was in the Frankland River area with Lane-Poole. On 10 November he collected yellow tingle, *E. guilfoylei* Maiden (#388 HH; 'Tree up to 100ft tall Frankland River'; 'Yellow Tingle') is written on one specimen sheet.[31] Many leaves collected had been eaten prior to collection. On 12 November 1920, he collected *E. megacarpa* F. Mueller (#391 HH), a 'Tree 80ft' and another at Big Brook (#242 HH), where he notes on the label: 'see photograph' (Y-346 AA; see Figure 5.10); *megacarpa* is sometimes called bullich or swamp karri, and has, as its name suggests, large fruit. These specimens in the Harvard Herbarium are now dark brown with age, still caterpillar-chewed, holed and twisted, with strong mid-ribs of the abaxial surface typical of eucalypts and the clear dark lines of their veins, with new growth stopped and caught forever. It appeared that Wilson often forgot whether it was 1920 or 1921; here each sheet has a different year but the specimens were collected on the same day. It was close to Christmas, it was summer, it was late November, it was getting warmer; all can be confusing to a visitor from the north.

In his notes from his travels published through the Arnold Arboretum, Wilson made an important point about the eucalypts to his American or northern readers. While he had criticised Australians for not knowing more about their eucalypts, he wrote:

> We of the north know little about the Eucalyptus and to us the Blue Gum (E. globulus Labill.) and one or two others do duty for the whole genus. We know that they are mighty trees which furnish valuable timber but I doubt if many of us realize the ornamental character and great beauty of the flowers of a number of the species.[32]

Here, Wilson was likely thinking of the bright flowers of *Eucalyptus ficifolia*, which he saw and collected (#388 HH) in the Frankland River area in these southern forests. He visited this place 'on purpose', specifically to see this

FIGURE 5.10: *Eucalyptus megacarpa*, known as bullich. A fire had been through this area some years before, and nearly all the trees seen in this picture had been scarred to quite high in the canopy, with several trees killed. Image: Wilson Y-346 AA.

impressive species.³³ This eucalypt, commonly referred to as 'ficifolia' is known as the scarlet flowering gum or crimson-flowered gum and has been reclassified as *Corymbia ficifolia*.

> It seems to have its home just in a very small area of country round the Bow River, just a few square miles—where it grows wild. There is very little of it, just small scrubby tree, tortured by fire. When the next fire goes through the country they may be destroyed altogether. People may say: 'What's the odds, we have plenty in King's Park.' But it seems a pity that such a thing should disappear from its natural habitat.³⁴

It was a keen observation of Wilson, or a comment on what he had been told by Lane-Poole, that too frequent fires can destroy eucalyptus groves and forests. Since Wilson's comment, ficifolia has become tremendously popular in gardens and is a staple of the horticultural trade in Australia due to its flowers. However, the wild tree seen by Wilson in 1920 (Figure 5.11) is much larger than the usual cultivated form of ficifolia today (Figure 5.12).

FIGURE 5.12: A usual form of *E. ficifolia* today in gardens, smaller than the wild tree photographed by Wilson.

FIGURE 5.11: An old and magnificent *Eucalyptus ficifolia*, red-flowering gum, in the Frankland River area, 1920. Desmond Herbert has climbed into the lower branches and appears to have a dog on his lap. Wilson described this species as 'of small size, quick growing, has large leaves and terminal masses of flowers from pale orange to crimson in color' and that it 'must rank among the most beautiful of trees'.[35] This photo is one of several used by Kessell and Gardner in their 1924 book *Key to the Eucalypts of Western Australia*. Image: Wilson Y-326.

At the time of Wilson's trip in 1920, botanists could only imagine where the eucalypts had come from, or how old they might be as a species. Wilson thought that they were 'modern, highly developed' trees that were juxtaposed with the 'Mesozoic growths' of the *Xanthorrhoea* and *Zamia* palm.[36] A hint of the great age of *Eucalyptus* is indicated by the astounding speciation of the genus. Recent genetic sequencing of *E. grandis*, a tall subtropical tree from eastern Australia, showed that eucalypts were living happily 109 million years ago.[37] If shown a fossil eucalypt of 52 million years ago found in Patagonia that showed flowers, fruits, and leaves, Australians who are not botanists, or not interested in botany at all, would still declare 'It's a eucalypt!'[38] (Figure 5.13). Pauline Ladiges, Emeritus Professor of Botany at the University of Melbourne, notes that these ancient fossils 'look just like the eucalypts down the road', mainly because the fruits of eucalyptus are so recognisable and distinct; they are 'the single most useful structure for differentiating species'.[39] These are Australia's 'gumnuts'.

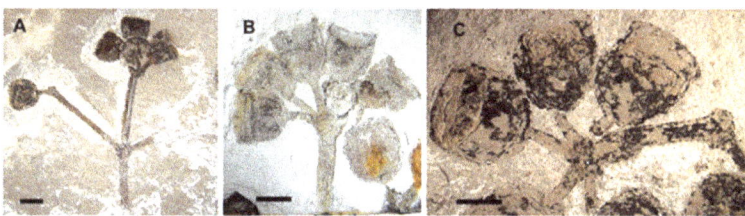

FIGURE 5.13: Fossils from South America, 52 million years old, showing eucalypt capsules that appear readily identifiable to modern eyes as eucalypts. Image adapted from Gandolfo et al., 2011 with permission from lead author and Creative Commons.

Because these fossils are so recognisable, it is thought that eucalypts must have been established in that tell-tale form even earlier than these dates, although fossils have not been found.[40] Eucalyptus fossils have been found in Patagonia, New Zealand, and Antarctica itself, all pieces of Gondwana, where they were in the company of dinosaurs for millions of years.[41, 42, 43] Wilson noted that the south-western landscape reminded him of the age of the dinosaurs — monsters of 'the remote Lizard Age'.[44] As a group, eucalypts are thought to have evolved in western Gondwanaland, 'somewhere in the Weddellian Biogeographic Province (which includes southern South America, western Antarctica, and south-eastern Australia)'.[45] Many other major Southern Hemisphere plant families evolved or diversified in this region, such as Podocarpaceae, the Nothofagaceae that so interested Wilson, Casuarinaceae, and Proteaceae.[46, 47]

Eucalypts evolved at very high latitudes near the South Pole. This seems startling, even given that the climate was very different to that of today and the temperature differences between pole and tropics nowhere near as great as today. Millions of years ago, when Gondwana was intact, subtropical and temperate conditions existed near the poles. Modelling shows that the mean annual temperatures millions of years ago at a latitude of approximately 60°S were a pleasant 17–21°C, and cycads, gingers, palm, and ferns were present.[48] These high temperatures and rich forests at high latitudes are 'astonishing' to us today.[49] There are no modern analogues to this situation; no forests grow at such high latitudes anywhere on the planet today.

These ancient high-latitude eucalyptus forests dealt with yearly cycles of extreme changes in light and dark.[50] At about 67°N and S, there are seasonal phases of continuous light, and months of dark. What morphological legacies might this experience of polar darkness have bestowed on the flora today, if any? Fossil eucalypts appear to have a very long and narrow leaf form. For living trees, this has long been taken as an advantage in high-light environments of woodland and open forests, and a likely adaptation to Australia drying over many millennia. However, that fossils also have this feature gives pause for thought. 'We can rule out a high sun elevation angle, since all these fossil locations were at relatively high latitudes when the fossils were alive',[51] and it is suggested that the long, narrow leaves that hang vertically, as Wilson had noted, capture light best in very low sun environments, as would have been found in the high latitudes near the southern pole so long ago. Today, there are concerns that there might be photic barriers to plants moving in response to climatic warming. While plants might shift toward the poles, 'tracking their optimal thermal environments', the 'daylength, the driver of daily and annual timing … is fixed by latitude'. Thus, trees that would benefit from cooler climates nearer the poles might struggle to move into new and different light environments compared with their ancestral history, with a 'thermal-photic mismatch'.[52] However, would eucalypts still carry resilience to long nights of polar darkness followed by long days with no night? It is an interesting question, because polar darkness is usually associated with cold and snow, but when eucalypts lived on Gondwana long ago, it was mild or temperate.

Plants moving north on the drifting continent of Australia would have faced a drying climate and increasing loss of nutrients from eroding soils as the millions of years passed. It has been proposed that as Australia drifted north and became more arid in the last 34 million years, 'reducing water availability (xeromorphy) in progressively higher light environments were the major drivers' of the evolution of another Australian family, the Casuarinaceae.[53] Wilson collected casuarinas at Bruce Rock (23 October 1920, #144 HH), York (a 65-foot *Casuarina glauca*, #207 HH), 20 miles from Coolgardie (12 feet, a casuarina with cones, #546 HH), on 10 December 1920 in the Adelaide Hills (6–7 feet, #612 HH), and recorded *C. torulosa* (now *Allocasuarina torulosa*) 40 miles west of Sydney (#634 HH) on 19 December, showing that Casuarinas range right across Australia.

That *Eucalyptus* speciated so spectacularly in this journey north, experiencing changing conditions of light, temperature, and rainfall, suggests that the genus has or had tremendous phenotypic plasticity. Stephen Hopper, former director of the Kew Gardens, noted in a major review that the outstanding capacity of *Eucalyptus* to respond to a range of environmental stresses could reflect an idea expressed by Ross Florence in 1981, that of 'prior adaptation of its progenitor to more specialized and disturbed niches within the Gondwanan rainforests.'[54, 55]

In the summer of 2019–20, south-eastern Australia began to burn. Scene after scene showed eucalypts on fire, often exploding in flames as the fire front arrived. Most eucalypts are superbly adapted to fire. It is extraordinary to imagine that a plant can survive an inferno and be burnt black, with all leaves gone, and then out comes green amid the black trunks, ash-grey soil, and silence of a stilled and hushed bush. Judging by several close-up photos that Wilson took of fire-scarred trees, he was deeply interested in this extraordinary capacity (Figure 5.14). Wilson's photos of fired trees included those killed, healthy old trees with fire scars, and younger trees that had at one stage been scorched. In Wilson's time, little was known about the strategies, the impacts of fire, or how eucalypts evolved to deal with fire – all intertwined issues.

Eucalypts survive fires by several mechanisms that were not understood in Wilson's time, though the 'wonderful vitality' and capacity to survive fire of the then two hundred known members of the group was known.[56] Dean Nicolle, an authority on the eucalypts, has now articulated four broad strategies that

FIGURE 5.14: Despite the burnt trunk from bushfire, this tree was still living. Image: Wilson Y-451 AA

eucalypts use to regenerate after fire, after the destruction of their crowns.[57] These are obligate seeding (relying on seedlings to regenerate), epicormic shooting from stems, sprouting from lignotubers, and a combination of the epicormic and lignotuber regeneration. Of these strategies, perhaps most extraordinary are those species that possess lignotubers. Lignotubers consist of a 'woody mass of dormant vegetative buds originating at the cotyledonary and early seedling leaf nodes.'[58, 59] With lignotuber-bearing trees, the crowns and branches of the trees are lost in a fire, thus new tops are only as old as the last wildfire. But the tree returns via the basal material held in their lignotuber, and some lignotubers have been radiocarbon dated to be thousands of years old. As examples, a *Eucalyptus argutifolia* was found to be more than 2,000 years old, a *E. curtisii* 3,970–9,000 years old, and a *E. phylacis* 6,380 years old.[60] A mallee, *E. oleosa*, in Mildura, in arid-zone Victoria, has been dated to 2,500 years old. Such longevity makes these trees almost immortal and phoenix-like. Other trees in the world have been dated to be ancient by their root systems, though their trunks are also much younger.

Another mechanism to escape death after fire is resprouting. Eucalypts are some of the most successful resprouters in the world. Resprouting is dependent on preventitious buds, which are bud-forming structures 'generally different from resprouting structures in many other plants.'[61] Wilson likely saw resprouting on his trip as they came across fire-damaged trees. Fossil evidence shows the existence of oil glands for more than 100 million years and this suggests that eucalypts have been fire-resistant since at least that time. This is a fascinating conception and relates also to Wilson's 'other planet', in that it has been hypothesised that the great rise of the angiosperms in the warm world of the Cretaceous (145–66 million years ago) was due to their rapid reproduction and growth – almost like weeds – with small seeds, which displaced the gymnosperms, and that angiosperms had 'fiery origins'.[62] In Australian conditions, the great angiosperms flourished, with the tallest in south-eastern and south-western Australia.[63]

Much literature today presumes fire to be the main driver of evolution in the eucalypts and the main arbiter of management practices. However, such a survival response to fire might have been almost incidental. The xeromorphic and sclerophyllous nature of much of the Australian flora is also representative of 'adaptation to low nutrients rather than to a dry climate, and that in Australia today, rainforest genera are more effectively excluded from an area by low nutrients than by a dry climate'.[64] Investigations in this complex and fascinating area continue. One concern today – that of continued logging of forests on low fertility sites – would no doubt have attracted Wilson's comment.[65]

Australia's climate has fluctuated over the last 100 million years, with wetter and drier epochs. For eucalypts, fluctuation of mesic to arid would have been an added driver of evolution, but the likely greatest driver has been soil. As Wilson noted, it is extraordinary that tall, large-boled, long-lived trees grow on poor soils. As an exceptionally well-travelled horticulturalist, he would have known that

> elsewhere in the world soils of similar status would be found to barely support tree growth, let alone tall, productive forest … In their subsequent evolution the attributes contributing to adaptation to low nutrient soils probably enhance the capacity of eucalypts to adapt to other environmental stresses, including a drying climate and an increasing frequency and severity of fire.[66]

Eucalypts have a distinctive smell. The scent is so strong it can adversely impact wine-making if grapes are grown adjacent to eucalypts. The aromatic compound in the leaves is a cineole-dominated terpenoid mixture that has antiseptic effects; it is a 'germ-killer' and was used as a disinfectant early in the nineteenth century.[67] For the tree, the terpenes deter insects, and attracts (for example, koalas) or repels vertebrate herbivores, among a host of other factors.[68] Even before Wilson's time, eucalypts fascinated biologists due to their oils and their scent. Oil was even considered as a method to group eucalypt species taxonomically,[69] but this idea was rebutted by Joseph Maiden, Director of the Royal Botanic Gardens, Melbourne.[70] Eucalyptus oil is poisonous, but a eucalyptus oil industry for medicinal purposes has existed since first established in Victoria in 1852.[71,72] While hundreds of species contain volatile oils, only about a dozen are used in the industry to produce throat lozenges and chest rubs.[73] These tried-and-true uses are as familiar to Australians today as to Wilson's companions in the 1920s. Indigenous Australians had used eucalypts for the same reasons; they contain eucalyptol, which has 'anti-bacterial, cough-suppressant, expectorant, nasal decongestant and respiratory anti-inflammatory properties', as well as a host of other active constituents that have medicinal benefits.[74] Indigenous people also used *Melaleuca alternifolia* to treat injuries and burns, and had an entire inventory of other plants used for medicinal purposes that has largely gone unnoticed. In 1890, a correspondent to the newspaper the *Sydney Morning Herald* scoffed at the 'discovery' of the medicinal properties of eucalypts by doctors and scientists, pointing out that the same knowledge had long existed and was commonly used by Aboriginal people, and called for 'honour to whom honour is due'.[75] That honour is slowly coming, with great interest from many.

Wilson was impressed with the eucalypts but not impressed with their use by settlers. He saw a great deal of waste and widespread lack of appreciation of the value that the trees and their timber might bring. There were three main issues that concerned him: the wastage of timber at the mills, the misuse of timber for inappropriate purposes, and the wrong value put on land with timber, with a preference of the Government for agricultural production such as dairy, compounded by a lack of any understanding of the value of managed timber. Of the mills, Wilson noted to the press:

> When visiting the Government Mill at Big Brook, I was shown a continuous fire which burns as long as the mill is running; at this mill they pretend to recover on 33% of the log, and probably really get less, the rest going into the fire. It is a shame to waste 66% of a karri log. I saw much good timber passing over the chute into this fire which is perpetual as long as the mills last, simply because it is not suited for particular purposes. Efforts should be made to find uses for it. In Japan with their best timbers they do not waste even the sawdust. They dry the sawdust for their fires. Contrast this with the waste of wonderful karri timber.[76]

Some of the southern forests of Western Australia had been destroyed in the years immediately following World War I, when returned servicemen and migrants from the UK were brought to the area under a soldier settlement scheme to 'open up' the land, for dairying in most cases, and potatoes in some. Both Lane-Poole and Kessell had to deal with ardent land settlement supporters in the government, for whom clearing meant development and votes.[77] Group Selection Schemes for assisted migrants followed the soldier settlement schemes, and opening up the karri forest for agriculture went ahead, with inexperienced settlers allotted virgin forest. But 'a lot of the settlers had come from impoverished U.K. cities and had never seen a cow in their lives'.[78]

The result was hardship. One of the problems that Lane-Poole saw was that the Lands Department assessed land as good if the trees were tall, presuming fertile soil – which is not the case. The increasingly university-trained foresters argued for managing timber and forests over agriculture's need for clearing and the subsequent waste of trees, with a very different vision to that which eventuated in most places. The tensions of rivalry and opposing ideas about the treatment of forests was to be seen again and again by Wilson as he crossed Australia in the early 1920s. His short trip to the southern forest giants in Western Australia was another foretaste of issues to come on this expedition and opinions being formed by Wilson of Australia's treatment of its woodlands and forests.

Towards the end of his time based in Perth, Wilson wrote to Sargent that

> this expedition has been quite successful in Western Australia. Thanks to the facilities placed at my disposal by the Government I have been able to cover about fifteen hundred miles of country to the east, south & southeast of Perth. I have seen all the different belts of vegetation in this part of the country & collected about five hundred specimens. I have seen all the important forest trees & taken photographs of them. In all I have taken eight dozen photographs. The Conservator of forests has presented me with nearly fifty photographs & specimens of about 450 species of woody plants. These include a complete set of the Eucalypts of Western Australia all of which have been authoritatively named by Maiden. The Government Botanist is making up a set of duplicates for us also. All this & also my own collections & photographic negatives will be shipped to you direct by the Conservator. Of books I have got hold of a number but about these I will write you separately.
>
> I leave here on Saturday for Adelaide staying over a couple of days in the eastern part of this state to see the Eucalypts peculiar to that region.
>
> I am too busy now preparing for my departure to write a proper letter but on the way across the continent I shall try to write you some account of the vegetation of this part of the world.
>
> My health is excellent & I have much enjoyed the forest wonder shown.[79]

A separate letter appears only partly preserved and is also on Palace Hotel stationary. In it, Wilson noted to Sargent:

I find there is quite a lot of literature on the forests etc of W. Australia which we have not got. This is being got together for us & I shall post it to you before leaving the State. Already I have […] on hand & also a number of photographs of forest trees from the Conservator. Lane-Poole knows his work & is popular & enthusiastic. He has a collector constantly in the field & intends to get together all the woody plants of the state. A set of duplicates is promised us. This is fortunate since it is utterly impossible for me to collect other than a few here and there. I wish I had my Chinese collectors here what a hand [ ] make!

The weather is glorious, the countryside gay with flowers nearly all of woody plants. My health is excellent & everything full of promise.

Having the good fortune that Lane-Poole had taken on the task of sending plant material from Western Australia home to Boston, so that they were not lost on the *Canastota*, Wilson set off from Perth on the train for the Goldfields of the arid Kalgoorlie region, 600 kilometres (nearly 400 miles) to the east.

Riding along on the train, he would leave the sandy coastal plain and its tuart and flooded gum and its banksia woodlands, rise through the heavily wooded Darling Scarp, where scatterings of white-barked Darling Range ghost gum (*Eucalyptus laeliae*) could be seen among the grey and blue jarrah and marri forest, and feel the train fall away into its rhythm as the country opened up into the Wheatbelt, with the flanks of majestic salmon gums glinting in the sun, along with the cleared country that so concerned him in its extent, and its revelation of the desultory attitude to nature that was so widespread and which was to become a theme of his entire visit to Australia. Soon, moving east, wheat fields would give way to the Great Western Woodland, today the greatest intact temperate woodland remaining in the world. Once more, Wilson was to be fully immersed in more of this other part of the planet's botanic garden.

PLATE 5.1: "Casuarina Fraseriana Tree 50 ft Albany 6 November 1920". No. 347. Now *Allocasuarina fraseriana*.

PLATE 5.2: "Eucalyptus marginata "Curly Jarrah" Tree 50 ft Albany." No. 351, 6 November 1920. A piece of bark is given to show the derivation of the common name of curly.

PLATE 5.3: "Eucalyptus ficifolia 40 ft Frankland River West Australia." No. 389. November 1920. Now classified as *Corymbia ficifolia*. The common name is the red-flowering gum. Likely to be from the tree in Figure 5.11.

PLATE 5.4: "Melaleuca Tree 40 ft. Frankland River W. Australia" No. 401. November 1920. Classified as *Melaleuca cuticularis*, the salt-water paperbark. Wilson has collected the paperbark (left).

PLATE 5.5: "Callistemon [(hidden by later text)] Albany W. Australia" No. 362. 6 November 1920. Classified as *Callistemon speciosus* and synonym as *Melaleuca glauca*, the Albany bottlebrush.

PLATE 5.6: "Leptospermum Bush 5 ft Albany W.A." No. 360. 6 November 1920. Classified as *Homalospermum firmum*.

PLATE 5.7: "Isopogon Bush 7 ft. Albany". No. 370. 6 November 1920. Classified as *Petrophile diversifolia*.

PLATE 5.8: Isopogon 3 ft Big Brook W. Australia". No 259. November 1920. Classified as *Isopogon sphaerocephalus*.

PLATE 5.9: "Hakea linearis? Bush 5 ft. Waterloo Western Australia." No. 234. November 1921 [sic]; the true date is 1920. Now classified as *Hakea varia*.

PLATE 5.10: "*Banksia littoralis* Harvey W. Australia" No. 233. November 1920.

PLATE 5.11: "Banksia coccinea 4 -12 ft. Albany" No. 348. 6 November 1920. The common name is Scarlet Banksia. *B. coccinea* is widely used in floriculture due to its stunning flowers.

PLATE 5.12: "Dryandra Bush 8 ft. Albany W.A." No. 371. 6 November 1920. Now classified as *Banksia formosa*; common name Showy Dryandra.

PLATE 5 13: "Adenanthos barbigera Busselton". No. 283. November 1920. Now classified as *Adenanthos obovatus*; common name basket flower or jug flower.

PLATE 5 14: "Persoonia 5 ft. Albany W. Australia". No. 361. November 1920. Now classified as *Acidonia macrocarpa*.

PLATE 5.15: "Boronia lanuginosa near Hamel W. Australia". No. 232. November 1920. Now *Boronia molloyae*, tall boronia. The boronias are highly scented.

PLATE 5.16: "Pimelea 3-4 ft. forests Big Brook W. Australia." No. 256. November 1920. Classified as *Pimelea lehmanniana*.

PLATE 5.17: "Pimelea 2 1/2 ft. Albany. W. Australia." No. 339. 6 November 1920. Classified as *Pimelea hispida*.

NOTES

1. EH Wilson, 'Notes from Australasia No. 1', *Journal of the Arnold Arboretum*, 1921, 2(3):160-3.
2. 'The State's forest resources. Visiting American scientist. Beauties of the Nornalup District', *Albany Advertiser*, 27 November 1920, p. 4.
3. J Dargavel, *The Zealous Conservator. A life of Charles Lane Poole*, University of Western Australia Press, Crawley, 2008, p. 59.
4. K Kullmann, 'The emergence of suburban terracing on coastal dunes: Case studies along the Perth Northern Corridor, Western Australia, 1930-2010', *Journal of Urban Design*, 2014, 19(5):593-621.
5. Government of Western Australia, *A Tuart Atlas: Extent, density and condition of tuart woodland on the Swan Coastal Plain*, prepared by the Department of Conservation and Land Management for the Tuart Response Group, Western Australia, 2003.
6. JJ Wentzel, MD Craig, PA Barber, et al., 'Tuart (*Eucalyptus gomphocephala*) decline is not associated with other vegetation structure and composition changes', *Australasian Plant Pathology*, 2018, 47:521-30.
7. Currently, the Ludlow Forest is being restored by volunteers, with a 45-year plan in place.
8. PG Nevill, D Bradbury, A Williams, S Tomlinson, and SL Krauss, 'Genetic and palaeo-climatic evidence for widespread persistence of the coastal tree species *Eucalyptus gomphocephala* (Myrtaceae) during the Last Glacial Maximum', *Annals of Botany*, 2014, 113:55-67.
9. Reported by traditional owner Karen Jacobs, describing walking to the island known as Wadjemup by Noongar people (now Rottnest Island), https://derbalnara.org.au/boodjar-a-changing-coastline.htm Wadjemup means 'place of the emus', but there are no emu there now.
10. I know of one tree that has appeared to be dying my entire lifetime.
11. EH Wilson, *Smoke that Thunders*, 1985, pp. 120-1.
12. Wilson, *Smoke that Thunders*, p. 115.
13. See M Bayly, 'Phylogenetic studies of eucalypts: Fossils, morphology and genomes', *Royal Society of Victoria*, 2016, 128:12-24.
14. RT Lange, 'Australian tertiary vegetation', in JMB Smith (ed.), *A History of Australasian Vegetation*, Sydney, McGraw-Hill, 1982, pp 44-89.
15. PY Ladiges, F Udovicic, and G Nelson, 'Australian biogeographical connections and the phylogeny of large genera in the plant family Myrtaceae', *Journal of Biogeography*, 2003, 30:989-98.
16. RH Groves, 'Present vegetation types', in AE Orchard (ed.), *Flora of Australia*, vol. 1, 2nd edn, Australian Biological Resources Study, Canberra, 1999, pp 369-401.
17. *Ballarat Star*, 21 April 1921.
18. SL Kessell and CA Gardner, *Key to the Eucalypts of Western Australia*, WA Forests Department, Government Printer, Perth, 1924.
19. Douglas fir was first described by the English botanist Archibald Menzies (1754-1842), from his visit to Vancouver Island during George Vancouver's 1790 voyage to the north-east Pacific Ocean. Archibald Menzies is also honoured in *Banksia menziesii*, 1754.
20. DYP Tng, GJ Williamson, GJ Jordan, and DMJS Bowman, 'Giant eucalypts—globally unique fire-adapted rainforest trees?', *New Phytologist* (Tansley Review), 2012, 196:1001-14.
21. A review of the largest trees in Australia is given in JL Williams, D Lindenmayer, and B Mifsud, 'The largest trees in Australia', *Austral Ecology*, 2023, 48(4):653-71.
22. From the New South Wales National Parks and Wildlife Service.
23. Numbers in Wilson's diary for the plant specimens are from #344 to #383.
24. *Albany Advertiser*, 27 November 1920, p. 4.
25. *Western Mail*, 25 November 1920.
26. CA Gardner, 'Trees of Western Australia—the red tingle', *Journal Dept of Agriculture, Western Australia, Series 3*, 1953, 2(3):407-16.
27. Wilson quoted in the *Boston Evening Transcript*, 30 September 1922.
28. Wilson, *Notes from Australasia*, p. 162.
29. *Albany Advertiser*, 5 March 1921, p. 3.
30. *Albany Advertiser*, 27 November 1920, p. 4. This appears to be a copy of the *West Australian*, 16 November 1920, p. 6.
31. The tingle trees likely get their name from the Indigenous word 'dingul dingul' for these trees.
32. Wilson, *Notes from Australasia*, no 1, p. 162.
33. *Smoke that Thunders*, p. 117.
34. *Western Mail*, 25 November 1920, p. 9. Note that Kings Park and Botanic Garden is in Perth.
35. Wilson, *Notes from Australasia*, no 1, 1921, p. 162.
36. *West Australian*, 21 October 1920, p. 7.
37. ME Jones, M Shepherd, RJ Henry, and A Delves, 'Chloroplast DNA variation and population structure in the widespread forest tree, *Eucalyptus grandis*', *Conservation Genetics*, 2006, 7:691-703.
38. MC Zamaloa, MA Gandolfo and KC Nixon, '52 million years old Eucalyptus flower sheds more than pollen grains', *American Journal of Botany*, 2020, 107(12):1763-71; see also EJ Hermsen, MA Gandolfo, and

MDC Zamaloa, 'The fossil record of *Eucalyptus* in Patagonia', *American Journal of Botany*, 2012, 99(8):1356–74.
39. MA Gandolfo, EJ Hermsen, MC Zamaloa, KC Nixon, CC González, P Wilf, et al., 'Oldest known *Eucalyptus* macrofossils are from South America', *PLOS One*, 2011, 6(6):e21084.
40. Comment by Professor Pauline Ladiges, University of Melbourne, 2023. See also PY Ladiges, 'Phylogenetic history and classification of eucalypts', in JE Williams and JCZ Woinarski (eds), *Eucalypt Ecology: Individuals to ecosystems*, Cambridge University Press, Cambridge, 1997, pp. 16–29.
41. EJ Hermsen, MA Gandolfo, and MDC Zamaloa, 'The fossil record of *Eucalyptus* in Patagonia', *American Journal of Botany*, 2012, 99:1356–74.
42. MA Gandolfo, EJ Hermsen, MC Zamaloa, KC Nixon, CC González, P Wilf, and NR Cúneo, 'Oldest known *Eucalyptus* macrofossils are from South America', *PLOS One*, 2011, 6:e21084.
43. M Macphail and AH Thornhill, 'How old are the eucalypts? A review of the microfossil and phylogenetic evidence', *Australian Journal of Botany*, 2016, 64:579–99.
44. *Smoke thatThunders*, p. 113.
45. RS Hill, YK Beer, KE Hill, E Maciunas, MA Tarran, and CC Wainman, 'Evolution of the eucalypts—an interpretation from the macrofossil record', *Australian Journal of Botany*, 2016, 64(8):600–8. Note that Western Antarctica is the smaller peninsula-like part of Antarctica today that reaches up to South America.
46. R Hill, 'Long term climate change and plant evolution in Australia', 2009, accessed 4 September 2021, media.adelaide.edu.au/institutes/enviroment/2009/cf1-bob.pdf
47. TL Dutra, 'Paleofloras da Antártica e sua relação com os eventos tectônicos e paleoclimáticos nas altas latitudes do sul', *Revista Brasileira de Geociências*, 2008, 34(3):401–10. With English abstract.
48. DG Greenwood and SL Wing, 'Eocene continental climates and latitudinal temperature gradients', *Geology*, 1995, 23(11):1044–8.
49. T Utescher T and V Mosbrugger, 'Eocene vegetation patterns reconstructed from plant diversity—A global perspective', *Palaeogeography, Palaeoclimatology, Palaeoecology*, 2007, 247:243–71.
50. DR Greenwood and JF Basinger, 'The paleoecology of high-latitude Eocene swamp forests from Axel Heiberg Island, Canadian High Arctic', *Review of Palaeobotany and Palynology*, 1994, 82:83–97.
51. RS Hill, YK Beer, KE Hill, E Maciunas, MA Tarran, and CC Wainman, 'Evolution of the eucalypts – an interpretation from the macrofossil record', *Australian Journal of Botany*, 2016, 64(8):600–8.
52. NP Huffeldt, 'Photic barriers to poleward range-shifts', *Trends in Ecology and Evolution*, 2020, 35(8):652–5. Today, this issue is of concern in the Northern Hemisphere.
53. RS Hill, SS Whang, V Korasidis, B Bianco, KE Hill, R Paull, and GR Guerin, 'Fossil evidence for the evolution of the Casuarinaceae in response to low soil nutrients and a drying climate in Cenozoic Australia', *Australian Journal of Botany*, 2020, 68:179–94.
54. SD Hopper, 'Out of the OCBILs: New hypotheses for the evolution, ecology and conservation of the eucalypts', *Biological Journal of the Linnean Society*, 2021, 133:342–72.
55. RG Florence, *Ecology and Silviculture of Eucalypt forests*, CSIRO, Collingwood, 1996, p. 52.
56. 'Field Naturalists Excursion', *South Australian Register*, 16 October 1896, p. 3.
57. D Nicolle, 'A classification and census of regenerative strategies in the eucalypts (*Angophora*, *Corymbia* and *Eucalyptus* – Myrtaceae), with special reference to the obligate seeders', *Australian Journal of Botany*, 2006, 54:391–407.
58. Ibid.
59. They are thought to be modified stem structures arising from the accessory meristematic tissue in the leaf axils.
60. For a detailed discussion of these features, see Nicolle, 'A classification and census of regenerative strategies in the eucalypts (*Angophora*, *Corymbia* and *Eucalyptus*—Myrtaceae), with special reference to the obligate seeders', 2006.
61. GE Burrows, 'Buds, bushfires and resprouting in the eucalypts', *Australian Journal of Botany*, 2013, 61(5):331–49.
62. WJ Bond and AC Scott, 'Fire and the spread of flowering plants in the Cretaceous', *New Phytologist*, 2010, 188:1137–50.
63. DYP Tng, GJ Williamson, GJ Jordan, and DMJS Bowman, 'Giant eucalypts—globally unique fire-adapted rain-forest trees?', *New Phytologist*, 2012, 196:1001–14.
64. RG Florence, 'The biology of the eucalypt forest', in JS Pate and AJ McComb (eds), *The Biology of Australian Plants*, University of Western Australia Press, Nedlands, 1981, pp 147–80.
65. Ibid, p. 175.
66. Ibid, p. 175.
67. For example, 'Eucalyptus', *Brisbane Courier*, 28 May 1886, p. 6.
68. C Külheim, A Padovan, C Hefer, ST Krause, TG Köllner, AA Myburg, J Degenhardt, and WJ Foley, 'The *Eucalyptus* terpene synthase

gene family', *BMC Genomics*, 2015, 16:450.

69 For example, see RT Baker and HG Smith, *A research on the eucalypts: Especially in regard to their essential oils*, WA Gullick, Government Printer, 1902.

70 AM Lucas, 'A nineteenth-century exploration of phytochemistry in botanical systematics: Joseph Henry Maiden and Eucalyptus kinos', *Historical Records of Australian Science*, 2017, 28(1):18–25.

71 BEJ Small, 'The Australian eucalyptus oil industry – an overview', *Australian Forestry*, 1981, 44(3):170–7.

72 First founded by Bosisto, a chemist and friend of Ferdinand Mueller: see *Sydney Morning Herald*, 26 July 1952, p. 8.

73 A book that discusses oil production in detail is John JW Coppen (ed.), *Eucalyptus: The Genus Eucalyptus*, CRC Press, 2002.

74 T Abbott and P Abbott, 'Technical information', 2015, cited in V Hansen and J Horsfall, *Noongar Bush Medicine: Medicinal Plants of the South-West of Western Australia*, University of Western Australia Press, Crawley, 2016.

75 JS Bray, 'Eucalyptus for fevers', *Sydney Morning Herald*, 25 December 1890, p. 6.

76 *Western Mail*, 25 November 1920, p. 9.

77 K Frawley, 'Visionaries to villains: The rise and fall of the foresters', in G Borschmann (ed.), *The People's Forest: A Living History of the Australian Bush*, The People's Forest Press, NSW, 1999, pp. 36–47.

78 K Campbell, *The Growing Up*, Hesperian Press, Carlisle, WA, 2004. Campbell recalls the hard life as a child in the soldier settlement scheme.

79 Letter from Wilson to Sargent, from Palace Hotel, Perth, dated 17 November 1920, AA Archives, W.XIV:A, box 21, folder 11.

CHAPTER 6

ARID, WITH WOODLANDS

In the Harvard Herbarium, specimens of the genus *Callitris* were found along a corridor in an old wooden cupboard that looked as if it might not have been opened often—perhaps not for decades. Inside a blue folder there was an old typewritten letter tucked in between specimen sheets. The letter was from the Conservator of Forests in Western Australia in 1923, Mr S L Kessell, Charles Lane-Poole having resigned in October 1921, very soon after Wilson had left Western Australia.

Stephen (Kim) Kessell was an Australian who had trained in Adelaide and Oxford and had served under Lane-Poole's direction. He was of like mind to Lane-Poole in his desire to see scientific forestry with well-trained staff. Soon after Lane-Poole's resignation, Kessell had written that Lane-Poole was 'a man of clear vision and high ideals who strove successfully to lay the foundations of a sound forest policy'.[1] Lane-Poole described Kessell as a 'first class sensible bloke'.[2] Kessell shared Lane-Poole's fury at the treatment of trees in Australia, writing in a report: 'the Anglo-Saxon settlers who populated Australia brought with them no traditions of forestry as a rural industry, and for 100 years or more the forests were looked upon as an enemy to be slaughtered'.[3]

Kessell wrote that 'Mr Gardner supplied the following notes':

> *Callitris glauca, R Br.* I have only seen this species at Black Flag Lakes, near Kalgoorlie, while in company with Professor Wilson, who has a similar set of specimens. I have good reason to believe that this species

extends over the Lake country to Lake Moore, where I have heard that there are trees of upwards of 70 feet, with a trunk diameter of 2 feet.[4]

Black Flag Lake is north of Kalgoorlie, the major town of the Goldfields following the 1890s gold rush in Western Australia. Gold had first been discovered in Coolgardie in 1892. By 1898, Coolgardie was the third-largest town in Western Australia, with a population of about 30,000, but by the time of Wilson's visit, Coolgardie was in steep decline, for gold had proved rich but fleeting. World War I had contributed to the loss of men and in the twentieth century Coolgardie's population went up and down until in the 1960s it was regarded as almost a ghost town. It has now made some resurgence through nickel and tourism and its population today is about nine hundred. The town is surrounded by small ghost gold-mining settlements in the bush and has a wonderful collection of late nineteenth century buildings in brick, granite, and sandstone, and some with stucco facades, or built of tin. From the old mine manager's house, you can see across the undulating land that had held so much promise for so many men.

Gold was found at Kalgoorlie in 1893, and this became the richest gold vein in the world. Soon the Kalgoorlie region was booming, leading to rapid population growth in Western Australia. The area attracted men from all over the world, largely because the earlier gold mining in eastern Australia in the 1850s had died away and eastern Australia was suffering a major economic depression at a time when there were no social services to assist the unemployed. The great gold strikes of California had also ended and the Yukon gold strike and stampede of 1896–9 to the Klondike had not begun. Kalgoorlie is still a thriving small city in the arid lands, with a population of approximately 29,000 (2020 figures).

Nothing is written in Wilson's diaries about how he got to Kalgoorlie, but the West Australian newspapers reported that Wilson was to take 'the Great Western Express for the Eastern States' on 20 November after several weeks in Western Australia, studying its timbers and flora in general. It had only been a few years before, in 1917, that the train line extended beyond Kalgoorlie to join the rest of Australia. At the time, the link—a tremendous engineering feat across waterless desert—was called the Great Western Express. It came to be known later as the Trans-Australian. Much later, in 1970, when a new standard gauge was laid, the train became the Indian-Pacific, joining Perth to Sydney without the tedious need to change trains, a point that Wilson had complained about. He noted that 'One of the strangest features of life in Australia … was the fact that State jealousy has resulted in a different railway gauge for every State, so that at the State lines all passengers have to change cars, and all freight has to be transhipped'.[5]

Before he left Perth for the Goldfields, Wilson said to the *Western Mail*, 'I have had an extremely interesting time in Western Australia, enjoyed every minute of it, and I wish the country well'. The route east to Kalgoorlie follows the line Wilson had travelled the previous few weeks into the Wheatbelt, along the route of Hunt's wells. The train took him beyond the Wheatbelt towns of Merredin and Westonia, where he had made collections. He broke his journey east for only a few days in the Kalgoorlie area.

As Wilson approached Kalgoorlie on the train from the west he would have seen, even in 1920, a curious thing as he headed east along the Goldfields Pipeline route. There are crops through the Wheatbelt, then suddenly, as if a line has been drawn across a page to denote some mysterious border, woodland from horizon to horizon. A major tree here is *Eucalyptus salubris*, one of the gimlets. These glorious lithe little trees, with glistening smooth red flanks that invite touch and caresses; gimlets are a favourite of many (see Figure 6.1). Wilson noted that the name comes from the screw-like shape of their trunks. He compared the gimlet to the salmon gum: 'The gimlet is much the same in appearance save that it is a smaller tree with duller trunk which is fluted & twisted like a screw hence the name gimlet' (see Figure 6.2).[6]

FIGURE 6.1: A stand of young gimlets north-east of Kalgoorlie in 2021, showing their fluted trunks and great beauty, whether singly or in a stand. Image: M. Grose 1147.

FIGURE 6.2: Desmond Herbert in 1920 with a stand of older gimlets. Image: Wilson Y-391 AA.

The salt lakes scattered throughout the arid areas around the Goldfields do not often fill with water. These are all paleo-systems of great age and signal previous fluvial times and are a major landscape type of the drier parts of south-west Australia. The age of this landscape has contributed to the high speciation. Black Flag Lake, White Flag Lake, and others north of Kalgoorlie are part of these strings of old lake systems (Figure 6.3). It is alleged that Black Flag was given its name because when alcohol arrived by dray from Kalgoorlie, the publican ran up a black flag to let the tired and thirsty miners know; the gold reef found had been rich. The lakes contain fresh water from time to time. The *Southern Times* reported in June 1896 that 'The lakes at Black Flag are full of fresh water, and are

FIGURE 6.3: Black Flag Lake near Kalgoorlie, WA. CSIRO Science image. Photographer: Willem van Aken; 18 January 1996. CSIRO science image Creative commons. Licence free to use. https://www.scienceimage.csiro.au/tag/lakes/i/5487/black-flag-lake-near-kalgoorlie-wa-/

navigable for a great distance. Timber is being conveyed by boat to the Devon Consols mine'.[7]

These are remote places, but Wilson's excursions in China and Formosa were to isolated areas rarely travelled by foreigners. Despite the remote and arid area, early botanists collected here due to the growing realisation of the great species richness in this region. At the same time, more was being discovered about Australia's fauna, and that Australia was the home of many of the world's most venomous snakes and a centre of global reptile biodiversity as well as botanical diversity.

Among men who arrived in this auriferous country looking for their fortune was a young Herbert Hoover, the future 31st President of the United States (Figure 6.4). Born on the 'nearly treeless eastern rim of the Great Plains',[8] he arrived at the age of twenty-two from Stanford's first graduating class of geologists, after working for the US Geological Survey in Colorado and California, then as a mine scout in Arizona and New Mexico.[9] In 1897 he took up a post as a mining engineer and in 1898 became manager of the Sons of Gwalia Mine at Kanowna, north of Kalgoorlie; this proved one of the richest gold mines in the world. Hoover was considered one of the most qualified mining engineers on the West Australian Goldfields, and one of the toughest and hardest negotiators, often pitting national groups of men against one another, but achieving remarkable successes and profits.[10] Though he described the region as one of 'black flies, red dust and white heat',[11] he also noted that 'after the sole annual rain the whole desert broke into a Persian carpet of different-colored immortelles which lasted for weeks'.[12] Hoover went back to the USA, returning briefly to Kalgoorlie in 1905 and in 1907, when he played a major role in rescuing a trapped miner.[13] In the Kalgoorlie Goldfields he had made the beginnings of the fortune that was to sustain him for the Presidency in 1929.[14] The years that Hoover spent in the Western Australian Goldfields are called his 'forgotten years' by American historians in much the same way as Wilson's years in Australia have been forgotten.[15]

Coolgardie was named after the Aboriginal word kurti-kurti, after a species of mulga from the vicinity, and advertisements in the local newspapers indicate that the town was often spelt Koolgardie in the early days. Neighbouring Kalgoorlie, 40km (24 miles) to the north-east, was named after an Aboriginal term, karlkurla, pronounced 'gull-gurl-la', used in the area by the local Wangkathaa (or Wongutha

FIGURE 6.4: Herbert Hoover, far left, at two mia mias or gunyahs (shelters). Hoover is holding in his right hand a spear and a nulla nulla or waddy (hunting stick) and in his left hand a woomera (spear-thrower), a boomerang, and a nulla nulla. The woman next to Hoover is holding a spear or waddy. Behind her is a small boy to the right and another older man to the left (obscured), who does not want to be in the photo, perhaps showing his distrust and dislike of the palpable and oncoming cultural change. The man next to her, sitting, also has a boomerang in his right hand and a spear in his left; the next man is holding two nulla nullas. The man on the far right is wearing a headband typical of both men and woman of the region. Their camp's mia mias are made of brushwood or sapling, with some canvas added. Wilson made no mention of meeting Aboriginal people in his travels. Mining wealth brought few benefits to them until recently. Image: Herbert Hoover Presidential Library, West Branch, Iowa. With permission.

or Wongi) people. It is the name of the fruit of the silky pear plant (*Marsdenia australis*).[16] This plant is also known as the cogla vine or bush banana and has a large edible fruit (Figure 6.5). There was pressure in the 1890s to give the new town a European name because it was thought by some that two similar-sounding names, Kalgoorlie and Coolgardie, would confuse people. However, John Forrest, the Premier of Western Australia at the time, was an explorer of inland regions and keen to keep Indigenous names. They never appear to be confused to this day.

FIGURE 6.5: Cogla vine (*Marsdenia australis*), known as karlkurla, silky pear or bush banana, near Black Flag Lake, north of Kalgoorlie. It grows by winding itself up other trees, often Acacia. The fruit fits neatly in the palm of the hand and is edible when young. Kalgoorlie is named after this species, which Wilson collected near Widgiemooltha on 22 November 1920 (#568 HH), writing 'Leichhardtia, flowers greenish-white [ ]'. Image: M Grose.

The history of preserving Indigenous names was strong in Western Australia. The concept of retaining local names had been espoused much earlier by the British geographer Captain Francis Beaufort, who wrote in 1818 that 'it appeared to me judicious to retain the vernacular names, wherever they could be distinctly ascertained, than to adopt those applied by other foreigners. The custom of inventing new names is still more pernicious to the true interests of geography'.[17] Many Western Australian towns bear indigenous names, as noted in the previous chapter. An early Governor of WA, John Hutt, instructed that naming policy, noting that retaining indigenous names for waterholes and springs 'in an irregularly watered country like Australia, is most valuable, besides that is only an act of fair justice to the first inhabitants or discoverers of any spot, to retain the name that they may have conferred upon it'. There was also a practical reason. New names would not be recognisable to Aboriginal people, who would then not be able to guide travellers to water.[18]

In 1891, the Royal Geographical Society noted that a standardised system should be adopted for British Colonies, at the same time as the United States was standardising native American names. There was comment in 1900, in regard to Aboriginal names, of 'recognising the advantage of having some definite system to work to, instead of leaving it to each explorer or surveyor ... some of whom seem to think that the principal thing is to get as many double vowels as possible'.[19] The writer might have been thinking of Kalgoorlie and Coolgardie, or Toodyay, Doodlakine, and Baandee, towns that Wilson passed by on the way to the Goldfields, or Widgiemooltha, which he was to visit, south of Coolgardie.

Around Kalgoorlie, the major regional centre for the Goldfields, many Aboriginal languages are spoken. They number more than twenty, each with their own vocabulary and grammar; some differences are overt and others subtle. They fall into two main language groups—the Mirning group (Mirning, Marlpa, Ngadju, Kaalamaya) and the Wadi group (Pitjantjantjarra, Yankunytjatjarra, Maduwongga, Cundeelee Wangka, Kuwarra, Wangkatja, Ngaanyatjarra, Ngalia, Tjupan, Manytjilytjarra, Pintupi) and others. Like many first nations languages all over the world, original Australian languages like these are under threat due to their small speaker population size, the small range where they are found, and the pressure of a major language such as English.[20] Research is being carried out to document and create dictionaries of these languages and to encourage the number of speakers to grow.

Though the letter from Kessell in the *Callitris* cupboard confirmed that Charles Gardner was in the Goldfields with Wilson, Wilson's diaries make no mention of him. Yet Gardner would have been a vital companion because he was the authority on the plants of the region. Gardner was a newly appointed Forest Department collector in 1920 and was no doubt the man being referred to when Wilson noted that Lane-Poole 'has a collector continually in the field'.[21] Gardner is in many images in the Wilson collection at Harvard. It is unclear as to where some of the images of this part of the Expedition are from, whether from Lane-Poole, or from Wilson, because many images held in the Arnold Arboretum and catalogued as Wilson's images are also seen in *The Primer of Forestry for Western Australia* compiled by Lane-Poole and published in 1922.[22] Wilson had noted that many images he had of the major trees in Western Australia were given to him by the ever-helpful Charles Lane-Poole; it is likely that all images were taken on the same trip.[23]

Charles Austin Gardner was an Englishman like Wilson. Born in Lancashire in January 1896, he had arrived in Australia in 1909 with his parents. Though without formal training in botany, he became the Government Botanist of Western Australia in 1929 and remained in that position until 1960. When Wilson met him, Gardner had already made a name for himself as a watercolour artist, specialising in landscapes and botanical paintings. He had studied landscape painting under James Walter Linton, the son of an English lord. Linton had studied at the Slade School of Fine Art in London, and gone to Western Australia to investigate issues related to his father's investments in gold mining.[24] In 1920, Gardner exhibited at the West Australian Society of Artists, mainly scenes of Perth and along Perth's Swan River. He had been encouraged by Desmond Herbert and the artist and journalist Emily Pelloe in combining botany and art.[25]

In *The Toxic Plants of Western Australia*,[26] by Gardner and the veterinary pathologist Dr H. W. Bennetts, published in 1956, many of the fifty-two watercolours were done by Gardner.[27] His watercolours include the very pretty Roe's poison, *Oxylobium spectabile* (named after John Septimus Roe, the first Surveyor-General of Western Australia, who had first recorded the deaths of stock due to poisoning from eating native plants in the early days of European settlement),[28] slender poison (*Oxylobium heterophyllum*), prickly poison (*Gastrolobium spinosum*), Champion Bay poison (*Gastrolobium oxylobioides*), Hutt River poison (*Gastrolobium propinquum*), spike poison (*Gastrolobium glaucum*), woolly poison (*Gastrolobium tomentosum*), wedge-leaved rattlepod (*Crotalaria retusa*), Birdsville indigo (*Indigofera enneaphylla*), whitewood (*Atalaya hemiglauca*), rusty poison (*Cryptandra leucophracta*), the frighteningly-named strychnine bush (*Strychnos lucida*), with its large fruits that look enticing to stock, caustic bush (*Sarcostemma australe*), and cabbage poison (*Velleia discophora*).[29] These poisonous plants intrigued Wilson in 1920, and Gardner might have been influenced by Wilson's keen interest in this group of plants all those decades before.[30]

In his role as a collector for the Forests Department, Gardner was expected to do an enormous amount of field work. He had been collecting in September, October, and November 1920 in the Goldfields area, was there collecting with Wilson in November 1920, and then remained in the Goldfields after Wilson's visit until March 1921—an entire summer. Gardner knew this region well and would have been an immensely knowledgeable companion.

Around the Kalgoorlie–Coolgardie Goldfields region, Wilson's plant collecting party appears to have travelled around in a Ford Tourer (Figure 6.6), with Wilson's chauffeur likely to have been William Finlay, who Wilson mentions as their driver in the Wheatbelt[31]. The motor car (or simply 'motor', as it was referred to by Wilson in his diaries) was key to getting about for a group of men, quickly and with less discomfort than by horseback. The Model T had been built by Ford in Detroit from 1908 and had quickly become popular in Australia; it was ideal. It was 2.9 litres, 4 cylinders, 20 horsepower, water-cooled, with thin tyres and a simple engine. Importantly for bush roads, it had good ground clearance, an ability to ride over stumps and was able to go through water, and, although of a lightweight design, it was rugged and tough enough for rough, unsealed roads.[32] It was also very economical on fuel, which was stored under the front seat and needed to be checked with a simple dipstick. The oil was checked by opening a valve under the car and just having a look.

With the motor, Wilson and party visited Black Flag Lake for *Callitris glauca* (now *C. columellaris*), a coniferous tree. Lane-Poole called this tree the Goldfields pine and described it as the 'largest of the pine shrubs of the Goldfields'. This small conifer attains a height at most of thirty feet, with a 'cedar-like appearance. The leaves are of a bluish-green, the bark almost black and fibrous. This tree ... occurs on the margins of salt lakes in open country. It is particularly

FIGURE 6.6: Here we have the typical botanical collector in the field in the Australian bush in summer in November 1920, complete with three-piece suit, tie, watch-chain, and a smart hat. The small trees are *Callitris glauca* (syn. *columellaris*), white cypress pine. This image was taken 12 miles west-north-west of Kalgoorlie and specimen #571 was collected. Wilson noted in his diary's specimen list 'see photograph'. This is arid country, with an average annual rainfall of <300 millimetres (<12 inches) and intense heat, yet the woodland has great species diversity. Note the Model T Ford Tourer in the background, with waterbags. The person is Desmond Herbert, identified by his masonic star. Image: Wilson Y-400 AA.

valuable as a fencing timber, since it resists white ants'.[33] Wilson had seen his first *Callitris* a few weeks earlier near Merredin.[34] He wrote in his diary that 'it is densely pyramidal in habit & this & its green colour make it conspicuous from a distance. The bark is rough & fibrous'.[35]

Today near Black Flag, huge, reticulated haulage trucks taking ore to be processed pass every fifteen minutes in the distance, then silence swallows them once more. In this landscape many cogla vines with their fat, green fruits are draped over prickly trees (see Figure 6.5 on page XXX). Here, I found Wilson's *Callitris* just as he must have seen them one hundred years ago; individuals stand out in the bush with their cedar-like appearance and their more chromatic green in a grey-green world (Figure 6.7). As Wilson noted, it is a brighter green than

FIGURE 6.7: *Callitris glauca* (centre) near Black Flag Lake, 2021, showing its very different form to the eucalypts in the distance. Image: M Grose.

other flora in the region (Figure 6.8). *Callitris* were in good company among family-like groves of mallets such as gimlets. The silence was outstanding. Even if with human company, chat would have seemed profane in this woodland. Little things are observed. Ants, nuts, ironstone bits, dead leaves, cogla vines full of smooth, green, handsome fruit, the wind making a different sound in each tree species, lizard-tracks, the glint of spiderwebs confidently spanning woody stems to woody stems from bush to bush across the red soil. Ants were busy and purposeful, though some seemed undecided and having to ask for advice and direction from others along the route, as I had because I missed the turn-off to Black Flag Lake amid the many roads that led to mines. The bushes and trees are scattered, as if each plant needs space and demands respect and separation, though I sensed, as I walked about as lightly and quietly as I could, that all is chattering underground in complex connections. What stories they might tell of changing landscapes and climates. Collectively, perhaps we are to them just a passing vapour that appears for a little time and then vanishes.

Callitris are members of the cypress family of the conifers, the Cupressaceae, a great and ancient family that existed on Pangea more than 200 million years ago before it broke up into Laurasia and Gondwana. That early beginning on the single supercontinent suggests why they look more like the plants of the Northern Hemisphere than eucalypts or banksias. Callitris-like trees spread widely across Gondwana.[36] *Callitris* survived the huge climatic changes from wet forests to dry aridity after Australia separated from Antarctica. They did so by adapting their xylem—the water transport system that keeps leaves connected to their roots—to more xeric conditions, to become Australia's most successful gymnosperm.[37] 'The current world title for the most water-stress tolerant xylem comes from a small tree species of *Callitris* that inhabits desert margins in southern Australia'.[38] It is *Callitris tuberculata*, from Western Australia, and its hydraulic conductance is maintained at pressures 'remarkably close to the practical limit of water metastability, suggesting that liquid water transport … has reached it operational boundary'.[39] In other words, the physics of water uptake gives a limit to coping with desiccation, one of the greatest challenges to plant life on the terrestrial planet, and this species operates at that limit.[40] This is why trees are usually excluded from the most arid ecosystems in the world, but not in Australia, as

Wilson noted with great interest (Figure 6.8). Wilson's biographer Roy Briggs noted that Wilson was known to call trees 'chaps'.[41] What respect we should show these 'chaps'.

FIGURE 6.8: A very green conifer in the arid regions, from Larter et al. (2015). 'The inset shows a close-up view of the tuberculate female cones of this species (photograph by A. Wesolowski)'. Note that *C. tuberculata* has been classified as synonymous with *C. preissii* and *C. columellaris*.[42] With permission.

Sargent had long been keen to have an excellent collection of conifers from around the world at the Arnold Arboretum. Wilson was later to write to Sargent about conifers planted in the Hobart Botanical Gardens. In a letter to a New Zealander who had a fabulous collection of conifers, Sargent wrote:

> We have here in our herbarium what we believe to be the finest collection of the conifers of the northern hemisphere in existence but we are still sadly weak in those from the region south of the equator. We have specimens of a great many of the New Zealand conifers but they are fragmentary and sometimes only from cultivated-plants. We very much desire to improve this collection and to obtain copious material of all the New Zealand conifers, including flowers leaves and fruits enough to make a good showing.[43]

Though the conifers are a wonderful soothing green, this is hard and arid country. It is fly country. In summer, when Wilson and his companions were here collecting, bush flies are at their peak. The flies (*Musca vetustissima*) crawl, buzz, and annoy eyes, ears, nose, and mouth. Bush flies get under sunglasses, get stuck in hair, crawl up noses, and hang like coats on the back. They cluster, swarm, niggle, irritate, and nip. The bush fly is a dung fly and noted for liking to walk over faces and faeces, in alternating order; they carry pathogens. Don't open your mouth.

These flies would have been like nothing Wilson had experienced before. Wilson noted in his diary of the 'plague of flies'[44] in the bush and that the flies were 'maddening'.[45] Herbert Hoover had thought the same thing about this region's bushflies. Hoover wrote to his cousin Harriette Miles about the flies:

> The Australian fly is much inferior, more vicious and less energetic than the American fly ... He always makes for one's eyes, so we always wear nets, which keep away what little air may be stirring. Yesterday, my cook made a bucket of cocoa and left it sitting by the fire while he went to the tent. When he returned, not three minutes later, he fished 391 flies out of it.[46]

Hoover might not have been joking.

Another pest of the region, particularly in Wilson's time, was the rabbit. European rabbits had been introduced in eastern Australia by early settlers but failed to survive in bushland. However, in 1859, a mere twenty-four wild rabbits were deliberately released in Victoria to provide sport for wealthy settlers.[47] They flourished and spread in plague proportions. The government of Western Australia tried to erect barriers to prevent rabbit movement westward, creating a series of 'rabbit-proof' fences, but to no avail; rabbits can burrow, and many fences were erected after rabbits had already arrived. They have been a disastrous pest of Australia's ecology, for plants, animals, and soils, with reduced agricultural production. Wilson described them as a 'bad pest'. In the Wheatbelt he had visited a section of the Number 1 Rabbit Proof Fence, which ran from Esperance on the south coast to Eighty Mile Beach near Broome on the north coast—an enormous distance, and he noted 'plentiful' rabbits near Number 2 Rabbit Proof Fence.[48]

The day after I went to Black Flag Lake, I sought out the site of my grandfather's hat shop in Hannan Street, Kalgoorlie's most famous street. Closed after my grandfather went to the Great War in 1916, it is now an Indigenous art gallery, and I was to be surprised by a rabbit story while there. A group who had come in from the desert were sitting on the pavement outside waiting for the shop to open to sell their paintings; we chatted. English was their second language; some spoke little but their own; I spoke none of theirs. We waited. I looked at my watch and kept looking. 'She'll come', a woman on the footpath said. I stopped looking and sat with these desert people, who knew patience better than I did. The curator, Monika Dvorak, arrived, and I asked about her surname. Yes, in the arid country of the goldfields of Western Australia works a direct descendent of the famous composer of the *New World Symphony*.

Monika told me a rabbit story related to the hat shop. There was a butcher next door, and the two shops shared back-of-shop premises; the hats were made from the pelts of rabbits. One day, a snobby woman asked the butcher for a rabbit. He showed her one, but she sneered that it was rather too small. Oh, said the butcher, I'll get another one from the back of the shop. Knowing that he only had the one rabbit left, he and his assistant pulled and stretched the rejected rabbit so that it appeared longer. He went back and presented the long version to the woman. Splendid she said, that is a much better rabbit! But, she said, I'll take both rabbits. I'll take the smaller rabbit too.

FIGURE 6.9: The quandong, *Santalum acuminatum*, also known as native peach, looks innocuous in a black-and-white photo, but is a distinctive little tree with reddish, round fruits that can be made into jam; they are high in vitamin C. Image: Wilson Y-396 AA.

Many plant species and animals are depicted in the paintings. One, by a Ngaanyadjarra woman from arid country north of Kalgoorlie, featured a large honeypot ant, a specialised worker ant, with grevillea (likely *Grevillea secunda*), quandong (*Santalum acuminatum*) (Figure 6.9), bush tomatoes (*Solanum centrale*),[49] and the cogla vine (*Marsdenia australis*) that had been growing in abundance at Black Flag, and the famous witchetty grub. Witchetty grubs, a high source of protein, are the larval stage of the wood-eating moth *Endoxyla leucomochla*. These moths are associated with the granite or Wanderrie wattle, *Acacia kempeana*, which is found all over arid and semi-arid Australia, and with other wattle species. Monika said that the flowers of the grevillea depicted were used by indigenous people to dip into water and suck the sweet nectar that exuded, as for many grevilleas and banksias. Some of the elderly ladies now in aged care homes love to receive these blooms as gifts from the bush, and they suck the sweetness, close their eyes, and remember the times of their youth in the desert, the scent of bush and red dirt, and the swish of grevillea blossom in the mouth and its sweetness on the tongue.

All Aboriginal people had been first-class botanists, ecologists, zoologists, and meteorologists, and learnt their lessons from earliest childhood from Elders and aunties and uncles. They learnt what to eat; whether it was eaten raw and, if not, how to process it; where it was likely to be found and in what season; what animals to follow to indicate plant availability; how to cook it; what was poisonous raw but edible when worked and cooked in what manner; what trees stored water in their roots; and what woods stored edible insects or small reptiles that were all good protein. This was an intensely refined and detailed knowledge of their local physical environment and its resources, which included the spiritual and ceremonial; these latter usually directed their manner of living in and with the land, and often supported a 'spiritual propagation' via maintenance ceremonies.[50] All could read what country was telling them through their personal long-term contact with the land, which we all once had but have now largely lost. Many old traditions and local wisdom gained from inherited knowledge passed on for hundreds of generations living in a particular region have been lost in the last years all over the world. It is a symptom of modernisation. What we have lost we can never reclaim, though we might try or even ache to remember. Such a widespread global loss of local knowledge of place reinforces the importance of assisting Aboriginal Australians to pass on their knowledge, wisdom, insights, spirit world, and language of their local and regional country, in all its heterogenous nature. Wilson appears to have had no contact with Indigenous people in his travels in Australia, though he is likely to have seen some tribal people on the edges of the arid lands of Kalgoorlie.

There is a long history of aridity in this flat, dry, tectonically old continent. The arid zone is Australia's largest biome today, with about 70% of the entire continent arid.[51] The north is tropical and only the eastern coast and south-east and south-western corners are mesic. The arid area is made up of a wide range of environments, including sandy and gibber deserts, ranges, and coastal plains such as the Nullarbor Plain that Wilson was soon to travel across. Vegetation varies. There are 'shrub woodlands, acacia and mallee-eucalypt shrublands, spinifex grasslands, tussock and hummock grasslands and chenopod shrublands'.[52] Wilson, being mostly interested in trees, was shown mainly woodlands in the arid zone.

This southern interior section of Western Australia that Wilson visited with Gardner and Herbert has large areas of semi-arid acacia and eucalyptus woodlands. This is part of the largest temperate woodland in the world, the Great Western Woodland. In this region of extraordinary biodiversity, '250 million years of continuous biological lineage have given rise to the red earth, the crusted granite domes, the blinding salt lakes and the gnarled tree trunks' so loved today.[53] It is a mosaic of ecosystems. There are about 13 million hectares (32 million acres) of acacia woodland called mulga (*Acacia aneura*), and 8 million hectares (20 million acres) of eucalypt woodland and forest.[54] Many of the trees are eucalypts, and many are quite small trees. All have attractive colours and grains, though they are 'very high density and timber is difficult to mill and dry'.[55]

Wilson said to the press that Western Australia needed 'some sort of bureau where our forest products can be chemically examined to see that proper use can be made of them'.[56] Now, members of the Goldfields Speciality Timber Industry Group (GSTIG) meet in Kalgoorlie twice a week in a large workshop adjacent to an enormous mound of old mining tailings. GSTIG aims to enable greater understanding of the less well-known timbers of Western Australia and their potential uses. I admired a gorgeous piece turned by Doug made from western myall (*Acacia papyrocarpa*), a small tree 3–8 metres (10–26 feet) tall. The piece was sensuously and silkily smooth, with some wood as tough and dark as ebony, then deepest brown with paler sections. Rightly, Doug would not part with it for anything.

Research is being done by GSTIG to work with the strong characteristics of these arid-zone timbers. There are hundreds of species of eucalypts that could be examined. As a student, and remembering my grandfather's life in the Goldfields, I had purchased Chippendale's 1973 book *Eucalypts of the Western Australian Goldfields (and the adjacent wheatbelt)*.[57] This veritable bible of eucalypts of the arid woodlands has now been expanded due to the great deal of research that has been done on Australia's eucalypts in the last few decades. Wilson had commented that Australians knew little about their own major genera the *Eucalyptus*, and he would surely have been fascinated to hear what is known today of their history, genealogy, geographical range, properties, evolution, and great diversity of form. Some of this diversity can be seen in the extraordinary range of fruits of the eucalypts (Figure 6.10).

FIGURE 6.10: A selection of eucalyptus fruits—some large, small, soft, tiny, spikey, rounded, knobbly, fat, narrow—seen in a sample box in the GSTIG shed in Kalgoorlie. Image: M Grose 1014. Trees under study by GSTIG for their timber properties include the redwood (*E. transcontinentalis*), the Goldfields blackbutt (*E. lesouefii*), the famous mallet known as the gimlet (*E salubris*) and salmon gum (*E. salmonophloia*), black morrel (*E. melanoxylon*) and red morrel (*E. longicornis*), the giant mallee (*E. oleosa*, called red morrel in Wilson's time) (Figure 6.11), Cleland's blackbutt (*E. clelandii*), merrit (*E. flocktoniae*), the salt gum of the salt rivers and lake systems (*E. salicola*), and the widespread grey mulga (*Acacia aneura*), gidgee (*A. pruinocarpa*), western myall (*A. papyrocarpa*), and minniritchi (*A. grasbyi*), named for its bark.

Several acacias have the unusual and beautiful minniritchi bark type. Typical minniritchi is red to red-brown and exfoliates in narrow curly strips that look like shavings that 'curl retrorsely from each end' (that is, they curl downward) on lateral branches and in some species the curling extends to the upper branches.[58]

In addition, GSTIG were investigating black oak (*Casuarina pauper*), the beautiful pixie bush (*Eremophila oldfieldii*), and turpentine bush (*Eremophila fraseri*), named after its scent, and various members of the Proteaceae—beefwood (*Grevillea striata*), corkwood (*Hakea suberea*), that does indeed have a thick corky bark, the delicately flowered native willow (*Pittosporum phylliyreoides;* now *P. angustifolium*) and the scented sandalwood (*Santalum spicatum*). The large range shows the complexity of the landscapes and the rich variety of flora in the Goldfields region's 'desert forest'.[59]

FIGURE 6.11: *Eucalyptus oleosa* var *longicornis*. This species occurs either as a tree or a mallee. Gardner discussed the many names for this species in a 1953 publication, *Journal of Agriculture* 2, 414. The image used was this one, accredited to 'E. H. Wilson, by courtesy of the Forests Dept' of Western Australia. It seems that Lane-Poole had copies made and they have been used by others.

FIGURE 6.12: Gimlet woodland north-east of Kalgoorlie, with an understorey of pearl bluebush *Maireana sedifolia*, showing the beauty in colour and form of this woodland. This bluebush lives for up to 300 years. Image: M Grose 1126.

These woodland trees have wonderful crowns that create a pillowing and billowing effect when seen with hundreds of trees (Figure 6.12). The wood of these arid-zone trees displays a great range of colours and textures. I asked George from GSTIG if he had a favourite tree to work with and, after only the briefest pause for thought, he said 'salmon gum' (*E. salmonophloia*). Salmon gum, the tree loved by Wilson, is one of Australia's glories. George, Baden, and Doug worked the following Saturday on a plank of salmon gum, passing the plank though a planer time and time again until very smooth. The tree itself has shades of pink in its trunk—hence its name—and the pink colours are seen throughout the timber (Figure 6.13). It is extremely pretty. Wilson had written that

> On the Goldfields areas where the rainfall varies from 5 to 10 inches this tree grows a hundred feet tall with a clean, polished trunk full ten feet in girth.... Indeed but for this tree and the Gimlet (E. salubris F.v.Muell) it is questionable if the goldfields of Western Australia, which have to date

yielded upwards of four hundred million dollars worth of gold, could have been developed.[60]

This is an interesting comment as it shows that Wilson was aware that the great problem of the early days of the Goldfields was the absence of water and how to get it. Wood was needed for the purification of water, for both domestic use and the mining companies.

FIGURE 6.13: The timber of salmon gum, showing a wonderful pink and darker grain. Image: M Grose, at GSTIG.

Many of the goldfield's species are very hard woods, and warp as they dry. Because of warping, these hard timbers are suited to speciality uses rather than larger furniture or floorboards (Figure 6.14). Neil Turner, who is a member of GSTIG and an advisor to them, has a specialist woodwork and design studio in Stratham, near the south-west town of Bunbury. Neil had been a wheat and sheep farmer but took to woodworking to ease the mental anxiety of droughts and poor seasons.

Sandalwood is one of Neil's favourite woods. Not only does it give a wonderful aroma, but the oils in the wood give a smoothness and ease of movement to any piece that he is working. It is oilier than eucalypts, which are

FIGURE 6.14: Wooden spoons from Western Australian timbers made by Neil Turner, who is based near Bunbury, Western Australia. The tree species from which the spoons are made are (from left) tuart (*E. gomphocephala*), wandoo (*E. wandoo*), two spoons from sandalwood (*Santalum spicatum*), red mallee (*E. oleosa*), lace sheoak (*Allocasuarina fraseriana*), where the lace is produced by medullary rays (cellular structures perpendicular to the growth rings and only found in some trees), and black mulga (*Acacia citrinoviridis*), also known as river jam and used experimentally in guitar and recorder-making. Wilson believed that the 'rightful purpose' of sheoak (second from right) would be for 'the making of the best furniture',[61] as it is now used. Image: M Grose.

famous for the oil from their leaves, which also makes them highly flammable. In Australia it is not the Indian sandalwood, *Santalum album*, which is a tropical tree of southern India and south-east Asia, but *Santalum spicatum*, an Australian native not found elsewhere.[62]

Two aspects of sandalwood from this region are perhaps surprising. First, Australian sandalwood (*S. spicatum*) supplies over half of the international sandalwood trade, and second, sandalwood from Western Australia has been traded since 1844 to eastern and southern Asian markets, where it is used for incense and joss sticks (joss is a type of glue made from the bark of *Litsea glutinosa*) for religious purposes, known as the agarbatti market.[63] Sandalwood is thus one of Australia's oldest exports. Wilson first saw sandalwood in the Wheatbelt and wrote that 'we saw a few small trees of sandalwood but all the large trees have been felled'.[64]

A very slow-growing and long-lived small tree, sandalwood is found in the central arid area of Western Australia extending towards the border with South

Australia. These gnarled and scraggy little trees, described by Wilson as 'rather ugly'[65] are often found on breakaway slopes ¾the scree slopes. While nothing to look at, even a tiny piece of wood gives a marvellous and cleansing scent. Sandalwood always needs to be treated with great respect. More than a century ago, however, sandalwood trees were ruthlessly harvested. Joseph Maiden, a botanist who Wilson was to meet in eastern Australia, wrote in 1903 that 'sandalwood is collected by pulling up the whole shrub or small tree by a horse and chain'.[66] Today, sandalwood cutting licenses are rare and highly controlled. Illegal harvesting and theft of sandalwood[67] is of major concern and there are large fines for either individuals or corporations.

Sandalwood trees are root hemi-parasites[68] and require a host tree for healthy growth. Their host tree in Australia is often an *Acacia*, a nitrogen-fixing plant. Nutrients and water provided to the sandalwood by the host are delivered by haustoria, which are root structures or organs unique to parasitic plants. A haustorium 'penetrates deeply into the root or stem of a host before establishing a physiologically effective contact with host xylem or phloem'.[69] The haustorium then extracts water and solutes for the parasite. The sandalwood tree itself has a small root system. This type of parasitism is highly successful and is common in nutrient-poor soils such as those in Australia.[70] It is many years before the heartwood, which is the source of the scent, is ready for harvest.

FIGURE 6.15: Desmond Herbert in the Goldfields, November 1920, with *Eremophila scoparia*, Broom bush (nguntu), and eucalypts in the background. Note the wide bright sky, the sharp foliage of the bushes, the grasses (likely *Aristida contorta*), and the sparse crowns of the eucalypts. Image: Wilson Y-392 AA.

Wilson travelled extensively in his short time on the Goldfields despite the heat of summer. In late November, he left Coolgardie to look at mallees. With Gardner and Herbert, he collected near Norseman (Figure 6.15), 187 kilometres (116 miles) south of Kalgoorlie, passing through Widgiemooltha, now an abandoned town with just a petrol (US 'gas') stop.[71] Around here, on 22 November 1920, Wilson found plant gold.[72] Many species of eucalypts were collected, including the two main gimlets *Eucalyptus salubris* ('50 feet', #538) and *E. campaspe* ('35 to 45 ft', #530), and '*E. calycogona* var. gracilis (Snap &rattle to 15feet', #530), '*E. Griffithsii* 30 to 60 feet. fl white filaments slightly pink', #535), '*E. Le Souefii*[73] (Goldfields blackbutt 30 to 45 feet', #536), *E. flocktoniae* ('White Gum filaments yellow 50 to 60 feet', #537), *E. stricklandii* ('30 feet fl greenish-yellow', #540), *E. torquata* ('fl red 20-30 feet', # 543 and #562, #563), '*Eucalyptus Ewartiana* 15ft', #545), *E. oleosa* ('Morrell 60ft', #561, and *E. transcontinentalis* ('White Gum', #564). This range highlights the diversity of the eucalypts that amazed Wilson.

Also collected were the bush banana, the cogla, which he annotated as 'Marsdenia Leichhardtiare flo greenish white [twinning] [ ] [ ]' (#568) (Figure 6.5 on page XXX) and a host of *Eremophila* species—'*Brownii*' (#521), '*Drummondii* (fl purple)' #522, '*Oldfieldii* (? Fl red 4 to 10 feet erect & spreading. Camel Bush', #541), another blue flowering *Eremophila* (#566), as well as *Acacia, Callitris, Cassia eremophila, Scaevola, Grevillea, Dodonaea*, and others not initially identified (Figure 6.16).[74]

In the Goldfields collections, some labels had under the Expedition title 'Africa', 'India' or 'Australia'. Wilson crossed out many and wrote 'W Australia'. I wondered if he had printed labels before leaving Boston and was running out of them in the 'veritable botanic garden' due to the numbers of specimens that he was collecting. For *E. ewartiana*, he had struck out 'Africa'. This is a very pretty tree, with slight, heart-shaped leaves. Unlike eucalypts from the southern forests, the leaves of this arid zone tree on the specimen sheet are not dark brown with age but light, as if just recently fallen and faded on the dry woodland floor. The fruits are like baby birds' nests, just a cup. The Harvard Herbarium houses a specimen sheet from Lane-Poole of the same species, collected in Westonia on 6 June 1919, though Lane-Poole's second daughter had celebrated her first birthday in Perth the day

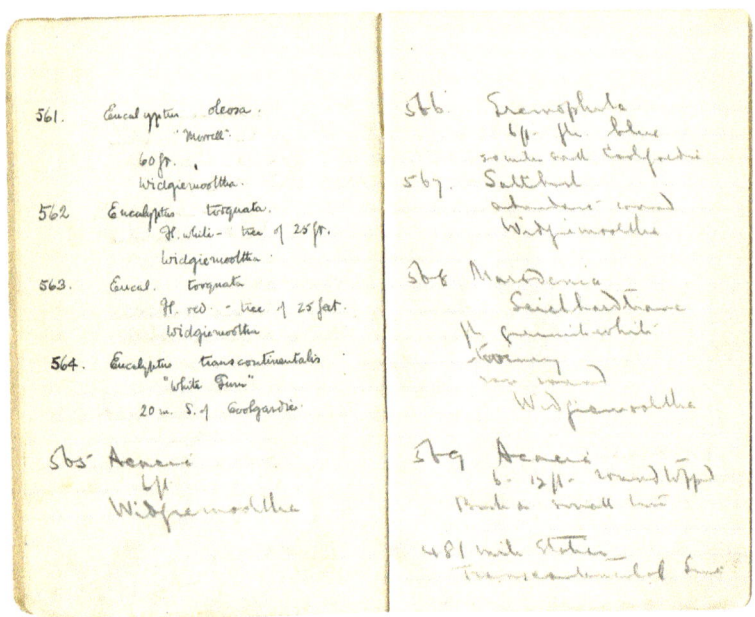

FIGURE 6.16: Wilson's diary showing another's hand (top left) and Wilson's own (lower left, and right) on the collecting trip to Widgiemooltha.

before. The longest period that Wilson had been at home with his wife Nellie and their daughter was less than two years;[75] it was only after his Australasian Expedition that this changed. One restless man with long absences from family had travelled in Australia with another man of the same ilk, it would seem.

Kalgoorlie in 1920 would have appeared very different to today. Wilson would have seen evidence that the land had been stripped of timber around Kalgoorlie and Coolgardie due to the need for timber to purify water. 'Nearly all the wooded areas north, west, and east of Kalgoorlie and within 50km south of Kalgoorlie were clearfelled' in the decades before Wilson's visit[76] (Figure 6.17). The timber had been used as fuel for domestic and industrial use. The problem of water was solved with the Goldfields Water Supply Scheme that Wilson had seen in Mundaring.

'Woodlines' were rail tracks that extended out from Coolgardie and Kalgoorlie into the surrounding woodlands and brought in these vast loads of timber. It was 'one of the largest industrial uses of timber for fuel anywhere in the world in the twentieth century'.[77] Almost 1.75 million tons of wood were cut from the region.[78] The land was so stripped of trees that by 1900, just seven years after gold had been discovered, the timber around Kalgoorlie had been cut out.[79] A woodcutter recalled in the 1920s: 'The bush boss, he used to make sure they cut every tree. There wasn't one tree left'.[80]

Concerns were raised by botanists about the extent of tree removal in the Kalgoorlie area as early as 1906. Dr Ludwig Diels (1874–1945), a German botanist based in Berlin, visited Australia some twenty years before Wilson. Diels was funded by the Humboldt Foundation for Nature Research and Travel, and he spent fourteen months collecting just in Western Australia.[81] Diels collected 4,700 specimens and his companion Ernst Pritzel 1,016, many from the arid inland of Western Australia.[82] Diels and Pritzel travelled along the route of the Goldfields Water Scheme Pipeline, as did Wilson on the train, and visited the areas around Kalgoorlie and Coolgardie, and collected to the north and south of those towns. Like Wilson, they were also collecting on the Goldfields in summer, when many eucalypts flower. Travelling through in 1901, Diels noted the extensive woodcutting of the regional woodlands. Between 1900 and 1960, clear-felling and selective cutting took out more than 2.8 million hectares (nearly 7 million acres) of woodland, a staggering toll that wrought havoc on the environment. 'From Kalgoorlie it progressively extended c.50km north, c.110km east, c.130km south-east and south-west, and c. 120km west'.[83] 'At any one time, there were woodlands yet to be cut, woodlands being cut, and woodlands regenerating after cutting'.[84]

Regeneration to restore the destroyed woodlands was begun as early as 1898, with clear-felled plots near Coolgardie. Lane-Poole set up experiments on revegetation in 1917 and 1919. Though there is only a record of plant specimens Wilson collected on the Goldfields, one image taken by Wilson appears to show

FIGURE 6.17: 'A typical woodcutters' camp', 1904. Farmers kept guns for stock and fox or dingo control. Here, the cutters have, ironically, set up their tent to gain some shade. There is a meat or food safe hanging from a tree, likely as far from ants as possible. The bottles with the camp oven are Hannan's beer (1895-1982) from Kalgoorlie; water was scarce in the bush. Identification of the beer bottles with thanks to the Beaten Track Brewery, Boulder-Kalgoorlie, and an anonymous Kalgoorlie bottle collector. Image: *Western Argus*, 13 September 1904.

this regrowth. Figure 6.18 shows Charles Gardner in front of a stand of 'thick regrowth gimlet around 10-15 years old' and thus too old to be Lane-Poole's experiments of only a few years before; there is fluting in the trunks, a feature of gimlets. But 'stands like this are very common in areas that regenerated after woodline clearing'.[85]

FIGURE 6.18: Charles Gardner in a regrowth stand of *Eucalyptus ravida*, the silver-topped gimlet, in the Kalgoorlie area in November 1920. This image also appears in Lane-Poole's book, where it is wrongly identified as *E. campaspe*, the silver gimlet (Kealley, personal communication, 2021). Image: Wilson Y-441 AA.

A great deal of resprouting has occurred since cutting along the woodlines stopped in the early 1960s.[86] After so many decades of abuse, it seems surprising that the woodland species did not reject human society but came back to live among us and show us their beauty, as if reminding everyone of what we destroyed. Ecological assessment of the woodlands as they are today suggests that the land appears to have largely recovered from the onslaught, but the structure of the revegetated salmon gum woodland differs from that of salmon gum in the original woodland. Revegetated trees have no large branches, with less bark and dead wood, with fewer large old trees and less structural complexity.[87] It is still unclear how old the felled trees were; some would have been hundreds of years old because eucalypts are known to live with multi-century timeframes, with individuals in the Great Western Woodland having great longevity, often more than four hundred years. With such great ages, it is no surprise to find that only half of the revegetated gimlets are reproductively mature after nearly fifty years.[88]

Histories of the woodlines do not say about the impact of destructive cutting of vast areas of fragile woodlands on the local indigenous people. This destruction is very contrary to their own sense of care for Country and the delicate balance of plants, soils, animals, and humans in spiritual connections with ancestors needed for the sustenance of the rhythm of the country.

Wilson was appalled by the destruction of this great temperate woodland. A core message on his departure from Australia was his concern for the 'wanton destruction' of forests. Always asked his thoughts by the press throughout his Australian trip, Wilson was asked his thoughts about the desert country around Kalgoorlie. But he said 'That is not a desert. There are trees growing there and other vegetation. It is an arid area certainly, but not a desert'.[89] Wilson was intrigued by trees in this region that grew tall although it was arid. He noted to the press that some of the country was *'all wrong'*, inasmuch as big trees flourished in dry areas. 'Those, it was thought, must have deep roots to get to water underneath, but that was dispelled on examination, for it was found that their roots were on the surface. It was a mystery.'[90]

From Kalgoorlie, Wilson continued his journey on the Great Western Express all the way to Adelaide and he complained of heat, dust, and boredom during the trip, commenting on the 'great nuisance' of having to change trains three times due to different track gauges between the Australian States at that time.[91] On his journey eastward, Wilson saw treeless desert when he travelled across the Nullarbor Plain, given the Latin name for exactly that. Once, long ago, the Nullarbor was a sea and its many limestone caves contain fossils of extinct animals such as the famous Tasmanian tiger, or thylacine. This railway line includes the longest straight stretch of track in the world, 478 kilometres (297 miles). Australian astronaut Andy Thomas remarked that the line is identifiable from space because of its unnatural straightness—a fine line, as if 'someone has drawn a very fine pencil line across the desert'.[92] The line crosses no permanent fresh water, or trees.

Wilson wrote:

> Thanks to facilities freely placed at the disposal of the Arnold Arboretum's Expedition by the Government of Western Australia and to the admirable arrangements made by the Conservator of Forests, Mr. C. E. Lane-Poole, I traversed some 2000 miles in the southern part of Western Australia. Alone I should have been completely lost among the extraordinarily varied and anomalous vegetation but the Conservator himself was my guide through all the important forest areas and through the sand plains and savannah regions I had the companionship of the Government Botanist, Mr. D. A. Herbert.[93]

Wilson felt that he 'was overwhelmed with courtesies and my heart is all of gratitude to Western Australia. Just everything possible has been or is being arranged for me'.[94] On arrival at Adelaide, he wrote to Sargent in Boston that he had an article for the journal that 'has been written hurriedly and under the difficulties inseparable from travel and needs a little polishing'.[95] He clearly liked to be working.

All through my stay in Roslindale, I lived on the edge of the Arnold Arboretum's grounds, which was a delightful location. Every night I admired the huge American beech out the window. Beyond it was a small cemetery housing men who had died during the Revolution and, dying in summer, heat had prevented their bodies from being taken the six and a half miles to Boston, and so they were buried in Roslindale, on a sloping hill. The site is marked by a flag, ever flying, flapping, draping, among the tall trees at the Arboretum's edge. It was seemingly neglected. But, perhaps, I thought, it gave that air because of the slippery path, patches of ice, piles of fallen leaves in late autumn, a scattering of lonely graves beside the busy peak hour road, the clapboard houses in pretty colours—some with Christmas decorations up as December came—and the bank leaking ice onto the footpath and along the street. And there, in my writing, is a lexicon different to American that Wilson, as an Englishman, would have been familiar with—we say 'footpath', while it is 'sidewalk' in America. From the sidewalk, the internment site is almost invisible. One must go up the steps to read the plaque on a puddingstone boulder: 'In Memory of the Soldiers of the Revolution Who Died in

Hospitals at Jamaica Plain and Were Buried in this Lot'. It is called the Walter Street Cemetery, and once housed forty men, who most likely died of smallpox.[96] Wilson, who was killed with his wife Hellen (Nellie) in a motor accident in 1930, was buried in Canada because his request was not to be buried in the United States, but in Commonwealth soil. He was an Englishman to the core. His last expedition to Australasia was almost entirely focused on British Commonwealth countries.

Back in Coolgardie, where Wilson had collected hard sclerophyllous grey-green vegetation, I visited the grave of another English Ernest, Ernest Giles (1835–97), who had died there of pneumonia at the height of the mining boom. It was a visit I had long wanted to make, to pay my respects to Giles, one of the greatest explorers of Australia in the nineteenth century. It was warm in the late autumn sun and the wind whispered softly in mallees and umbrella-like mallet eucalypts in the cemetery surrounding his grave, but they gave little shade. There was red, rusty, iron-clad dirt and sharp stones and small bush flies, the grave of a nurse murdered at the Coolgardie Hospital in 1898 and the man who had murdered her and then turned the gun on himself in the next suite of graves, with more than a quarter of the graves those of infants, mostly unmarked; some were of Afghans, who were cameleers across the arid interior and successful businessmen on the Goldfields.

Ernest Giles had, like Wilson, gone to isolated places on many expeditions. Giles, unlike Wilson, was largely unrecognised in his time, except by other explorers, who knew what trials he had endured and what country he had revealed. Giles travelled by camel and horse in five extraordinarily hard expeditions in extreme heat through 'heartless deserts' in 'useless searchings for water' (his own words), with dying horses, in glorious visual country later painted by artists and photographed by tourists. He did not find fine regions or land that might be improved for pasture or cropping but land of little pastoral value known for aridity, sand, stone, scrub, spinifex, flies, ants, heat, remoteness, and red dusty dirt. Giles wondered at fame, writing that: 'The explorer does not make the country, he must take it as he finds it; and … to the discoverer of the finest regions the greatest applause is awarded …' [97] Likewise, it seems, Wilson did not find in Australia pretty lilies or glorious deciduous trees to take home to the United States; he received no applause for his long trip to Australia and this journey was largely lost and then hidden as remnants in cupboards and herbarium stacks.

A person who found finance and supported Giles was the botanist and director of the Royal Botanic Garden, Melbourne, Baron Ferdinand von Mueller (1825–1896), who first described many eucalypts and other Australian species. Wilson wrote that von Mueller 'had correspondents all over the world and distributed seeds and specimens far and wide. During his lifetime he was the dominant figure in Australian botany'.[98] Giles's journals give information about trees, other plants, small gum creeks and 'pine-clad hills' of *Callitris*.[99] He noted details of plants and soil in the landscape of 'anomalous vegetation' that astounded Wilson, that 'fine looking oak trees … grow to a height of 20 to 25ft of barrel without a branch and then spread out into a fine and shady top. They … appear to inhabit the poorest region as far as soil is concerned, for they grow out of pure red sand'.[100] Like Wilson a few decades later, he was intrigued at how trees grow in this environment. How fitting it would be to plant some *Callitris* at Giles's grave, or one of the sheoak trees (*Casuarina* sp.) that he and his party took shelter under from the heat, though their shade is light, as Wilson had noted of Australian trees. Giles might have found comfort in the shade of the deeply leafy American beeches of the Walter Street Cemetery. But that is another place in a very different part of the planet.

NOTES

1. SL Kessell, 'Report of the Forests Department for the year ended 30 June 1922', Western Australian Parliamentary Papers, 1922–23, no. 14.
2. Cited in J Dargavel, *The Zealous Conservator: A Life of Charles Lane Poole*, University of Western Australia Press, Crawley, 2008, p. 60.
3. M Roe, 'Kessell, Stephen Lackey (Kim) (1897–1979)', *Australian Dictionary of Biography*, National Centre of Biography, Australian National University, vol. 15.
4. Recorded in a letter sent by the Conservator of Forests in 1923, then SL Kessell, dated 28 July 1923, from the Forests Department, Western Australia. The letter was not in the AA Archives but was found in the specimens of Callitris in the Harvard Herbaria. Lake Moore is far to the north of the area travelled by Wilson. Ian Kealley (personal communication) noted that *Callitris glauca* grows taller north of Kalgoorlie. An earlier letter from Kessell mentioned 'Mr Gardner, who has been able to pack up for you specimens …', and a booklet on Western Australian conifers had been sent to the AA in 1923.
5. Recorded in the *Boston Evening Transcript*, 30 September 1922, magazine section, p. 1.
6. Wilson, diary, 22 October 1920.
7. 'Black Flag Lakes', *Southern Times*, 13 June 1896, p. 3.
8. G Jeansonne, *The Life of Herbert Hoover, Fighting Quaker 1928–1933*. Palgrave Macmillan, 2012, p. 1.
9. W Coughlin, *Stanford Magazine*, March/April 2000, stanfordmag.org/contents/into-the-outback; also noted in *Coolgardie Miner*, 22 May 1897, p. 7.
10. Hoover had trained at Stanford, where he received his degree in 1895 in geology. He spent six months at the Sons of Gwalia Mine before being appointed Chief Engineer for the Chinese Bureau of Mines.
11. 'What did Hoover do in Western Australia?', Herbert Hoover Presidential Library and Museum, 2012, accessed 29 August 2022, web.archive.org/web/20120118014403/http://www.hoover.archives.gov/info/faq.html#Australia
12. H Hoover, *The Memoirs of Herbert Hoover: Years of Adventure 1874–1920*, Macmillan, New York, 1951, p. 32.
13. Information given at the Palace

PLATE 6.1: "Euc. campaspe Widgiemooltha". No. 539. 22 November 1920. *Eucalyptus campaspe* is the silver gimlet.

PLATE 6.2: "Eucalyptus calycogona var. gracilis 'Swamp Rattle' 15 ft tall 20 miles from Coolgardie." No. 530. November 1920. Now classified as *Eucalyptus celastroides*, known by the Noongar name of mirret.

PLATE 6.3: " Euc. Ewartiana 15 ft Widgiemooltha". No. 545. 22 November 1920. Now *Eucalyptus websteriana*, heart-leaved mallee.

PLATE 6.4: "Eucalyptus flocktoniae Widgiemooltha". No 537. November 1920. Known as merrit, the Noongar name.

PLATE 6.5: "Euc. torquata flo white Widgiemooltha". No. 562. 21 November 1920. *E. torquata* normally flowers pink or red-coral, and hence its common name of coral gum, but white floweres are also found, as here by Wilson.

PLATE 6.6: "Euc. Stricklandii Widgiemooltha." No. 540. 21 November 1920. Now *Eucalyptus stricklandii*.

PLATE 6.7: "Melaleuca Tree 15 ft flo white 20 miles from Coolgardie." No. 533. November 1920. Classified as *Melaleuca sheathiana*; common name is boree or borree, the Noongar name.

PLATE 6.8: "Hakea mulitlineata 20 miles south of Coolgardie". No 550. 22 November 1920. Common name is grass-leaved hakea, a glorious large shrub that flowers bright pink in winter and spring; thus the flowers not seen by Wilson.

PLATE 6.9: "Melaleuca 6 ft 20 miles south of Coolgardie". No. 547, 22 November 1920. Species not identified.

PLATE 6.10: "Grevillea 25 ft flo. Greenish and pink 15 miles south of Coolgardie." No. 525. "22 xi 1920". Classified as *Grevillea nematophylla*. The common name is water bush.

Hotel, Kalgoorlie. Found also in newspaper reports, such as: 'Hoover is on the spot' (at the mine) in February 1907 in Kalgoorlie; 'Mining', *Sun* (Kalgoorlie), 24 February 1907, p. 10. Hoover had arrived only a few days before, on 18 February, on the *Mooltan* (ship) from England, as reported in 'Mainly about people', *Daily News*, 18 February 1907, p. 4.

14 Herbert Hoover Presidential Library and Museum, accessed 24 November 2020.

15 See G Blainey, 'Herbert Hoover's forgotten years', *Business Archives and History*, 1963, 3(1):53–70; and A Heintz, *Herbert Hoover's Forgotten Australian Years*, Herbert Hoover Presidential Library and Museum, 1966; J W Kirwan, 'Hoover in W.A. Some Goldfield's Memories', *West Australian*, 6 November 1928, p. 16. Kirwan was president of the Legislative Council. Kirwan notes that Hoover's association with the Coolgardie Chamber of Mines 'may be regarded as Mr Hoover's first introduction to public affairs'. This comprehensive article on Hoover's life in the WA Goldfields gives a clear date of arrival of 1896, while other sources give 1897. Kirwan consulted with those who knew Hoover.

16 This information from the Karlkurla Dreaming Cultural Tour notes, 2020.

17 N Green, 'Explorers and the Aborigines', p. 424, in M Hercock and S Milentis with P Bianchi (eds), *Western Australian Exploration 1836–1845: The letters, reports and journal of exploration and discovery in Western Australia*, Hesperian Press, 2011, pp. 419–37. Green is citing F Beaufort, *Karamania, or, a Brief Description of the South Coast of Asia-Minor and of the Remains of Antiquity. With plans, views etc. Collected during a survey of that coast, under orders of the Lords Commissioners of the Admiralty in the years 1811–1812*, 2nd edn, R Hunter, London, 1818, p. 11.

18 Ibid. Green discusses this issue.

19 R C Clifton, 11 September 1900, in *The Spelling of Native Geographic Names*, issued by direction of the Honourable Minister for Lands, T Bryan Printer, Perth, 1901.

20 T Amano, B Sandel, et al., 'Global distribution and drivers of language extinction risk', *Proc. of the Royal Society B*, 2014, p. 281, doi:10.1098/rspb.2014.1574.

21 Letter from Wilson to Sargent from Perth, 17 November 1920, AA Archives, W.XIV:A, box 21, folder 11.

22 Mutual images are, from Wilson's list: Y-335 Jarrah; Y-340 Blackbutt; Y-326 Crimson flowing gum, *E. ficifolia*; Y-? WA Peppermint; Y-401 Salmon gum showing motor car; Y-391 Gimlet; Y397 Morrell, *E. longicornis*; Y-443 Goldfields blackbutt, *E. Le Souefii*; Y-440 Grey gum, *E. Griffithsii*; Y-438 Goldfield redwood, *E. transcontinentalis*; Y-444 Goldfields whitegum, *E. Flocktoniae*; Y-445 Goldfields yellow flowering gum, *E. Stricklandii*; next p. 85; Y-384 York gum, *E. foecunda* var. *loxophleba*; Y-450 Kurrajong, *Sterculia Gregorii*; Y-448, Goldfields pine, *Callitris glauca*; Y-405 Raspberry jam, *Acacia acuminata*, with camera case; and Y-393 Sandalwood, *Santalum cygnorum*.

23 Suggested by Ian Kealley in 2021.

24 A Gray, 'Linton, James Walter Robert (1869–1947)', *Australian Dictionary of Biography*, 1986, vol. 10.

25 N Stewart, 'Emily Harriet Pelloe (1877–1941)', *Australian Dictionary of Biography*, 1988, vol. 11.

26 Gardner is the author (appellation C.A Gardn.) of many of the poison plants in the book.

27 The majority of the watercolours in the book were done by Edgar Dell.

28 John Septimus Roe recorded the death of stock in his diary, 23 October 1835. Quote from p. 80, M Hercock, *The Western Australian Explorations of John Septimus Roe 1829–1849*, Hesperian Press, Carlisle, WA. The death of these animals is now considered almost certainly due to ingestion of the poison pea, a species of *Gastrolobium*-like box poison, prickly poison, or York Road poison, as noted by Alex George in Hercock (above, p. 80, footnote).

29 Gardner never married and bequeathed all his original paintings to the Benedictine Monastery at New Norcia, which was founded by Spanish monks in 1847; it is Australia's only monastic town. It was from the Abbott that Gardner learnt Latin, to assist in naming species.

30 Stephen Kessell and Charles Gardner published the *Key to the Eucalypts of Western Australia*, a larger work, in 1924. Gardner also published *The Wild Flowers of Western Australia* in 1959. He had intended to work on more volumes, but this work was never completed. He became well-known across Australia for his tremendous contribution to the collection and naming of species, with thousands of plant specimens housed in the herbarium of Western Australia. During his lifetime he described eight new genera and around two hundred new species. *Eucalyptus gardneri* is named after him. In 1937 Gardner became the first Australian Botanical Liaison Officer at the Royal Botanic Gardens, Kew, where Wilson had studied and worked in 1897 to 1898, forty years earlier.

31 Wilson, diary, 26 October 1920.

32 M Simpson, 'Henry Ford's Model T and its impact in Australia', Museum of Applied Arts and Sciences, Sydney, 2015, https://maas.museum/inside-the-collection/2015/07/30/henry-fords-model-t-impact-in-australia/

33 Lane-Poole, *A Primer of Forestry*,

With Illustrations of the Principal Forest Trees of Western Australia, Government Printer, Perth, 1921, p. 89. Note that 'white ants' is a term for termites.

34 Wilson, diary, 22 October 1920.
35 Wilson, diary, 25 October 1920, near Bruce Rock.
36 The history of the Gondwanan subfamily the Callitroideae is discussed in M D Crisp, L G Cook, et al., 'Turnover of southern cypresses in the post-Gondwanan world: Extinction, transoceanic dispersal, adaptation and rediversification', *New Phytologist*, 2019, 221(4):2308–19.
37 M Larter, S Pfautsch, et al., 'Aridity drove the evolution of extreme embolism resistance and the radiation of conifer genus *Callitris*', *New Phytologist*, 2017, 215(1):97–112.
38 TJ Brodribb, 'Progressing from "functional" to mechanistic traits', *New Phytologist*, 2017, 215(1):9–11.
39 M Larter, TJ Brodribb, et al., 'Extreme aridity pushes trees to their physical limits', *Plant Physiology*, 2015, 168(3):804–7.
40 Ibid.
41 R Briggs, '*Chinese Wilson*.' *A Life of Ernest H. Wilson, 1876–1930*, HMSO, London, 1993, p. 7.
42 A Farjon, *A Handbook of the World's Conifers*, Brill, Leiden, Boston, 2nd edn, 2017, pp. 238 and 243.
43 Letter from Sargent to Mr TW Adams, Esq., Greendale, Canterbury, New Zealand, AA Catalogue 967.
44 Wilson, diary, 27 October 1920. This is from his visit to the Wheatbelt with DA Herbert.
45 Ibid.
46 WJ Coughlan, 'Into the Outback', *Stanford Magazine*, March–April 2000.
47 F Fenner, 'Deliberate introduction of the European rabbit, *Oryctolagus cuniculus*, into Australia', *Rev. Sci. Tech*, 2010, 29(1):103–11.
48 Wilson, diary, 25 October 1920, near Quairading.
49 Bush tomato is also known as desert raisin or kutjera. Bush tomato is now being harvested by Aboriginal pickers commercially and grown commercially in selected regions. It is an arid lands plant.
50 See, for example, a discussion of values in P Sutton and K Walshe, *Farmers or Hunter-Gatherers? The Dark Emu Debate*, Melbourne University Press, 2021, pp. 24–45. See also WH Edwards, *An Introduction to Aboriginal Societies*, Social Science Press, 1988, p. 43.
51 H Lambers, 'Introduction', in H Lambers (ed.), *On the Ecology of Australia's Arid Zone*, Springer, Switzerland, 2018.
52 M Byrne, L Joseph, DK Yeates, JD Roberts, and D Edwards, 'Evolutionary history', in H Lambers (ed.), *On the Ecology of Australia's Arid Zone*, Springer, Switzerland, 2018.
53 A Watson, S Judd, J Watson, A Lam, and D Mackenzie, *The extraordinary nature of the great western woodlands*, April 2008, accessed 31 January 2023, https://www.researchgate.net/publication/236335929, 2363.
54 GR Siemon and IG Kealley, *Goldfields timber research report*, Department of Commerce and Trade, and others, 1999, p. 3.
55 Ibid.
56 'A study of trees: Professor Ernest H Wilson's visit. Our wonderful flora', *Western Mail* (Perth), 25 November 1920, p. 9.
57 GM Chippendale, *Eucalypts of the Western Australian Goldfields (and the adjacent wheatbelt)*, Australian Government Publishing Service, Canberra, 1973.
58 BR Maslin, 'Acacia (Leguminosae-Mimosoideae): A contribution to the flora of Central Australia', *Journal of the Adelaide Botanic Garden*, 1980, 2(4):301–21. See also BR Maslin, 'Four new species of *Acacia* section *Juliflorae* (Fabaceae: Mimosoideae) from the arid zone in Western Australia', *Nuytsia* 1999, 24:193–205. Retrorsely means turned backward or downward.
59 I Kealley and M Clews, 'Western Australia's desert forest', *Landscope*, 1995, 10(4).
60 Wilson, 'Notes from Australasia, No. 1', *Journal of the Arnold Arboretum*, 1921, 2(3):162.
61 *Albany Advertiser*, 27 November 1920, p. 4.
62 Wilson gives this as *Santalum cygnorum* Miq. in 'Notes from Australasia, No. 1', p. 163.
63 In the early years of settlement of Western Australia, sandalwood brought in 30% of the State's export revenue, accessed 25 January 2023, www.wa.gov.au/organisation/forest-products-commission/western-australian-sandalwood
64 Wilson, diary, 22 October 1920.
65 Wilson, 'Notes from Australasia, No. 1', p. 163.
66 Applegate, 1990. Cited in C Williams, *Bush Remedies*, Rosenberg Press, New South Wales, 2020.
67 'The harvesting of sandalwood is managed under the *Biodiversity Conservation Act 2016*. It has replaced the *Sandalwood Act 1929* and the *Wildlife Conservation Act 1950*, and it significantly improves the State Government's ability to protect native species and important biodiversity assets. The Act increases the penalties for the illegal harvesting of sandalwood and under the new legislation, the maximum penalties for illegally harvesting wild sandalwood are now $200,000 for individuals and $1 million for corporations.' https://www.wa.gov.au/organisation/forest-products-commission/western-australian-sandalwood
68 Hemi-parasites have functional chloroplasts and can perform photosynthesis. An excellent description of Australian

hemi-parasites is given in H Heide-Jørgensen, 'Hemiparasitic Santalales', in *Parasitic Flowering Plants*, Brill, Leiden, 2010, pp. 25–144.

69 J Pate, T Bell, and B Verboom, 'Unravelling the secret lives of plant root systems: Voyages of discovery "Down Under"', in *Biodiverse Vegetation of South West Australia with Implications for Cultivation and Conservation of Plants*, University of Western Australia Publishing, Crawley, 2020, p. 218.

70 MC Press, 'Dracula or Robin Hood? A functional role for root hemi-parasites in nutrient poor ecosystems', *Oikos*, 1998, 82:609–11.

71 Widgiemooltha had been previously spelt Wigimoola, Wijimoola, and Widgemooltha, showing a homing in on the English rendition of an indigenous place name.

72 The page now has the emendation of SGM Cann ? *Eucalyptus websterianna* Maiden, 11 July 1966.

73 *Eucalyptus lesouefii*.

74 I have kept the species names as Wilson wrote them; when named after a person, the tradition at the time was to capitalise the species name.

75 Noted in TM Holway, 'History or Romance? Ernest H Wilson and plant collecting in China', *Garden History*, 2018, 46(1):3–26.

76 I Abbott, 'Fauna and ecology of the Great Western Woodlands', in Bianchi, *Woodlines of Western Australia*, Hesperian Press, WA, 2018, pp. 329–34.

77 WA Museum notes about the woodlines.

78 G Smith, cited in B Bunbury, *Timber for Gold: Life on the Goldfields Woodlines 1899–1965*, Fremantle Arts Centre Press, South Fremantle, 1997, p. 15.

79 M Sharp, *The Goldfields Woodlines*, 2016, accessed 4 December 2021, www.outbackfamilyhistory-blog.com/the-woodlines/

80 Reported in Bunbury, *Timber for Gold*, 1997, p. 82. However, Ian Kealley, who spent thirty-four years researching goldfields trees, believes that the woodliners left the sandalwood.

81 Much of Diels's extensive collection was lost in the bombing of Berlin during World War II.

82 'In 1906 Diels published his observations in a book *Die Planzenwelt von West-Australien südlich des Wendekreises* (The plant world of Western Australia south of the Tropic). This was of 416 pages and a landmark, publishing new information on the flora, descriptions of the vegetation, and a small coloured vegetation map of Australia'. From JS Beard, 'The botanists Diels and Pritzel in Western Australia: A centenary', *Journal of the Royal Society of Western Australia*, 2001, 84:143–8.

83 I Abbott, 'Fauna and ecology of the Great Western Woodlands', in Bianchi, *Woodlines of Western Australia*, 2018, p. 329.

84 Bianchi, *Woodlines of Western Australia*, 2018, p. 336.

85 Ian Kealley, personal communication, 2021.

86 Bianchi, *Woodlines of Western Australia*, 2018, pp. 251–6.

87 See comments by botanist and ecologist Ian Abbott, in Bianchi, *Woodlines of Western Australia*, 2018, appendix.

88 CR Gosper, CJ Yates, GD Cook, JM Harvey, AC Liedloff, WL McCaw, KR Thiele, and SM Prober, 'A conceptual model of vegetation dynamics for the unique obligate-seeder eucalypt woodlands of south-western Australia', *Austral Ecology*, 2018, 43(6):681–97. Obligate seeders are those that, after fire, can only regenerate from seed.

89 Ibid.

90 *Observer* (Adelaide), 4 December 1920, p. 37. This comment was made by Wilson in Adelaide, where he had gone by train. It is a little surprising, since he had crossed the Nullarbor Plain, where there are no trees at all for hundreds of miles.

91 Letter to Sargent, from Grand Central Hotel, Adelaide, 28 November 1920, AA Archives, W.XIV, box 21, folder 11.

92 Penelope Debelle, 'Adelaide's own spaceman Andy Thomas', *Advertiser* (Adelaide), 10 October 2014.

93 Wilson, 'Notes from Australasia, No. 1', 1921, p. 160.

94 *West Australian*, 21 October, p. 7.

95 Letter to Sargent, from Grand Central Hotel, Adelaide, 28 November 1920, AA Archives, W.XIV, box 21, folder 11. The journal he refers to is the forerunner of the current Arnold Arboretum journal *Arnoldia*, known as the *Bulletin of Popular Information*, commenced in 1911.

96 See WH Marx, 'Revolutionary War burial site near arboretum', 2020, first published in the *Jamaica Plain Citizen*, 26 May 1988, accessed 16 November 2020, www.jphs.org/colonial-era/revolutionary-war-burial-site-near-arboretum.html

97 Written at the end of Giles's fourth expedition, which covered 4,000 km (around 2,500 miles) in 1875. Quotations from *Ernest Giles's Explorations, 1872–76*, South Australian Parliamentary Papers 1872–76, Friends of the State Library of South Australia, Government Printer, Adelaide, 1875, p. 332.

98 Wilson, *Plant Collecting*,1985 edn, p. 192.

99 From various notes in *Ernest Giles's Explorations, 1872–76*, introduction by V Hankel, 2000; 'pine-clad hills', 14 September 1872, p. 29; 'useless searchings for water', Monday 22 December 1873, p. 187.

100 Ibid., diary entry 26 August 1872, p. 7.

CHAPTER 7

'STOP MAKING HAVOC OF YOUR HERITAGE OF NATURAL FOREST'

Wilson's diary entries are scant for the entire Expedition, and scarcely exist for eastern Australia, although he spent months there. Of his time in Adelaide, little is said. From Adelaide he travelled on the train, briefly to Melbourne, then to Sydney, then sailing across to New Zealand in late January 1921, where he stayed for February and March. After returning to Sydney, he sailed to Hobart. Back in Sydney, he then went up the eastern seaboard of Australia to Brisbane by train, and visited Bribie Island (near Brisbane) and Fraser Island by launch. After seeing his collection off on the ill-fated ship *Canastota*, he sailed for India (see Table 7.1).

While Wilson stated to the press in Melbourne that 'I am anxious to see your forest plants in Gippsland',[1] there is no record of this visit to see the tall eucalypts of that area, which include *Eucalyptus regnans*, one of the tallest angiosperms.[2] Evidence of his travel itinerary and plant specimens are almost absent, except for a few pages of very short trips in New South Wales, which were partly to see some tourist sites and towns as well as trees.[3] It seems that after that early visit in New South Wales in late December 1920 and the first fortnight of January 1921, he did not keep a diary, or it was lost at sea. Information about eastern Australia is thus fragmentary and confined to his *Plant Hunting* articles and interviews he gave to the press. Newspaper records suggest that he was extremely busy, and this might be the reason he had little time for writing.

TABLE 7.1: Wilson's travels in eastern Australia and New Zealand

1920	November	Arrives Adelaide, South Australia (late November)
	December	Collecting in South Australia; departs for Sydney via Melbourne, arrives 20 December; Christmas 1920 in Sydney, collecting north of Sydney near Gosford and Narara
1921	January	'during Christmas holidays', Manning district NSW, Taree, Bulga Plateau Dubbo, NSW (13 Jan) to look at *Callitris*; Tumut
	February	31 Jan arrives Auckland, New Zealand; February in New Zealand
	March	New Zealand
	April	In Tasmania and Victoria 21 April Ballarat and Creswick, Anzac Day (25 Apr) in Melbourne Vic; Stanley, Tasmania. Hobart (27 April, Easter Sun).
	May	Brisbane, Queensland (arrived by train, 13 May), Bribie Island 19 May Maryborough, Fraser Island
	June	Townsville, collecting 7 June. 21 June on Thursday Island, Torres Strait, collecting; Darwin; on to Singapore and Java, then India

NOTE: The table gives an approximate timetable of Wilson's travels in eastern Australia and his collecting localities or visits. Some of these are taken from newspaper reports that give the dates of his arrival or departure; diaries, if they did exist, have been lost. Many travel dates were determined from the dates given on plant specimen sheets. Like other herbaria, the Harvard Herbaria contain more plants collected in Western Australia than from eastern Australia. Though Wilson lost the specimens from the eastern states, a few remain or were sent by other collectors. These dates are closer than those given by Howard.[4]

When Wilson arrived in Adelaide, he immediately attracted attention, with many visits from local dignitaries and botanists. The newspapers introduced him to their readers. The *Observer* newspaper grumbled:

> The average Australian has few scruples about felling trees with an axe or 'grubbing' them, and then using the wood for fuel. Too often good timber goes to waste in smoke. There is yet comparatively little system practised in the forests by timber getters. This fact is deplored by those who have made a hobby of afforestation, both local men and visitors from abroad. It has not escaped the attention of an American who arrived in Adelaide a few days ago. He landed in Western Australia in October to begin a big undertaking. He is here for the express purpose of taking a census of our native trees. Far from being dismayed by that immense task, Mr. Ernest H. Wilson … apparently revels in it. For more than twenty years he has been doing similar work. The first work he engaged in was participation in obtaining a census of American trees and shrubs. Later he went abroad and spent 11 years in China going through the Boxer rebellion, the Russo-Japanese War and other stirring times.[5]

Wilson visited Sir William Sowden, who was a journalist, and chief lead writer and editor of the newspaper, *South Australian Register*, and was interviewed for that paper at Sowdens' home. From there, it seems that other papers copied material, and Wilson's story was widely circulated. Sowden was a 'keen field naturalist and forest conservationist',[6] and on this occasion, Sunday 28 November 1920, Wilson met many botanical authorities from South Australia. Subsequently, the *South Australian Register* had a substantial article about Wilson, and, judging from its contents, the writer had more of an opportunity for a long chat with him than most journalists.[7]

On 11 December 1920, the *Observer* (Adelaide) noted that 'one of the foremost men in the English horticultural world is at present in Adelaide – Mr. Ernest H. Wilson, who is assistant director of the Arnold Arboretum on the Harvard University'.[8] He was always referred to in Australia as the Assistant Director of the Arboretum, and as 'Professor'. Around Adelaide in mid-December 1920, Wilson collected on a trip that took him from Adelaide and its Hills to Gawler, 42 kilometres (26 miles) to the north of Adelaide, and then to the Mount Crawford Forest Reserve, in the Mount Lofty Ranges east of Adelaide. He travelled with Mr Forman McLean, an American plant pathologist and forestry expert based in the Philippines who happened to be in Adelaide. Mount Crawford is a historic gold-mining district. Wilson was greatly impressed by the beautiful scenery, with its ancient red gums, 'recalling an old English park in its verdure, growth and peacefulness'. In the Hills, Wilson collected *Callitris propinque* 'tree 20-35ft [ ] common in Hills' (#527), a Hakea, '3-5 ft Hills common' (#573), *Banksia marginata* '3-15 ft flo yellow on Hills common' (#574), and near Gawler an *Acacia* '5-15 ft' (#575). He added to the newspapers that South Australian trees are mainly evergreen and their colouring 'rather gray-green as compared with the deep green foliage of trees of the old worlds'.[9] At Mount Crawford, and the Mount Lofty Ranges, he again collected a small suite of mainly shrubs, such as *Pimelea*, *Baeckea*, *Dodonea*, *Hibbertia*, and *Darwinia*, but also an '80-100 ft flo pink common' *Eucalyptus* (#589, but he left the name of the common species as a question), and other eucalypts that he was able to identify at Mount Lofty, as well as those he could not.[10]

The article that reported this little trip in South Australia was prefaced with:

> Afforestation is a live subject, and unless it is taken in hand on an adequate scale, those who come after us will be embarrassed. Much more

could have been done than has been done in this State. Sufficient has been accomplished, however, to show that with advantage leasehold land in poor country, with a good rainfall, should be resumed for afforestation purposes. The growing of timber is not a private job – it is one for the community.[11]

Clearly, the writer had a point of view.

Wilson was pushed to give his impressions about this live subject to the newspaper men. They reported:

Dr Wilson was candid. 'You know,' he said, 'I cannot tell you much about my impressions until I have had a chance to look around … What I can say is this: I am collecting specimens of Australian flora. I already have many specimens, and I do not expect to complete my investigations before June next. I say to you, right now, that it is equally as important to preserve records of trees and plants as it is to record the history of the long extinct dodo or any prehistoric monster. I will tell you this,' said the visitor with a burst of confidence. 'The wheat belt I saw in the vicinity of Gawler is one of the finest in the world. I saw miles and miles of golden grain awaiting harvest, and it gladdened the heart.[12]

This comment shows that although Wilson had criticised the extensive clearing of the Wheatbelt in Western Australia, he was not simply against agriculture. Instead, he was concerned at the poor use of forest lands, and appealed against the wastage of timbers. As newspapers reported, 'The appeal is not for the preservation of alluvial valleys that are suitable for agriculture, it is against the ruthless stripping of the land of forests which are really more valuable in their natural state than the land can ever be for agricultural or grazing purposes.'[13] This theme was key to Charles Lane-Poole's ideas about agriculture and forestry, and no doubt Wilson would have discussed this issue while travelling with Lane-Poole in Western Australia. Lane-Poole was later to write of a rational use of land, 'where it can be made to produce food crops profitably, it should be farmed; where it will produce a higher yield in timber it should be devoted to forestry'.[14] He believed that the agronomist and the forester should 'stand shoulder to shoulder'.[15]

On 4 December 1920, the *Observer* in Adelaide had a large article about Wilson's visit. To put it in its time, in the same edition was news of an assassination in Germany, members of the Sinn Fein arrested, assistance for blind returned soldiers, the League of Nations, a gloomy recession in England and Lloyd George's response, the boredom of the ex-Kaiser exiled in Holland among cows, Mesopotamian Oil possibilities for American companies, and an air race in New York where the fabulous speed of 191 miles per hour (over 300 kilometres per hour) was attained.

Wilson spent Christmas Day 1920 in Sydney, and noted to Sargent that 'the heat is great' and everything was closed for the holidays and some people were away, but that he was nevertheless 'enjoying excellent health'.[16] There were many Christmas days that Wilson was alone at a great distance from his family, both in his long travels in China and again in Australia. His next Christmas was to be in India, after being alone for all of 1921. Wilson had commented on his loneliness in a letter to his mother, stating that on his trip to Japan, Korea, and Formosa [Taiwan] in 1917–18, his wife and daughter accompanied him, based relatively close by while he collected, making the trip 'much pleasanter and infinitely less lonely for me'.[17] In Australia, he would have suffered the emptiness of hotel rooms in cities and country towns, among people but without closeness or familiar ties.

While in New South Wales, Wilson travelled with Richard Dalrymple-Hay, who had been appointed as the first Commissioner of Forests for New South Wales, in 1916. Dalrymple-Hay was another pioneer of Australian forestry who strengthened 'state control and policy over forest resources against the ongoing traditional development ethos of the lands and agricultural bureaucracies and forest exploitation by the sawmilling industry'.[18] Of their journey to the Castlereagh Forest reserve 40 miles west of Sydney, Wilson wrote that

> it was a pleasant ride over poor roads & we had many fine views of Sydney & its suburbs. The forest was acquired 4 years ago & is nearly all pole-growth of Iron bark (Euc. Siderophloia) with a little white [fern] mixed. It is a splendid example of what can be obtained in a few years by keeping out fire.

They lunched at Windsor, 'the 2nd oldest town in N.S.W.'[19]

While in Sydney through the Christmas period in 1920, he also met Joseph Maiden, a British-born and educated botanist who had gone to sunny Australia at the age of twenty-one due to ill-health and the recommendation for a warmer climate. Maiden was the foremost botanist of his day in Australia; he had founded the National Herbarium of New South Wales in 1901, was the world authority on the Eucalyptus, had been awarded the gold medal of the Linnean Society in 1915 and made a Fellow of the Royal Society; he was 'active in the movement to retain large areas of native forests', and was a kind man revered by his staff.[20] Wilson described Maiden as 'an enthusiastic and kindly gentleman. Unfortunately he is suffering from some bone disease which prevents him walking other than a few hundred yards.'[21]

During the Christmas holidays of the summer of 1920–21, Wilson travelled into rural New South Wales and visited the Manning River region near Taree. He was accompanied on a three-day tour by Dalrymple-Hay, and joined by local forestry officers. These men appear in several images that Wilson took (Figure 7.1). On the first day, they went to Cooperhook and Mitchell's Island State forests, where he saw re-afforestation experiments. Wilson noted pure stands of 'blackbutt (*E. pilularis*) pole growth up in 150ft like ships masts under-growth chiefly *Casuarina torulosa* very little [ ] which is kept down by fire. En route saw fine stands Water Oak, Casuarina glauca, a tall species.'[22] He went on the second day to see 'a fine specimen of red cedar' on a pastoral property at Burrill Creek, and to look at 'Wingham Brush'.[23] Newspapers later reported that:

> Professor Wilson was much interested in this brush, and took a number of photographs of the remarkable fig trees, ferns, and undergrowth. On the third day the Bulga Plateau was visited, and the Professor Wilson was shocked at the waste of valuable timber that had been burnt off. There, also, he took a number of photographs of trees, timbers and bark. Professor Wilson's tour concluded with a trip to the Tuncurry prison plantation, where pines are growing, and the State forest reserve between Failford and Nabiac. He will proceed to Queensland and Tasmania.[24]

His forest visits in NSW seem to be limited to forestry sites. In late December, with Dalrymple-Hay, Wilson saw

> mostly pole growth of Grey Gum (Euc. Punctata), iron bark (Euc. Paniculata), spotted gum (E. maculata) & Blue gum (E. saligna) & White mahogany (E. acmenoides). The spotted gum with its spotted bark and white [ground] is the Lady of the Forest of this part of N.S.W. In open

pasture I saw a fine tree of Red Cedar (Cedrela australis) and was told that formerly much good cedar grew here.²⁵

Near the prison, Wilson saw some specimens of *Livistona australis*, the cabbage tree palm, and other tropical palms (Figure 7.2). These indicate that he was in warmer climes, and he noted in his diary that he was now seeing 'dense jungle' and that the weather had been 'oppressively hot with thunderstorms in evening'.²⁶

FIGURE 7.1: An unknown forester with tall timber in the state forests of New South Wales. Wilson Y-293, AA.

FIGURE 7.2: Tall palms and *Eucalyptus* in the moist sclerophyll forest and rainforest margins, New South Wales. Wilson Y-295, AA.

FIGURE 7.3: Snow gum, *Eucalyptus pauciflora*, in the High Country of south-eastern Australia, is famous for the colours of its trunk. It is one of the few areas that one species dominates the landscape. Unlike many parts of Australia, this region did experience glaciation in the last Ice Age. Image: M Grose 1625.

In the first week of January, he was still in NSW and had some 'easy' days collecting 'Euc. rubrida (Red Gum) E. stelluate (Black Sally) & E. Dalrymphona & e. cariacea (Snow Gum) … The Snow Gum & Black Sally reach the attributed limits of the forest in the district & evidently withstand considerable ice & snow. The day was perfect. Clear sunshine with cool wind & not the least bit hot.'[27] An example of a snow gum is in Figure 7.3. On January 4, 1921, he collected a large number of eucalypts, now apparently lost. They were found between Picton and Mossvale: (#661 to #687) – narrow-leaved ironbark, white gum, Sydney peppermint, broad-leafed messmate, narrow-leaved messmate, grey gum, grey ironbark, coast ash, scribbly gum, red gum, white stringybark, and red mahogany. After noting that he saw Mountain Ash, '*E. Gigantea*' on 8 January in the Jounama state forest, and the Jillabenan cave the next day, he has no more diary entries for the entire Australian trip, and I have relied on newspaper reports.

During his travels in January, Wilson clearly began to consolidate his opinion of what he saw in Australia's treatment of its forests, and he was not impressed.

Wilson spent February and March in New Zealand, and was back in Melbourne, the capital of Victoria, in late April 1921. His first trip out of the city was to visit the Victorian Forestry School in the small country town of Creswick, not far from Ballarat; the school has now been incorporated into the University of Melbourne. Wilson noted to a representative of the newspaper *The Ballarat Star* that 'good work was being done … that the equipment was small, and therefore insufficient, but within the limits that were possible very fair results wer [*sic*] being obtained'.[28]

> The people of Australia should set themselves to help the forestry services to prevent a recurrence of the errors of the past that denuded this fine country of much of its primeval timber wealth. We do not want all Australia to become a desert. There is enough of that now. Australia, too, should have a forestry school of its own as soon as possible; it is advisable for the training of your officers under Australian conditions, and I think the delay in founding this school is deplorable indeed.[29]

Charles Lane-Poole had been urging a national forestry school since he had arrived in Australia in 1916 as Conservator of Forests for Western Australia. Though Victoria had the Creswick school, and New South Wales had a school, both were small, had few students, and were poorly equipped, as Wilson noted. Lane-Poole saw a 'need to train foresters at the university level and apprentices in a training school'.[30] The 1920 State Premier's conference had approved a national forestry school to be established in New South Wales, and this must have been very much on Lane-Poole's mind when he travelled about with Wilson a few months later. However, it was not until May 1925 that the Australian Government approved the establishment of a national forestry school in Canberra, the new national capital. The move of the proposed site to Canberra had been Lane-Poole's suggestion, and the school was very much his vision, for which he is remembered. Wilson said to the press that 'the school would answer requirements as far as forest rangers wer [*sic*] concerned, but a great central institution was essential if Australia's forests were to be propertly [*sic*] looked after, and its supplies were to be made the best use of. He emphasised the necessity of educating the public to a realisation of the value of a national asset, which was of the greatest importance, and said that it was only by securing individual interest in this matter that this could be done.'[31] Wilson commented on this issue many times as he travelled around eastern Australia.

In Creswick, Wilson made strong criticism of their little arboretum, which he described as 'a wreck of a thing'. The reporter from the *Creswick Advertiser* noted that Wilson was 'emphatic' about his opinion. This was a very long article from a conversation between the reporter and Wilson. Wilson expounded on the need for a better arboretum, and how an arboretum could be used to test trees for production. At the time, no testing was done before planting, and Wilson was both appalled and amazed. 'At present they had no data at all to go on. There was even a lack of knowledge of the eucalypts, their own native trees, whilst they had data about pinus insignis, an imported tree.'[32]

Wilson commented that in this district, three crops of *Pinus insignis* and Corsican pine (*Pinus nigra*) could be grown each century, something that could not be done in the northern hemisphere. He also felt that farmers, though they were planting windbreaks, could grow much wider breaks and 'could go in more for planting trees'.[33] This was a prescient comment, and one that farmers have been taking up in the last twenty years, but slowly.

He travelled to the nearby small city of Ballarat, which lies 110 kilometres (68 miles) north. There, he inspected Ballarat's Avenue of Honour, planted between 1916 and 1919 to honour the men and women from the district who enlisted in World War I. Wilson would have seen the original plantings of twenty-three species, mainly ash, oak, maple, alder, birch, lime, poplar, and elm; many of these did not flourish in the area and were replaced with *Quercus* and *Fraxinus* species. Every tree has a memorial plaque for a soldier – 3,771 trees in total, over a length of 22 kilometres (14 miles).[34] The Melbourne *Herald* reported that Wilson 'remarked that the only similar avenue of which he knew was the 27 mile road in Japan. The Ballarat Avenue, he added, was an example of what could be done in the way of tree-planting in Australia.'[35, 36] But Wilson was not done with criticism. He once again noted to the local papers the need for better education of the public for a wider appreciation of the value of forests.[37]

Waiting in Melbourne to get a ship to Tasmania, Wilson wrote to Sargent of the many hold-ups in his travels plans due to strikes and holidays.[38] I noted the date on his letter, 25 April; this is Anzac Day in Australia, a public holiday. First observed in 1916, Anzac Day is somewhat the equivalent of Memorial Day in the United States; it marks the day of the first major military action of an independent Australia in 1915, and is a day of remembrance of the Australian and New Zealand Army Corp (Anzac). Due to delays caused by the Spanish influenza epidemic of 1918–20, the Melbourne 1921 event was the first major march on Anzac Day held in Australia; crowds were large, and World War I veterans were out in full to re-join old mates, march with their units, and lunch together at the pub.[39] It appears that Wilson did not leave his grand Victorian hotel, being a little removed from the Melbourne march route.

Wilson was in Tasmania days later, and wrote an article for the Arnold Arboretum's journal about the Hobart Botanical Gardens, which he posted back. He noted that many of the northern species were incorrectly named in the botanical garden, and that 'on the whole the deciduous trees of the northern Hemisphere do not thrive', although conifers were 'more at home'.[40] Wilson felt that the Southern Hemisphere conifers in the Hobart Botanical Gardens were the 'most interesting', and picked out *Araucaria excelsa* (Norfolk Island Pine), *A. bidwillii* (bunya-bunya), *A. cunninghamii* (hoop pine), and a South American and a South African pine. He went into some detail about the excellent specimens of northern hemisphere pines found in Hobart. To the press, he was scathing, describing the garden as 'the most burnt-up gardens he had ever set his eyes upon' due to the lack of water provision. As a result of 'starving the gardens of water, you are letting this tree [a rare Mexican Pinus patula] die from the top.'[41] 'Disastrous results were accruing.' Wilson's questioner pushed – as 'his last thrust' – for a new subject, American politics and President Harding, but Wilson made no comment, and replied with a 'conclusive declaration' that 'I believe in sticking to my leather'.[42] The reporter wished him bon voyage.

While in Tasmania, Wilson did extensive tours of forests of the south, north-west, and west coasts. He felt that the future of Tasmania should be largely based on forestry.[43] There, he must have seen some of the world's tallest trees once more, but there is no record in diaries of his response of viewing this colossal flowering forest. Mountain ash (*E. regnans*) is the tallest tree species in Tasmania. Today, these giants are still being logged despite their capacity for massive height and girth, for great personality, and huge carbon storage; older trees store more carbon than younger because, unlike us, trees increase their metabolism as they age. In many cases, the only trees saved are the largest trees with girths over 5 metres (16 feet), or 85 metres (279 feet) tall, or more than 280 cubic metres in stem volume. This foolish policy leaves us with only youth and elderly. Once the elderly come to the end of their lives, youth will take centuries to be dubbed as some surviving today are – Centurion (100.5 metres, 330 feet), Icarus Dream (97 metres, 318 feet), or Gandalf's Staff (84 metres, 276 feet), which are all in the later stages of their long lives of perhaps five hundred years.[44] El Grande (79 metres, 259 feet) was burnt in 2003 due to unnatural causes. And perhaps we will forget how tall younger trees can grow when they are left alone, and not caught in such havoc. When one sees clear-felling of large trees, and great trees left as patches without their family groups, we see the loss of an entire community.

Tasmania is an island with a small population, and often beset with economic issues and fights between loggers and environmentalists. Yet tourism to see some of the world's giant, hefty trees in these great forests needs to be pushed more strongly, to be more widely known around the world.

Wilson might have seen *Nothofagus* in Tasmania, and perhaps in autumn colours, since he was there in April. *Nothofagus* is the only deciduous tree species

FIGURE 7.4: *Nothofagus gunnii*, deciduous beech (in colour, right), on the Tarn Shelf, western Tasmania. Image: Gumboots Photography (Fiona Walsh), with permission.

in Australia. Today *Nothofagus* is usually found off the beaten track, with most now found in the Tasmanian World Heritage Area (Figure 7.4). Western Tasmania is the stronghold of diversity in *Nothofagus*, with other centres of diversity being in north-eastern Tasmania.[45] These lovely trees dominated south-eastern Australia 35 million years ago at about the time the continent was making its final break with what is now Antarctica. Wilson was photographed with *Dicksonia antarctica*, the Tasmanian tree fern, a relic of Gondwana (Figure 7.5).

FIGURE 7.5: Wilson in Tasmania against a backdrop of *Dicksonia antarctica*, the Tasmanian tree fern. Behind him is a *Dicksonia* of great age, as these are slow-growing (3.5-5 cm per year) and do not produce spores until they are about twenty years old. Image: Wilson Y-518 AA.

In Tasmania, he was accompanied by Mr LG Irby, who had been appointed Forest Conservator in Tasmania in 1919 and had carried out an audit of Tasmania's forests. Wilson was 'enthusiastic on the immense forestry possibilities of the Island State. The visitor was particularly interested in the indigenous conifers, and from his great experience of this type of tree all over the world, was able to throw fresh light on certain of the problems affecting their regeneration and perpetuation.'[46] Wilson was not, however, the forest saviour that other reports might have indicated, but also saw that many trees could be millable 'twice in a century' and could thus support a Tasmanian forestry industry. His idea was to manage existing eucalyptus forest. He said to a reporter:

> but to achieve this it is not enough to cut out all the best trees and leave the rest to look after themselves. The areas must be cut right through, big and little, and a fire put through afterwards to clean things up. Then the seed will come up everywhere, and the youngsters properly thinned out and protected from subsequent fire will have their chance.[47]

Irby's appointment as Conservator had led immediately to talk of 'waste lands' in Tasmania – that is, land seen as unproductive for agriculture or forestry. Such areas were sand dunes and button-grass plains on poor peatlands. Many saw a need to turn these areas into trees to assist Tasmania to grow into a large population and thriving economy.[48] Wilson was taken to 'some of the great waste lands of the State … in the north-west and the button-grass plains and vast stretches of sand dunes on the west coast between Strahan and Zeehan'.[49] The newspaper reported that Wilson investigated the depth and nature of the soil, and that the 'celebrated dendrologist was deeply impressed' about the 'latent possibilities for the growth of exotic conifers, and in fact expressed the opinion that in much of our waste lands Tasmania possessed the finest tree-milling propositions in the world'. This impression, the eager journalist said, by so great an expert, strengthens the opinion 'given by numerous foresters that Tasmania holds the key to the main supply of soft woods for the whole Commonwealth eventually'.[50] Wilson was also able to be of service to the local industry, and gave advice and suggestions as to the most suitable species, from an economic point of view, these being pines.[51]

The arguments and passions put forward by Irby are a lesson. Irby voiced the opinion in 1922 that 'the possibilities were immense, and if they could solve the problem of planting trees on the waste lands of the State, their names would go down in history'. Indeed. In 1922, Irby had further developed this theme and suggested that 'children starving in Europe and Asia' could be brought in as immigrants to Tasmania because the 'planting out of forests could be done by children over 10 years of age' to fix the 'treeless wastes' with introduced conifers to 'wipe out the national debt of Tasmania'.[52]

Historians and the rest of us now can write that 'mercifully, the "waste lands" did not cooperate, and neither did the Commonwealth Government, while the depression saw a lack of funds for silviculture planting'.[53] The iconic button-grass plains, a sedge (*Gymnoschoenus sphaerocephalus*) that is seen as a signature species of these moorland regions, are now renowned ecologically, and are a major tourist drawcard in the Tasmanian Wilderness World Heritage Area. Wilson had clearly walked into a new and very political situation where his comments might have been taken beyond his expectation.[54] Curiously, at the same time and at the same meeting in 1922, the committee reiterated the 'vital question of the prevention of the destruction of the vast heritage which lies in our forests'.[55] Yet landscape-scale destruction has continued. Policies and words might exist, but the truth is written on the landscape.

For all his encouragement of planting pines in Tasmania, Wilson stressed that one of the foremost planks of Australian forest policy should be conservation of existing forests. Wilson was aware that forestry was often political, but in Australia he spoke freely. Professor Wilson, said the newspapers, 'is acutely observant. He thinks that our forestry laws are ample, and he approves of the officers of the Forestry department, but he blames political influence for the prodigious waste of our timbers. It was notorious, before Professor Wilson found it out for himself, that some of the best forest country in Victoria was being sacrificed for the benefit of timber-cutting concerns and against the advice of the forestry officials.'[56] This was a recurring problem, and a recurring theme that reflected not only Wilson's idea of the main purpose of the Expedition, but his growing concern at the worldwide destruction of forests that he had witnessed again in Australia. As someone widely travelled in his day, which was rare, he was well placed to make a global assessment. Both he and Sargent were concerned at the loss of large trees across the world. Visiting Melbourne, he 'expressed great consternation at the destruction of timber representing millions of pounds in the indiscriminate, ruthless manner of "improving" the country'.[57]

Few truly appreciate how many of our old and large trees, how much of our forests we have lost. It has not been a sudden loss that has all occurred since the industrial revolution as many might imagine, though there was rapid loss in Australia. Rather, it has been a gradual cutting down for thousands of years. Two thousand years ago, when Julian, the last pagan Roman Emperor, camped with his troops on the islet in the Seine now known for Notre Dame de Paris, he was nearly surrounded by tall forests where wolves, bears, and lynx roamed. But surely and steadily, we cut our fuel for cooking and warmth, housing, and room for agriculture and towns. As each generation passes away, each succeeding generation loses the memory of what had once been over much of our world. 'Once upon a time', WG Sebald wrote, 'Corsica was entirely covered by forest. Storey by storey it grew for thousands of years in rivalry with itself, up to heights of fifty metres and more, and who knows, perhaps larger and larger species would have evolved, trees reaching to the sky, if the first settlers had not appeared and if, with the typical fear felt by their own kind for its place of origin, they had not steadily forced the forest back again.'[58] The phenomenon of the loss, generation by generation, has been called both shifting baseline syndrome and environmental generational amnesia. Amnesia is the telling and far more powerful word, as it suggests that we forget and perhaps even disbelieve what once was.

Wilson had noted the slow cutting and loss of trees that had occurred in China. Of the Chengtu Plain, he commented that 'In such a highly cultivated area the natural flora has, of course, been destroyed. The few indigenous shrubs and herbs that remain are relegated to the sides of streams and graveyards'.[59] It might well be that Wilson was less alarmed by loss of trees in China as the initial action had long passed away in time. However, in Australia, he saw the active process, the destruction, the empty wastes, the clear pillage, the stunned wildlife, and could feel the change being hammered on the environment as the forest was forced back. Wilson warned us. We have largely forgotten that many people warned us long before it became more popular to do so.

Wilson had a great deal to say about what was widely reported as the 'waste of timber' in Australia. He said that 'the Americans … were perhaps the greatest offenders in the world in regard to waste of timber, but from what he had seen Australians were 'jolly good seconds'.[60] He said more about forest destruction the longer he stayed in Australia, into 1921. After visiting the Bulga Plateau in northern New South Wales, he lamented the state of forests and their destruction. The Bulga is a beautiful area. The *Sydney Mail* described 'the vivid impression upon the mind of the visitor … at the stupendous panoramas of mountain and gully. Standing on the wooded edge of the great bluff that drops sheer about a thousand feet … one gets the impression that he is indeed surveying the world at his feet.' It is 'densely covered with big timber'.[61] But the destruction of this area was known and written about, and it might well be that Wilson was taken there to have an esteemed international visitor add his weight to the condemnation of what had been taking place.

Earlier in 1920, the *Sydney Morning Herald*, one of the nation's foremost newspapers, had highlighted the destruction of forests on the Bulga. They described the Bulga plateau, which rises to an elevation of over 610 metres (2000 feet), and its region. The article is, however, a far cry from today's conservation of forests or Lane-Poole's concept of forest preservation. It is worth giving a substantial portion of this piece to show, as one reads along, what men like Lane-Poole and Wilson were up against – a solely timber mindset, not a forest one, with management of the timber. It begins critically:

> Bulga is late in the field, and, unfortunately for its prospects of immediate settlement the very value of its asset in timber is likely to retard its growth in the way that the Dorrigo and Comboyne plateaus have developed. That may seem unfortunate for the Bulga, but it is all to the good of the State. Both, the Dorrigo and the Comboyne have been developed only at the cost to the State of prodigious wealth in the timber gone up in smoke. The settlers cannot be held blameable as they were allowed to take up their blocks in the heart of forests of magnificent hardwoods and in the valuable softwood brush. They could do nothing until the timber was disposed of, and in the absence of good roads, timber mills or railways, there was only one way to dispose of it – by cutting and firing. It was only another case of settlement preceding railways, but the destruction of millions of feet of

the best timber in the Commonwealth was the price paid for it. The Bulga timber, valued at hundreds of thousands of pounds, narrowly escaped a similar fate. Even now one can see on the Bulga magnificent tallowwood, blue gum, and brushbox standing bare and stark in a forest of dead trees.

The article goes on, and ends with the value of timber in feet:

> The Bulga State forest comprises about 10,000 acres of magnificent timber, and 12,840 acres of Crown lands adjoining were sub-divided for settlement into farms for returned soldiers. About this time, however, the State was taking stock of its resources in timber, and the result showed that these, instead of being almost unlimited, as was anticipated, were really much smaller than they should be in a State of its extent. This forest area had been subdivided, and intersected with roads at great expense in readiness for the proposed settlement, when the Government, realising the huge sacrifice of timber which must follow, decided to call a halt. Over 50 blocks of land had been made available, but eventually only about a dozen were allowed to go to ballot. It was mainly by reason of the resolute determination of the then Minister for Lands (Mr. Ashford) to save most of the Bulga timber, and thus prevent a re-petition of the folly which sacrificed so many of the coastal forests, that this decision was final. The assessment of timber on this area marked down for settlement was 210,000,000 superficial feet of present commercial value, or an average of over 16,000 feet per acre. About two-thirds of this timber was hardwood, principally blackbutt, tallowwood, blue gum and brush box, and its quality was perhaps higher than any in New South Wales.[62]

It then describes the potential of the timber and how to remove it, if only a railway were to be there to aid its greater extraction; there was more money to be made.

Wilson saw the destruction of the Bulga Plateau first-hand, visiting there in early February 1921. He was 'positively shocked' by what he saw in the waste of trees. He said:

> You people, – or, rather, most of you – call this improving the country. Well, that word "improvement" used to describe the ringbarking and burning-off of most valuable timber in this indiscriminate, ruthless way

has not the same meaning to me that the lexicographers give. The waste I saw on the Bulga area alone was appalling; it was wicked. I have no hesitation in saying that the destruction, to speak only of those parts of the Bulga Plateau that I myself traversed, represent millions of millions of pounds. For heaven's sake stop making havoc of your heritage of natural forest. Stop this destruction which is going on at Bulga, and doubtless at other places, too. It is a most urgent matter. The world today is suffering from a shortage of timber, and if your resources had not been wasted you would have been in a position to become suppliers of a commodity that is becoming more and more valuable every day. You have been destroying your own timber, and importing timber from Canada, the United States, and New Zealand. Lots of your own timber could have been used for the very purposes for which you are importing it from other countries. There seems to be a sort of arboricidal mania here.[63]

This was a theme that Wilson never lost in Australia. His pleas about 'arboricidal mania' were repeated across a score of newspapers in eastern Australia, including the *Armidale Chronicle*, the *Richmond River Herald and Northern Districts Advertiser*, the *Toowoomba Chronicle*, the *Cairns Post*, the *Don Dorrigo Gazette and Guy Fawkes Advocate*, the *Daily Examiner*, and the *Sydney Morning Herald* – large and small presses. A press article from Melbourne noted that 'There are tens of thousands of thoughtful Australians in all parts of the Commonwealth who will agree with him.'[64] The article finished with 'But the advice which the Professor summed up in the sentence – "For heaven's sake stop making havoc of your heritage of natural forest" – is sound and needs reiteration.'

Havoc is a powerful word. The plant collector William Purdom, collecting in China in 1912 for the Arnold Arboretum, had used the same term to describe the 'wanton destruction of what must have been fine old forests' and the 'havoc' of soil degradation in the Northern Provinces of China.[65]

Wilson was appalled that much of the land on the Bulga had been cleared for dairying, which brought a 'paltry' return compared to the judicious harvesting and management of fine timbers. He stated that this entire area should have been left as a forest reserve. He also decried the loss of trees in agricultural areas:

little of no thought is given to the future, and it is soon found that many districts are so denuded of trees that the timber, firewood, shade and shelter supplies are exhausted. In this respect New South Wales in common with all new countries, has been a sufferer. But the point I want to make is that there is a great deal of land that you have stripped of its forests for agricultural and grazing purposes that is really not adapted for those purposes at all. That is where you have made a mistake.[66]

Wilson arrived in Brisbane, Queensland, from Sydney on 13 May 1921, by the Southern mail train, and was met at Brisbane Central Station by the Queensland Director of Forests, Mr EHF Swain. Wilson's stated aim in Queensland was to examine, in particular, softwoods such as the 'hoop, bunya, and kauri pines, and the trees from which the varied woods suitable for cabinet making are produced'.[67] He visited the Imbil Forest with Swain and the Queensland Government Botanist, Mr Cyril T White (Figures 7.6 and 7.7).

Wilson's host in Queensland, Edward Swain (1883–1970), was an early advocate of setting aside national parks for forests, had strong ecological views, protested the cutting of tropical forests, and attempted to save them from development. Swain was to be Director from 1918 to 1932 and during this period he had an 'almost mythical hold … on the imagination of many foresters of the period'.[68] He was a complex and brilliant man of a humble background from the Sydney suburbs.[69] Outspoken and visionary, Swain saw forests as the spiritual base of a modern society. Like Wilson, he lamented the waste of turning forests into farms and had seen that 'farming families and towns went through unnecessary hardship' as a result of government policies that pushed for agriculture in forested areas that were not suitable for farming. He bluntly referred to them as 'hillbilly settlements', and his vision was to replant degraded farms and save 'the old growth wilderness as a salve for the psyche'.[70] Swain believed that in most of human history, we had lived 'in harmony with forests before the age of agriculture'. Swain's sentiments about forest and human psyche stressed that forests play an important part in spiritual well-being and identity; he saw forests as 'spiritual reservoirs' deeply connected to a person's capacity to 'remain spiritually healthy'.[71]

Cyril White, who had collected hundreds of plants in Papua in 1918, was praised by Wilson for 'his ability and enthusiasm'. The newspapers reported that although White was 'a young man, the American knew him by repute before coming to Queensland, and he had "measured up" beyond expectations'. Wilson had added that 'Queensland should be proud of the fact that it had the one Australian-born Government Botanist, and that one a young worker' who Wilson thought would make his international mark.[72]

In Queensland, Wilson visited Benarkin State Forest, north-west of Brisbane and due east of the Bunya Mountains and Imbil, further north on their way to Fraser Island. While the *Araucaria* were his focus (see Figure 7.8), he would have seen the rainforest ecosystem, with blackbutt (*Eucalyptus pilularis*), tallowwood (*E. microcorys*), white mahogany (*E. acmenoides*) and other gums, and ironbark (*E. siderophloia*). Wilson also saw 'a most curious vegetable feature in the bottle tree (*Brachychiton rupestris*), which not only looks like a vast bottle as it stands on the ground, but yields drinkable water when tapped like a maple tree'.

FIGURE 7.6: Edward Swain (right) and another man (possibly Cyril White) on horseback, with kauri pine, *Agathis robusta*. Image: Wilson Y-300 AA.

The bottle tree, Wilson noted, 'possesses wood that is almost totally useless', being very soft.[73]

FIGURE 7.7: Cyril White, the Queensland Government Botanist, with *Banksia integrifolia* (determined from its height). Image: Wilson Y-277 AA.

FIGURE 7.8: This image was published in the *Observer* (Adelaide), with the following annotation: 'In the Bunya Mountains, Queensland, the Bunya Bunya Pine (Aricaria Bidwillii [sic]) grows to a height of 200 ft. This is a specimen growing in the virgin forest. Note the bird's nest fern growing on the butt. The photographs on these pages were taken by Professor E. H. Wilson, of the Arnold Arboretum, Harvard University, America.'[74] The man in the photo is Captain Samuel White, who was in the Wilson party. Image: Wilson Y-309 AA.

On 19 May, Wilson travelled to Maryborough, Queensland, accompanied by Swain, Cyril White, Mr Dalrymple-Hay – the Chief Commissioner for Forestry in New South Wales, who had travelled with him to Queensland – and Captain Samuel White, an Australian authority on birds.[75] They 'proceeded by motor launch … to Fraser Island, where Professor Wilson wishes to see the kauri pine in its native habitat'.

Fraser Island is the world's largest sand island. Its Indigenous name from the Butchulla people is K'gari, meaning 'paradise'. Here, Wilson saw the extraordinary growth of large trees on infertile, white-grey sand, as well as banksias (Figure 7.9); the island is noted for rainforest and eucalypts, heathland, marsh, mangroves, and perched dune lakes surrounded by forest. Wilson was particularly struck to see tallowwood, blackbutt, turpentine (*Syncarpia glomulifera*), scrub box (or brush box or Queensland box, *Lophostemon confertus*), flooded gum (*E. grandis*), and red stringybark (or red mahogany, *E. pellita*) – all Myrtaceae – on these sands.[76] He stated that 'the timber grown at Fraser Island was of excellent quality. One would scarcely expect to find such magnificent trees growing on pure sand. It only went to show what could be done with a regular rainfall.'[77]

Captain White described their visit:

> You may travel for miles under a canopy that hardly lets the sun through. This is held aloft by huge butts up to seven and eight feet in diameter, and as high as 100 feet without a branch. Fraser Island was a great eye-opener to me. One would expect to see nothing there excepting low scrub, but the centre of the island is covered by magnificent forest. I counted hundreds of trees up to eight feet in diameter, and spaced only 12 yards apart. These are growing in what appears to be loose white sand. Professor Wilson, of the Harvard University … said to me, 'Everything is wrong but there it is. One would never expect timber to grow on this soil. Everything in Australia appears to be upside down'. Professor Wilson has taken photographs of typical forest trees, and has gathered a vast amount of data. Financial provision has been made for his department to carry on for 1,000 years.[78]

FIGURE 7.9: *Banksia aemula*, Wallum banksia, in Queensland, showing details of infloresence and serrated leaf margins. This photo was likely taken at Fraser Island. Image: Wilson Y-284 AA.

On Fraser Island, Wilson also saw bunya bunya pine (now *Araucaria bidwillii*), hoop pine (Figure 7.10; now *Araucaria cunninghamii*), and kauri (*Agathis robusta*). The two pines are relicts from an earlier age, when Araucariaceae (such as monkey puzzle trees) dominated much of Gondwana – members are found across the southern hemisphere, in Norfolk Island, New Zealand, New Caledonia, Chile, southern Brazil, and Argentina. The family now has its greatest speciation in New Caledonia.[79] It is comprised of species in the genera *Araucaria*,

Agathis, and *Wollemia*, the latter (*Wollemia nobilis*) having been discovered in a deep, narrow canyon among precipitous sandstone escarpments in the Blue Mountains of New South Wales in 1994, to great excitement in the botanical world. The Wollemi pine is regarded as a living fossil because it closely resembles 120-million-year-old fossils from western Queensland, and thus this new pine's location has been kept a secret, due to its low numbers. It is now being grown in the nursery trade in Australia, using *in vitro* propagation.[80]

The Araucaria family of great, tall trees was once common in the northern hemisphere, but died out there at the time the dinosaurs did, approximately 65 million years ago. The Petrified Forest National Park in Arizona, USA, contains the remains of Araucariaceae that grew on a tropical flood plain 200 million years ago.[81] The great American naturalist John Muir, who worked for the beginnings of the National Park movement, thought that Araucariaceae were the most interesting trees that he had seen, a wondrous sight, and the South American species those he most wanted to see before he died.[82]

The bunya bunya pine is a well-loved tree, a 'denizen of the scrub'[83] in Queensland. It is famous for the size of its cone – up to 10 kilograms (22 pounds), though most are 4.5 kilograms (10 pounds) – and the need to never stand under the tree's canopy in case a seed cone drops on your head (Figure 7.11). Its name is an Indigenous one, and was variously given as bunya, banza-tunza, boonya, bunyi, banya bunya, bunnia, bonyi-bonyi, and banua-tunya.[84] It grows up to 50 metres (164 feet), and this height was a reason that Wilson was keen to see it for himself, in association with the hoop pine. It is a sacred tree to Indigenous people, and bunya feasts were once held. They were 'times of great spiritual significance, when Aboriginal people gathered to receive strength from Mother Earth. They were also times for arranging marriages, settling disputes, and for trading goods and sharing dances and songs.'[85] Because this strong association was recognised, the bunya was protected very early, in 1842, with no licences to cut timber being given. However, this wisdom, respect for Indigenous ideas, and foresight did not last long, and in 1860 that earlier legislation was rescinded and 'Unoccupied Lands' freely logged, leading to the end of Aboriginal feasts in about 1900. Great damage was done very quickly, and finally this controversial logging ended when the Bunya Mountains National Park was created in 1908.

FIGURE 7.10: The annotation of this image by Wilson in an Adelaide newspaper reads: 'THE WELL-KNOWN QUEENSLAND PINE. Capt. S. A. White, C.M.B.O.U., admiring a young Hoop Pine (Aricaria Cunninghami) [sic] at Fraser Island, Queensland. This is one of the most valuable white softwoods utilized for indoor work.'[66] Captain Samuel White (1870–1954) was a wealthy South Australian racehorse owner, who had fought in the South African wars, and was a conservationist and ornithologist. He explored through self-funded travels, examining birds in central Australia and the Nullarbor Plain (South Australia) between 1910 and 1921. He was President of the Royal Australasian Ornithological Union (RAOU) 1914–16. Image: Wilson Y-288 AA.

FIGURE 7.11: A giant Bunya pinecone, dropped in 2016 at the Australian Botanic Garden, Mount Annan, New South Wales, with permission from the Botanic Gardens of Sydney. The size of this cone does remind us of the great age of this species, which goes back to the Jurassic and Cretaceous and is a part of Wilson's sense of seeing in Australia living fossils harking back to the dinosaurs. The Araucariaceae family was once widespread in both hemispheres, but is now mainly found in the Southern Hemisphere.

Captain White noted that timber waste and tree loss would impede the huge potential he saw in Fraser Island for tourism in future years, which was a prescient comment. He described Wilson as follows:

> I had the good fortune also to meet again Professor Wilson, of the Arboretum of the Harvard University, who passed through Adelaide recently on an investigation trip of the timbers of the Commonwealth. He was excellent company, and is a scientist who brings method into his work. He will take back to America some of the finest photos of forest trees yet taken in Australia.[87]

Not long after his visit to Fraser Island, Wilson spoke, on 30 May 1921, 'at a Royal Society of Queensland function at the Brisbane University'. He said that 'in regard to trees, the white man was the most destructive animal that ever walked the earth'. Continuing, he emphasised the interest and importance of the Australian flora, and said that, 'inasmuch as trees in general were steadily disappearing from the world, conserving measures were very necessary.'[88] The week before he had given a short and much milder address to the Brisbane Field Naturalists' Club, where he was elected a member. The report on that night noted that 'the beautiful flower-fields of West Australia impressed him equally with the magnificent forests of Queensland – he could not understand why these were slightly termed "scrubs" … Professor Wilson said he was particularly interested in the mountain plant known as the Queensland waratah.'[89] Here, in the naturalists' club, Wilson was once more the horticulturalist.

His criticism of the term 'scrubs' was well-founded (Figure 7.12). It was a term used for the rainforests of the eastern coast, and was meant to be 'slightly off-hand, even derogatory'. The rainforest was to be relentlessly exploited, almost to the end of the twentieth century, when it was miraculously realised to be World Heritage rainforest with the 'outstanding universal values'[90] that Swain and others had seen so long before. Wilson considered that the 'State forests at Benarkin had been largely ruined by indiscriminate felling and burning, and it would take many years of careful nursing before they would be productive.'[91] Even the forest workers lived to change their minds, though most were stuck for money at the time they were involved in logging. One logger commented decades later: 'when you think of the kauri pines and maples and crowsfoot and beautiful trees like that, that were felled and burnt, it was a terrific waste. I've often wondered whether it would have been better to have kept the whole of this region for timber and not let one stick of it be felled for land.'[92]

Clearing, that 'arboricidal mania', went on even in the thick tropical rainforest of Queensland. The feeling of the times is perhaps shown by this press description, given after an accolade to the glorious natural environment that few in Australia knew of, and it reveals both pride and concern at the clearing of this land for sugarcane:

FIGURE 7.12: The caption reads 'clearing scrub', near Cairns, Queensland in 1925. This is rainforest, with a complex understorey under large, mature trees. The derogatory term 'scrub' was widely used; 'this thick jungle scrub' and 'glorious jungle scrub' seem incongruous descriptions.[93] Image: *Queensland illustrated 70 exquisite views*, c.1925, collection of John Young, permission needed. Via www.qhatlas.com.au/photograph/clearing-scrub-babinda-c1925, accessed 11 January 2023 with permission, Centre for the Government of Queensland.

But as one sees the advance of civilisation in the jungle country there cannot be anything but conflicting emotions. The great attempt to populate the North with white people must compel unstinted approbation; but progress can only be made at the expense of country, the beauty of which makes an unfailing appeal to the most exacting traveller. Many a proud boast can be heard by owners of cultivated acres that, where the soil was now yielding riches to the tiller, there stood only two, three, or four years ago, thick jungle scrub, which, at first appearance, looked as though it would need an army of men to clear. Vast tracts of magnificent scenic territory have yet to bow before the sweep of the woodman's axe; and a long period must elapse before the glorious panoramic spectacle is lost for ever to the sightseer.[94]

It was lost. Queenslander Judith Wright (1915–2000), one of Australia's most well-known poets, lamented the loss of forests in the brigalow (*Acacia harpophylla*) belt that runs along the eastern areas of the tropical forest west of the Great Dividing Range due to clearing by dragging cables, tree crushers, and aerial spraying, among a host of assaults on this now endangered ecological community. What remained of this landscape by 1970, fifty years after Wilson's visit, was in stark contrast to that remembered from her youth:

> Suffer, wild country, like the ironwood
> that gaps the dozer-blade.
> I see your living soil ebb with the tree
> to naked poverty.

Clearing of the land was so severe that less than 3% of the forest that Wilson saw now lies in protected areas.[95] The naked poverty that Wright wrote about, and the 'arboricidal mania' that Wilson spoke of, continued.[96] Wilson took several images of clearing but few show identifiable trees; the hoop pine left after clearing in Queensland can be seen in Figure 7.13.

FIGURE 7.13: Cyril White with the remains of (likely) *Araucaria cunninghamii*, the hoop pine, after clearing. The hoop pine, *Araucaria cunninghamii*, is even taller than the bunya, growing to 60 metres (197 feet). Wilson Y-302 AA.

Of his travels in Queensland, Wilson noted to the press his thoughts – they appeared to be always asking for his comments. *The Week* (Brisbane), reported thus:

> After a visit of inspection to the State forest areas at Benarkin, Imbi., and Fraser Island, Professor E. H. Wilson … who is visiting Australia investigating forest trees of the continent, and inquiring into the question of co-operation between American and Australia in the exchange of timbers,

states that he has been struck with the necessity for embarking on reafforestation on a large scale, and for removing the Queensland Forestry Department from political control.

This was an interesting remark, because Swain, who had become Director of Forestry in Queensland in 1918, had great plans for conservation. When Swain had joined the Forestry Department in Queensland, it had been described by a visiting South African forester as 'consisting of six men and a girl, a man for each day of the week and a girl for Sunday', due to lack of funds and government support.[97]

Yet Wilson regarded Queensland as the most favoured State for forest wealth, with 'magnificent forests' and a fine range of trees – softwoods and hardwoods.[98] Wilson said he was

> impressed with Queensland's vast timber resources. There was not a country in the world that was in a better position than Queensland; it was the queen of all lands, and should be able to supply not only its own timber needs, but a large proportion of the world's requirements ... What was badly needed was extensive reafforestation and conservation in order to cater for the requirements of future generations. At present the timber of Queensland was being rapidly exhausted, and the little money that was being spent on developmental and experimental work was provided by loans from the Federal Government. Despite the fact that £150,000 revenue was derived annually direct from the forests, none of this was earmarked for reafforestation purposes. It would take fully half a century to develop a forest crop![99]

Wilson believed that because of the time needed for forestry and conservation of forests, 'many Governments would come into being, and there was always the danger of one nullifying the good work of the other'.[100] Wilson's comments about removing political control were met with a response from an affronted Minister for Lands.[101] But the press joined in, attacking the Government, who took all the money from forestry, and, as Wilson said, 'not a penny went back into forestry'.[102]

Wilson was also asked about the problem of insect pests, and he linked these to the 'destruction of native birds' – since birds were the chief agents in the suppression of insect pests, they needed to be protected. It was all due, Wilson said, to an upset in the 'balance of nature'. Wilson 'advocated the strictest protection of native birds as the chief agents for the suppression of insect pests', and he went on to predict that 'in time, you will have to have plant-doctors, just as you have for the human family'.[103] He felt that in America, the protection of native birds had come too late, and

> it was for Australia to take the lesson to hear and before it was too late profit by the experiences of other countries. Already a tremendous amount of damage had been done in Queensland. Valuable timbers had been burned or used for such work as fence posts, and it was high time a halt should be called. Australians must remember that their magnificent heritage of forest timber would not last for ever.

The reporter had noted to Wilson that certain State forests were still being alienated for farming. This action, the reporter added 'annoyed him greatly. He remarked bitterly "Let the game go on. It is the old story – burn, burn, burn! Leave nothing for the future but do not wonder if future generations curse your shortsightedness".'[104]

The reporter noted to his readers that 'it is hardly necessary to add that Professor Wilson does not agree with the Government going into the timber business. Wilson pointed out that Queensland, the richest timber state, was the only one without a modern Forestry Act. He felt that 'at least one-third of the money earned by the forests should revert to them to ensure future supplies'. 'The Professor' – said the writer – 'evidently felt keenly on the subject and forcibly added' a very strong statement. Wilson said:

> What is happening is that you are not only prostituting your forests by allowing vast areas to be burnt off, but you are exploiting them in the name of the Government. Why will people not see that these great gifts do not belong in their entirety to the present generation? They are only ours on trust.[105]

Wilson's words here of prostituting forests are a far cry from the sedate horticulturalist describing pretty, new species for cultivation in California. He is angry. He was very widely reported across Queensland, in all the regional newspapers up and down the coast and inland, often in long articles repeating his pleas, warnings, and comments. This was a very different kind of expedition.

PLATE 7.1: "Eucalyptus acervula Southport Tasmania" No [(missing)]. 1 April 1921. Now *Eucalyptus eugenioides*, thin-leaved stringybark.

PLATE 7.2: "Eucalyptus Tree 40 ft x 8 ft girth Mt Crawford S.A." No. 604. 10 December 1920. Classified as *Eucalyptus obliqua*, messmate.

PLATE 7.3: "Eucalyptus alba Tree 40–60 ft x 3–6 ft Townsville Queensland." No. not given. 7 January 1921. *E. alba* is known as white gum.

PLATE 7.4: "Melaleuca squarrosa Smith Bush 10 ft flo yellowish Thickets around Strahan W. Tasmania." No. not given. 9 April 1921.

PLATE 7.5: "Melaleuca Bush 1-4 ft. flo rose-purple open country Windora N.S.W." No. not given. 2 May 1921. Classified as *Melaleuca thymifolia*, thyme honey myrtle.

PLATE 7.6: "[ ] Shore of Great Lake Tasmania." No. not given. 13 April 1921. Classified as *Orites revolutus*.

PLATE 7.7: "Hakea 3-5 ft common around Millbrook S.A." No. 573. 10 December 1920. Classified as *Hakea rostrata*, beaked hakea.

PLATE 7.8: "Banksia marginata Cav. Bush 4–12 ft common around Stanley. W. Tasmania." No. not given. 7 April 1921. This is the silver banksia.

PLATE 7.9: "Banksia integrifolia [Linn f] Tree 20-45 ft x 2-7 ft. Stradbroke Island near Brisbane, Queensland." No. on tag is 711. 3 June 1921. *B. integrifolia* is the coast banksia.

PLATE 7.10: "Banksia 3 ft flo golden-yellow open Windsor New South Wales." No. not given. 2 May 1921. Classified as *Banksia spinulosa*, the hairpin banksia.

PLATE 7.11: "Banksia aemula R.Br. Tree 15-45 ft x 2-7 ft sand coast Bribie Island Near Brisbane Queensland." No. 782. 25 May 1921. *B. aemula* is the wallum banksia.

PLATE 7.12: "Banksia latifolia R.Br. Tree 6 ft Stradbroke Island Near Brisbane Queensland." No. on tag is 781. 3 June 1921. *B. latifolia* is known as broad-leaved banksia.

NOTES

1. 'Man who loves trees', *Herald* (Melbourne), 14 December 1920, p. 14.
2. Thought to be the tallest flowering plant in Wilson's time but now contested.
3. Notes of extant diary entries are given in Appendix 2.
4. RAEH Howard, 'Wilson as a botanist', *Arnoldia*, 1980, 40(4):154–93.
5. 'Australian timber. An American's impressions', *Observer* (Adelaide), 4 December 1920, p. 37.
6. C Bridge, 'Sir William John Sowden (1858–1943)', *Australian Dictionary of Biography*, volume 12, 1990.
7. 'Australian timber. An American's impressions', *Register* (Adelaide), 29 November 1920, p. 9.
8. 'A famous horticulturist', *Observer* (Adelaide), 11 December 1920, p. 8.
9. 'Our timber. Impression of two visitors. W.A for hardwood. S.A. for pine', *Daily Herald* (Adelaide), 3 December 1920, p. 2.
10. These species specimen sheets have not been found in the Harvard Herbaria, and thus were lost at sea.
11. 'Our timber. Impression of two visitors. W.A for hardwood. S.A. for pine', *Daily Herald* (Adelaide), 3 December 1920, p. 2.
12. Ibid.
13. 'Australia's natural heritage', *Daily Examiner* (Grafton), 25 January 1921, p. 4.
14. Cited in J Dargavel, *The Zealous Conservator. A Life of Charles Lane Poole*, 2008, p. 118. Here, Lane-Poole was writing in 1926. In Australia following World War I, there had been soldier resettlements, with subdivisions made on forest land; many of these failed due to the poor soil once the trees had been felled.
15. Ibid.
16. Letter to Sargent from the Australia Hotel, Sydney, 20 December 1920, AA Archives, W.XIV, box 21, folder 11.
17. Letter to his mother, 5 February 1914, cited in RW Briggs, *'Chinese' Wilson. A Life of Ernest H Wilson*, HMSO, London, 1993.
18. I Bevege, 'The elusive Richard Tycho Dalrymple-Hay', *Australian Forest History Society Inc. Newsletter*, 2017, 71:8–9.
19. Wilson, diary, 19 December 1920.
20. M Lyons and CJ Pettigrew, 'Maiden, Joseph Henry (1859–1925)', *Australian Dictionary of Biography*, volume 10, Melbourne University Press, 1986.
21. Letter to Sargent from the Australia Hotel, Sydney, 20 December 1920, AA Archives, W.XIV, box 21, folder 11. The bone disease mentioned is rheumatoid arthritis, which Maiden suffered from after a collecting accident in 1911.
22. Wilson, diary, 28 December 1920. *Casuarina glauca* is now named *C. obesa*. It is a tree 6–15 metres tall and is known as swamp oak, inland swamp oak, and other names, today.
23. These notes from *Daily Telegraph* (Launceston), 15 January 1921, p. 8.
24. Ibid.
25. Wilson, diary, 29 December 1920. Spotted gum is now classified as *Corymbia maculata*. It has a powdery bark.
26. Wilson, diary, 28–31 December 1920.
27. Wilson, diary, 7 January 1921.
28. *Ballarat Star*, 22 April 1921, p. 4.
29. 'Timber resources', *Daily Standard* (Brisbane), 17 May 1921, p. 4.
30. Dargavel, *The Zealous Conservator*, p. 49. See the index in Dargavel's book for details of the school.
31. *Ballarat Star*, 22 April 1921, p. 4.
32. 'Australia's timber resources', *Creswick Advertiser*, April 1921. This article was only found at the AA Archives. No date or page number found.
33. *Creswick Advertiser*, April 1921. As above, no date or page number found.
34. vwma.org.au/explore/memorials/2179, accessed 21 November 2020.
35. 'Ballarat's honor [sic] avenue', *Herald* (Melbourne), 21 April 1921, p. 1. The article notes that Wilson had said this after a visit the previous day. This news was also repeated in the *Weekly Times* (Melbourne), 30 April 1920, p. 30, under country news.
36. Wilson was likely referring to the Nikko-Cryptomeria Avenue near Imaichi City, Tochigi Prefecture, that has a total length of 22 miles (35 kilometres). It was planted in 1628–48, with *Cryptomeria japonica* trees.
37. *Ballarat Star*, 22 April 1921, p. 4.
38. Letter to Sargent, from the Menzies Hotel, Melbourne, 25 April 1921. Wilson also wrote about writing the letter to catch the mail service to the United States before it closed, reminding us of how the mail used to operate.
39. A Dawn Service was not held until the 1930s. It is now a feature of Anzac Day. Sixty thousand Australians and 18,000 New Zealanders died in World War I. In 1918 Australia had a population of 5 million and New Zealand 1.15 million. In one letter, Wilson noted that the population of Perth in 1920 was about 33,000.
40. Wilson, 'Notes from Australasia, III. The Hobart Botanical Gardens', *Journal of the Arnold Arboretum*, 1921, 3(1): 51–5.
41. 'A modern Garden of Eden', *Mercury* (Hobart), 5 April 1921.
42. Ibid.
43. *Mercury* (Hobart), 18 May 1921, p. 4.
44. Comments from The Tasmanian Big Tree Hunters.
45. J Worth, G Jordan, G McKinnon and R Vaillancourt, 'The major Australian cool temperate rainforest tree *Nothofagus cunninghamii* withstood

Pleistocene glacial aridity within multiple regions: Evidence from the chloroplast', *New Phytologist*, 2009, 182:519–32.
46 *World* (Hobart), 7 December 1921, p. 6.
47 'A modern Garden of Eden', *Mercury* (Hobart), 5 April 1921, p. 4.
48 Gwenda Sheridan, 'Insights into Tasmania's cultural landscape: The conifer connection', *Australian Garden History*, 2011, 22(4):6–12.
49 *World* (Hobart), 7 December 1921, p. 6. This was the conservator's annual report to the Tasmanian Government, presented to the House of Assembly, and it described Wilson's visit; it was also noted in the same way in the *Examiner* (Launceston, Tasmania) on the same day, p. 3.
50 Ibid. Note that the reference to the Commonwealth is likely to the Commonwealth of Australia.
51 *Advocate* (Burnie, Tasmania), 25 May 1921, p. 2.
52 'Forestry in Tasmania', *World* (Hobart), 2 March 1922, p. 3. It was estimated that 1,000 boys per annum could be brought into Tasmania to plant spruce, larch, and pine.
53 Sheridan, 'Insights into Tasmania's cultural landscape', 2011.
54 For more insight concerning the political debate about forests and government attitudes, see 'The need for a forest policy', *Brisbane Courier*, 6 June 1921, which includes comments on the tensions between forestry men and the Lands Department in Queensland.
55 *World* (Hobart), 2 March 1922, p. 3.
56 'Australia's timber lands. Folly of denudation', *Argus* (Melbourne), 26 April 1921, p. 4. This article was repeated in regional newspapers outside the Victorian capital: *Maitland Daily Mercury*, 28 April 1921; *Australasian* (Melbourne)

(1864–1946), 30 April 1921; *Maitland Weekly Mercury*, 30 April 1921; *South Eastern Times* (South Australia), 6 May 1921.
57 'Horticultural notes', *Australasian*, 26 March 1921, p. 8.
58 WG Sebald, *Campo Santo*, Penguin, 2005, p. 36.
59 Wilson, *A Naturalist in Western China with Vasculum, Camera and Gun*, 1913, chapter IX, The Chengtu Plain: 'The garden of Western China', p. 109.
60 *Observer* (Adelaide), 4 December 1920, p. 37, in an article entitled 'Australian timber: An American's impressions', with a subtitle 'The waste of timber'. This story was copied from and repeated in other newspapers: *Recorder* (Port Pirie, South Australia), 1 December, p. 3; *Daily Commercial News and Shipping List* (Sydney), 8 December, p. 5; *Daily Commercial News and Shipping List*, 3 December, p. 4; *Morning Bulletin* (Queensland), 17 December, p. 8.
61 *Sydney Mail*, 9 February 1921, report on Wilson's visit.
62 'On the land. The Bulga Plateau. Valuable timber resources', *Sydney Morning Herald*, 29 April 1920, p. 5. The eucalypts mentioned are tallowwood, blue gum, brush box, and blackbutt.
63 *Northern Herald* (Cairns, Queensland), 16 February 1921, p. 34. First reported in the *Herald* (Melbourne), 2 February 1921, p. 6, then the *Sydney Mail*, 9 February 1921, p. 28.
64 'Forest wealth', *Weekly Times* (Melbourne), 12 February 1921, p. 33.
65 F Gordon, *Will Purdom: Agitator, Plant-hunter, Forester*, Royal Botanic Gardens, Edinburgh, 2021, p. 213.
66 Ibid.
67 'Studying tree life', *Brisbane Courier*, 14 May 1921, p. 3. The article gives a short summary of Wilson's career and the Arnold Arboretum – 'the arboreal museum at Harvard'.

68 A summary of Swain's life is given in *South-East Queensland forests: A pictorial history*, Private Forestry Service, 2013, pp. 54–60. 'People and Trees'.
69 For a detailed assessment of Swain's attitudes to forests and forestry and his belief in their importance to human spirituality see GA Barton and BM Bennett, 'Edward Harold Fulcher Swain's vision of forest modernity', *Intellectual History Review*, 2011, 21(2):135–50.
70 'The forest man: EHF Swain', accessed January 2021, www.abc.net.au/radionational/programs/archived/hindsight/new-document/4265230
71 These comments from Noongar lands from south-western Australia made by Mr C Humphries, a senior Ballardong culture and language Elder. Quotes from *Ballardong Noongar Budjar 'Healthy Country – Healthy People'*, Avon Catchment Council, Government of Western Australia, accessed 2 September 2021, www.wheatbeltnrm.org.au/sites/default/files/knowledge_hub/documents/ballardong_noongar_budjar.pdf
72 'Cordial visitor. Queensland botanists appreciated', *Daily Mail*, 6 June 1921. Wilson 'added a word of appreciation of the interest taken in forestry and natural history' by that newspaper. Wilson also became a life member of the Queensland Field Naturalists' Club, which had a keen interest in botany and birds.
73 Quotes from *The Boston Evening Transcript*, 30 September 1922; this is a report of his entire trip.
74 *Observer* (Adelaide), 12 November 1921, p. 5.
75 *Telegraph* (Brisbane), 20 May 1921, p. 6.
76 'Queensland's Forest Wealth', *Brisbane Courier*, 24 May 1921, p. 6.
77 *The Week* (Brisbane), 10 June 1921, p. 3. The scientific key to the success of these large species is

in the mycorrhizal fungi present. Fraser Island is now a World Heritage site.
78 'Birds and forests', *Register* (Adelaide), 14 June 1921, p. 5. The trip is further explained in a later newspaper report: 'Timber waste. Glories of Fraser Island. Captain White describes visit to Queensland', *Daily Herald* (Adelaide), 27 June 1920, p. 6. Captain White was there to comment on the relationship of birds and control of an insect, a pest of cedar.
79 Peter Kershaw and Barbara Wagstaff, 'The southern conifer family Araucariaceae: History, status, and value for paleoenvironmental reconstruction', *Annual Review of Ecology and Systematics*, 2001, 32:397–414.
80 For a discussion of the Araucariaceae, and American perspectives and leaf forms very different from pines, see WP Armstrong, *Pacific Horticulture*, 2010, 80(2), and 'The Araucaria family: Past & present', accessed 30 January 2021, www.pacifichorticulture.org/articles/the-araucaria-family-past-present/. The Wollemi pine is expensive to buy, but widely available. The Wollemi site was even kept a secret during the severe bushfires of 2019–20, when secret water-bombing was done to protect the area.
81 SR Ash and GT Creber, 'The late Triassic *Araucarioxylon arizonicum* trees of the Petrified Forest National Park, Arizona, USA', *Paleontology*, 2003, 43(1):15–28.
82 See J Muir, *John Muir's Last Journey – South to the Amazon and East to Africa: Unpublished journals and selected correspondence*, MP Branch (ed.), Island Press, 2001.
83 J Huth, 'Introducing the bunya pine, a noble denizen of the scrub', *Queensland Review*, 2002, 9(2):7–20.
84 Ibid.

85 Huth, *Introducing the bunya pine, a noble denizen of the scrub*, 2002. Huth gives a full account of the bunya and the story of its use and then protection.
86 *Observer* (Adelaide), 12 November 1921, p. 4.
87 'Timber waste', *Daily Herald* (Adelaide), 27 June 1921, p. 6. It is a pity to hear of the fine photos taken being lost at sea.
88 'Men and matters', *Worker* (Brisbane), 9 June 1921, p. 10.
89 The night of 23 May was reported in the *Daily Mail* (Brisbane), 25 May 1921, p. 8, under the title 'Natural history: Its value and interest'.
90 G Borschmann, *The People's Forest. A living history of the Australian bush*, 1999, p. 223.
91 'Forestry', *Queenslander*, 25 June 1921, p. 2.
92 Borschmann, *The People's Forest*, 1999, p. 223.
93 'Little known: What southern tourists miss', *Brisbane Courier*, 5 August 1920, p. 9.
94 Ibid.
95 Information on Expedition National Park, Queensland Government Parks and Forests, Department of Environment and Science, accessed 16 February 2021, parks.des.qld.gov.au/parks/expedition/about/culture
96 The Queensland Government is now developing policies to bring back the brigalow. www.qhatlas.com.au/content/brigalow, accessed 11 January 2023.
97 Recorded in J Powell, *People and Trees*, 1998, p. 56.
98 *Daily Mail*, 3 June 1921, p. 7.
99 *The Week* (Brisbane), 10 June 1921, p. 3.
100 'Forestry', *Queenslander*, 25 June 1921, p. 2.
101 *Queenslander*, 25 June 1921, p. 2; 'Tree farming: The state's part. Minister's reply to criticism', *Daily Mail*, 5 June 1921, p. 7.
102 *Morning Bulletin* (Rockhampton, Queensland), 10 June 1921, p. 8,

in a long and passionate article titled 'Queensland forestry'.
103 'Our timbers. American visitor's impressions. A plea for conservation', *Daily Standard* (Brisbane), 3 June 1921.
104 Ibid.
105 Ibid. This statement was reported widely and the article repeated.

CHAPTER 8

WILSON MORTIFIED BY AUSTRALIA'S 'HABIT OF DESTRUCTIVENESS'

Wilson extolled the value of trees wherever he went: 'Oh, I should like to see a real enthusiasm for trees. Are they not, leaving out man himself, Nature's noblest creations?'[1] He was highly aware of tree loss by removal or neglect, all over the world:

> Some species are as extinct as the dodo. There are 18th century records of trees in America that no longer exist. Maybe it is the same in Australia. There are several species in American and China that today are reduced to a few individuals ... in Australia, the scarlet flowering gum (ficifolia) occurs in a wild state only in one small patch near Denmark (West Australia), though there are of course, plenty of cultivated examples.[2]

He noted that there were several tree species in China and Japan that were no longer seen in the wild. Wilson also spoke of the strong need in Australia for a change in public attitudes to trees, and the need to foster such a change. He said:

> The creation of a public sentiment for trees is what is needed, in this country. It behoves all who realise the seriousness of the evils of forest destruction to lose no opportunity in educating the people. Much may be done in the schools. Few seem to sufficiently appreciate what an important factor trees are in the wealth, health and beauty of a country. When you destroy the forests you upset the balance of nature. Besides the aspects

of utility and beauty, trees play an important part in health. They live on the very gas (carbon dioxide) that is poisonous to the animal kingdom, and if the vegetable kingdom were destroyed the animal kingdom would automatically die too.[3]

He lamented the destruction of trees that was so prevalent across the continent, and believed that the 'habit of destructiveness' might have been a legacy of seeing the forest as a menace and an enemy to be subdued. He said that 'we might have expected the passing of these ideas and conceptions' over time, 'but if the tendency to destroy is not so active, there has come in its stead an apathy which is little better than destructiveness. The present generation must remember that it only holds the forest in trust for future generations.'[4] These comments still ring true today: that there might have been a shift from an 'appalling and wicked waste'[5] and destruction of trees in parts of Australia, but it has moved to often a simple apathy towards them. There remains a lack of will and motivation to conserve and cherish; the science is there but the values placed on our landscapes are often still lacking true care.[6]

Wilson lamented that there was a great need 'of educating the public to a realisation of the value of a national asset, which was of the greatest importance, and said that it was only by securing individual interest in this matter that this could be done'.[7] He was, like Lane-Poole, largely speaking to deaf ears. Like Lane-Poole, Wilson spoke very strongly while in Australia of the need for better protection and kinder use of forests, and his voice grew more strident the more he saw. But the 'arboricidal mania' that staggered Wilson was to continue for decades, if it has ceased at all. Around Australia, examples of arboricidal mania can still be seen today in many streets and towns despite modern ambitions for greening – trees savagely pruned, adventitious growth taken off, tops lopped off, limbs left without foliage, trees removed often on the whim of a householder, complaints of a tree being 'too tall' or that 'messy' fine leaves have dropped on roses or turf, and the lack of planning of roads with trees as an integral part of the street. We have a long way to go. In 1921, Wilson said that 'he hoped the people of Australia would think seriously about forestry. The majority of them had not started to think, but if every man would plant a tree … he would be doing something which the children of the future would thank him for.'[8] He also said that 'the great thing was to conserve such trees as nature had given, and to plant those … which are necessary'.[9]

The second requirement that Wilson thought important was for a national forestry school and education of foresters for Australian conditions and species. He acknowledged the 'general agreement' among foresters for such a school, and felt that the delay in establishing such a school was 'wicked'.[10] Part of his feeling about this issue was that, because of the absence of proper training and insights, 'this and preceding generations have thoughtlessly destroyed much of the forest wealth of the country'.[11]

Wilson was also disappointed in the poor use of native trees in parks and gardens all over Australia. He was exasperated at Australia's lack of appreciation of its native trees, and felt that more eucalypts should be planted in public parks. In a talk given on his return to Harvard in 1922, Wilson noted that 'the flowers to be seen in Australia and other countries settled largely by English-speaking people are mostly the same flowers which are to be seen in the home land. While some of the native flowers are grown, many of them are looked upon as too common for cultivation.'[12] The idea that they are common is not an excuse now for lack of planting, given the loss of native vegetation from most Australian cities, particularly in the major cities. The spotted gum, which has a beautiful white trunk with dun-coloured markings, he regarded as 'the lady of the woods'.[13] He said:

> I like your eucalypts – your gum trees, as you call them. I don't think Australians appreciate them as much as they should, judging from the number of imported trees one sees in your parks and gardens, instead of the gum itself. If you used your own native trees, your parks would have a more distinctively Australian appearance, and that I think, should be your aim.

The availability of native trees and shrubs to the public in nurseries has expanded enormously since Wilson's visit one hundred years ago, particularly since the 1970s, but there is still lack of native plantings. The issue of native versus exotic flora is still of concern, and a matter of debate for landscape architects and botanists in Australia today. Wilson's comment about a more distinctly Australian appearance is echoed in concerns that have been raised since the 1980s on the homogenisation of plantings – or 'biotic homogenisation', as it has been termed –

where exotic species are replacing unique native species and everywhere looking somewhat the same.[14] Some Australian states have had more success at planting native trees in public places, and have more nurseries that specialise in indigenous plants to encourage householders to plant natives. Generally, however, there remains a mixture of exotic and native trees in Australian public parks, with older public parks reflecting the sentiments that Wilson criticised. Still lacking in many suburbs and towns is what Wilson described as a 'real enthusiasm for trees'.[15]

Wilson also believed in a great need for a handbook on the Australian flora, one that 'you can take out into the bush and turn it up and get acquainted with the different plants. To begin, it might be a manual of the trees; they are not so numerous, but intensely interesting. Later on there should be one of the shrubs, illustrated.'[16] Today, Wilson would be impressed by the many volumes available for plant identification. However, many are for the initiated, and more needs to be done to educate young people in schools about the unique nature of the Australian flora and why it needs to be celebrated.

Wilson stated to the press that Australia burned its valuable timbers, with a misunderstanding of their value. From the Wheatbelt came this story:

> Being told of a farmer who was ruined in improving a farm, one of the improvements consisted of ring-barking 3,000 acres of wandoo, he [Wilson] said: 'He deserved to lose his money. I am glad he lost it. It is one of the terrible sights that one sees here, the way you burn your valuable timbers. You are no worse than others. It has happened in every part of the world, and the older countries have found out and regretted their mistake. But here in a new colony you had all the benefit of the mistakes that others have made, and you should profit by it.[17]

It appears that we did not. Wilson was one of the voices that cried a warning one hundred years ago.

In the Arboretum archives was a curious article from the *Australian Forestry Journal* of 1925 that Wilson might have kept as an essay of interest for himself. It was titled 'Forest devastation in Australia and Tasmania' and was by Dr Arnold Heim (1882–1965), a geologist from Zurich, who had travelled to Australia in 1921.[18] It was a resounding condemnation of Australian practices of ringbarking and burning good forests. 'It is like a battle field,' Heim wrote.

> Not only is the future being deprived of the splendid forest, which is vanishing from off the face of the earth on every side, but in consequence of the short-sighted materialism of our day the entire country and its splendid climate is being ruined. In vain have distinguished Australian scholars raised their voice in protest.[19]

Although the editor of the *Australian Forestry Journal* claimed that the situation Heim described was now out of date, Wilson's keeping of this article might suggest that he thought that a watch was needed on the state of forest consideration in Australia, for his own observations suggested that Heim's comment was not out of date. Wilson believed that the conservation of forests should be 'one of the foremost planks of Australian policy' to address the very things that appalled Heim.[20]

Wilson made another remark about Western Australia, but he felt this about the whole country:

> I go home very favourably impressed with what I have seen generally … I do not wish to criticise, but if I was out to criticise, all I would do would be to take photographs of these areas where every tree has been destroyed, and write underneath it 'Improved Land in Western Australia'. The pity is that it is so general. One realises that people have got to live, have got to win sustenance from the soil, and trees and other things must disappear to give room for this sustenance. But a lot of the land will not grow anything as good as trees.[21]

Wilson noted to the press that 'There was the cry to get the land cleared. And when it was cleared what was the result? A few cows were put to graze on the country. That was, of course, necessary, but it opened grave questions for the future … To the present age the forests were an inheritance for use … they should be handed on to the future generations, (Figure 8.1).

FIGURE 8.1: Extensively cleared woodland with few of the original trees left, likely in New South Wales, 1921. Judging by the remnant trees, many of the trees cleared must have been large and old. Image: Wilson Y-425 AA.[22]

In Victoria, he said:

> The trouble is that you do not realise how great an asset timber is. Timber resources are about the richest national asset that Australia possesses. You may think a lot about your agricultural lands and vine lands, and your mines but, strictly speaking, your best asset is your forest lands. Mines will peter out, but forests are, or should be, for ever. Trees do not belong to one generation; they belong to this generation, and to the future also.[23]

As he was about to leave Australia, he was reported as saying, under a large newspaper headline of 'SAVE THE TREES. A VISITOR'S PLEA': 'In quitting Queensland my last words are to plead with Queenslanders for better protection of their forest wealth. It is richer than they dream, and of untold value to themselves and to the world.' This was the chief of the impressions that he gained, the article said.[24]

Another final comment on Australia to journalists?

You have got one of the richest lands in the world. You are a most fortunate people. You get so much for doing so little, compared with conditions in other countries. You have one of the most beautiful countries in the world to live in, and the easiest conditions. I have been here some months now and it seems to me that Australians do not appreciate their good fortune. There should be no discontent here at all. This place is a paradise, and you don't seem to realise it because the conditions in many of the older countries have never been brought home to you.[25]

Harsh indeed.

Having read so many comments about trees and forests made by Wilson, and noting the questions and focus of the press and others, it is clear that the extent and knowledge of ecology and conservation in the 1920s was considerable, and that we appeared to have missed an opportunity one hundred years ago to forestall or prevent much of the later destruction of forests and woodlands. 'If trees only had votes,' Wilson said, 'they would not be treated as they now are'.[26] The dates of birth of many of these men who spoke for and worked hard for conservation of forests and forest preservation for future generations suggest something strongly: botanist Cyril White (1890), Charles Lane-Poole (1885), Edward Swain (1883), Wilson himself (1876), Captain Samuel White (1870), and William Sowden (1858) – all were born in the 19th century, and were mostly or entirely educated in that century. Yet today, we are often told that the environmental movement only began in earnest with Rachel Carson's *Silent Spring* of 1962, or, for landscape architects – particularly those in the United States – after Ian McHarg's *Design with Nature* of 1969. But forest conservationists like Lane-Poole, now heroes for keeping forests with their ecosystems, lived and, in most cases, died, well before those books and publicity, and the more widespread knowledge of ecology. The move towards conservation had a longer history. Influences on Wilson's generation included the natural history writings of John Muir and his ecological advocacy, the rise of National Parks as 'America's best idea', with the first national park (Yellowstone) in 1872 in the USA as protection against extractive uses of outstanding natural or wilderness lands, and the National Parks Service in the USA beginning in 1916, which has been copied globally since. Long and detailed debates occurred in the US about codifying the uses of National Parks and forest land all through the first two decades of the twentieth century, and these would have been followed closely by men concerned with the use of timber, such as Wilson. In Australia, John Ednie-Brown was Western Australia's first Conservator of Forests in 1896, and, like Charles Lane-Poole after him, was concerned with the preservation of forests. Ednie-Brown wisely wrote:

> The forests are nature's gift, and should be looked upon and dealt with accordingly, as an inestimable inheritance of great commercial and climatic value; besides much of the land upon which the best timber grows is, as a rule, of little or no value for agricultural purposes, and I maintain, without fear of logical contradiction, that what is now upon it is the very best kind of crop that will ever be seen there. To destroy it therefore, for the sake of a few more blades of grass, is suicidal and reprehensible in the extreme.[27]

However, it appears that Wilson was typical of his time, in that there was both 'an aesthetic and a utilitarian response'[28] to the idea of natural resources to 'benefit future generations'. This was seen in his strongly pragmatic ideas about the commercial potential of timber in Australia, while at the same time strongly lamenting the loss of natural forests for mere blades of grass, and speaking of the need to preserve them for future generations. There is a certain ambiguity in Wilson's position. It was likely a difficult path for him to tread, between utilitarian need and what his heart as a lover of trees might have told him, and this can be seen in his horror of wasting timber, whether burnt or for minor and trivial uses that were (and often remain) so widespread in Australia. The practical need for timber and the aesthetic and ecological desire for preservation is still a core tension today in our treatment of our forests and woodlands.

Yet none of this comes out in Wilson's only writing on this Expedition to Australasia, *Plant Hunting* in 1927, later published in 1985 as the book *Smoke that Thunders*.[29] This record is in part travelogue, part horticulture and plant descriptions, and part history of places, with a note about early botanical collectors in Australia, a group to whom his 1927 book was dedicated. Only in places did I find that Wilson drew on his scant diaries or letters for this publication; perhaps he had other notes, now destroyed or lost, because the detail of species is considerable, particularly given that he was a visitor from another part of the planet. He also had a formidable memory, described as 'prodigious', and was seen as a 'walking encyclopaedia'.[30]

Nowhere in Wilson's 1927 book do we sense his anger at the clearing and waste of timber, his 'for God's sake stop', his pointing out the 'wicked' nature of 'appalling' habitat lost,[31] his contempt and dismay for clearing rich and beautiful timbered forest for dairy cows that gave poor financial gain, his dismay at the 'indiscriminate ruthless manner of 'improving' the country',[32] his worry that nature would one day exact her payment in now-salted land in Western Australia, and his constant concern for what trees and forest might be left for future generations. It seems a curious thing that he concealed these issues from his readers. However, Wilson had done the equivalent when he wrote his best-selling books about plant hunting in China – he had removed the dirt and squalor and 'much of the mental and physical strain that the journeys imposed'.[33] Nowhere do we find in Wilson's China exploits angry comments about deforestation in China. Wilson's biographer Roy Briggs noted that Wilson 'softened and lightened' the stories for the public.[34] We can only conjecture as to why. Briggs was able to examine extensive journal notes for Wilson's travels in China. In contrast, there are scant journal notes made by Wilson in Australia, and I have thus relied on newspaper articles to point out the contrast between his thinking as reported by journalists and what he wrote in *Plant Hunting* and everywhere else.

Over the course of his travels in Australia, Wilson had moved from his amazement at the colours, forms, and sheer diversity of species he found in the 'veritable botanic garden' to a wonder at forest giants still largely unknown outside Australia, to a dismay at the extensive clearing of large trees with the extraordinarily rare capacity of thriving on nutrient-poor soils, to a growing and definite anger at the false idea of 'improving' country by denuding it, and the widespread 'arboricidal mania' and general apathy to trees that he saw so clearly across the entire country. In this Australian sojourn – now in an English-speaking country and meeting dozens of major players in forestry, foresters, lumbermen, politicians, and followed about by newspaper men – he was able to fully comprehend the state of Australian forests and woodlands through his frank dialogues with local experts and to use his wide knowledge of place and policies. His comments were widely reported and taken note of by governments, even if they did not act upon them or if government officials denied the state of forests that he reported. Given the size of Australia, this was a quick trip, but long enough to gain a perspective on the treatment of landscape across the continent.

Wilson was a witness to what was happening to Australian forests and woodlands, and he viewed all that he saw with a critical eye. Nowhere in all his years of global travel did he more clearly speak his mind in public. The Australasian Expedition paints Wilson less as a plant collector, for which he was already famous, and more as an astute and immensely knowledgeable man speaking strongly at the beginning of a greater environmental awareness.

NOTES

1. *Herald* (Melbourne), 2 February 1921, p. 6.
2. *Herald* (Melbourne) and *Northern Herald* (Cairns, Queensland), 16 February 1921, p. 34.
3. 'Millions wasted. Professor's plea. To save forests. Wonderfully rich Australia', *Sydney Morning Herald*, 20 January 1921, p. 8. Repeated in various newspapers, including *Toowoomba Chronicle*.
4. Ibid.
5. Ibid.
6. A point made in A Leavesley and J Mallela, *Desert Channels – the Impulse to Conserve*, CSIRO 2014.
7. *Ballarat Star*, 22 April 1921, p. 4.
8. *Ballarat Star*, 21 April 1921. This is a report from what appears to be a public seminar.
9. 'Australia's timber resources. A great asset. Visiting expert's views', *Creswick Advertiser*, April 1921.
10. 'Our timber wealth. Prof. Wilson's views. Growth of American trees', *Daily Telegraph* (Sydney) 12 May 1921, p. 4.
11. 'Forests and forest products', *Daily Telegraph* (Sydney), 14 March 1921.
12. Article, undated but likely 1922, titled 'E. H. Wilson returns', AA Archives, W.XV, box 26, folder 12.
13. 'Horticultural notes', *Australasian* (Melbourne), 26 March 1921, p. 8.
14. JD Olden and N LeRoy Poff, 'Toward a mechanistic understanding and prediction of biotic homogenization', *American Naturalist*, 2003, 162(4):442–60.
15. Ibid.
16. *Western Mail*, 25 November 1920.
17. 'A study of trees', *Western Mail*, 25 November 1920, p. 9.
18. Heim is noted as being in Australia in 1921 in *ETH Zürich*, https://doi.org/10.3929/ethz-a-000646758
19. A Heim, *The Australian Forestry Journal*, 13 February 1925, AA Archives, W.XV, box 26, folder 11. This is a reprint from the *Swiss Forestry Journal* of 1924. An image of Wilson is on the front of the Australian journal.
20. 'Millions wasted. Professor's plea to save forests. Wonderfully rich Australia', *Sydney Morning Herald*, 20 January 1920.
21. This was a major article: 'A study of trees: Professor Ernest H. Wilson's visit. Our wonderful flora', *Western Mail*, 25 November 1920, p. 9.
22. 'Australian timber. An American's impressions', *Observer* (Adelaide), 4 December 1920, p. 37.
23. 'Our timber reserves. Australia's great asset. American expert's views', *Ballarat Star*, 21 April 1921.
24. 'Save the trees. A visitor's plea. Forestry act needed', *Daily Mail* (Brisbane), 3 June 1921, p. 7.
25. *Ballarat Star*, 22 April 1921, p. 4. This was next to an advertisement for the wonders of asbestos in homes.
26. *Ballarat Star*, 21 April 1921.
27. MAC Fraser, *Notes on the natural history, etc. of Western Australia, being extracts from the Western Australian Year Book for 1900–1901*, Government Printer, Perth, 1903.
28. RW Sellars, *Preserving Nature in the National Parks: A History*, 2nd edn, Yale University Press, 2009, p. 43.
29. EH Wilson, *Plant Hunting*, volume 1, Stratford Company, Boston, Mass., 1927; Wilson, *Smoke that Thunders*, Waterstone, London, 1985.
30. RW Briggs, *'Chinese' Wilson. A life of Ernest H Wilson, 1876–1930*, HMSO, London, 1993. Foreword by Roy Lancaster.
31. 'Millions wasted. Professor's plea to save forests. Wonderfully rich Australia', *Sydney Morning Herald*, 20 January 1921.
32. 'Horticultural notes', *Australasian* (Melbourne), 26 March 1921, p. 8.
33. Briggs, *'Chinese' Wilson*, 1993, p. 2. Wilson had commented on the unstable political situation in China in *A Naturalist in Western China with Vasculum, Camera, and Gun*.
34. Ibid.

CHAPTER 9

EXTENDED JOURNEYS

In a letter dated 18 April 1921, Sargent suggested to Wilson that he might extend his journey, and travel from Mombasa to Nairobi, then by train to Cape Town. Exhaustive new preparations were needed. Sargent wrote:

> I … haven't yet been able to find out about the sailing of the steamers from India … I find that the place to go first by rail from Mombassa [sic] to Nirobi [sic] where there is a firm of outfitters, Newland, Tarlton & Company, who can supply boys and what outfit you need. The place from which to start for the forest appears to be Kijobi which is on the railroad three hundred miles west of Mombasa. This is only one day's march from the Aberdare Mountains … According to forestry reports the best trees in East Africa are on these mountains … It appears to me that if you get to Mombasa it would pay to go south to Durban which is the end of a railroad which would take you to Victoria Falls and down to Capetown, and then come home by a Capetown-England steamer. This would enable you to see a great deal of additional country and probably would not take more than three months longer than from Mombasa direct to Europe.[1]

On 3 June 1921, a year after he had left Boston, Wilson departed from Brisbane by steamer, bound for India via Singapore and Java on the ship *Montoro*. Table 9.1 gives his route. The eternal plant hunter and tireless worker,[2] when the steamer called in to the remote and tropical Thursday Island in the Torres Strait on 21 June, he collected specimens (Figures 9.1 and 9.2). This was his last collection in Australian waters. From there, he travelled to Port Darwin, described by him as 'rather primitive and roughly tenanted',[3] then across the Timor Sea to Java. In Java, and again in Malaya, he saw extensive destruction of forests for rubber plantations. While in the Malay states, he noted that he

> rode for seven hours by railway train, or more than three hundred miles, through what had not many years ago been a continuous and infinitely varied tropical forest, but which is now one vast rubber plantation ... This condition illustrates the general tropical tendency to the substitution of cultivation for forest life, and the consequent extirpation of species.[4]

Today, such loss goes on, with the addition of palm oil as the reason for destruction.

TABLE 9.1: Wilson's travels beyond Australia

1921	June	Townsville, departed 3 June on the ship *Montoro*[5] 21 June on Thursday Island, Torres Strait, collecting; sailed from Darwin for Java 25 June
	July	Departed for Singapore, Penang, and Malay States,[6] Rangoon; arrived Calcutta 19 July
	August	Northern India and (now) Pakistan: Lucknow, Dehra, Simla (to see 'great deodara', *Cedrus deodara*, Kashmir (garden of Nazimbagh, giant oriental plane tree, *Platanus orientalis*), Agra, Assam (*Pinus khasya*)
	September	Southern India: Madras, Nilghiri Hills, sacred bow tree (Buddha) (*Ficus religiosa*)
	October	India to Ceylon 6 October
	November	India, departed Bombay 4 November for Mombasa, arrived 11 November
	December	Kenya and Uganda
1922	January	Kenya and Rhodesia (visited Victoria Falls)
	February	South Africa (Transvaal, Natal, Zululand)
	March	South Africa; Port Elizabeth to Cape Town
	April	South Africa; departed from Cape Town for Southampton 7 April 1922[7]
	May	Arrived London 31 May 1922. Back at Harvard September 1922

After four months on the Indian subcontinent, Wilson followed Sargent's suggestion–or was it a directive?–of a route in eastern Africa for three more months, and arrived at Cape Town, finally sailing for England in April 1922. He arrived back at Harvard in September 1922 after an absence of more than two years, having departed Boston on 8 July 1920.[8] He was interviewed by the

FIGURE 9.1: Wilson collected a *Melaleuca* on Thursday Island in June 1921. A 'Tree 20ft'; the species name is hard to discern, but likely *M. viridiflora*, the red-flowering paperbark, which is found across the tropical north of Australia and into southern Papua New Guinea.

FIGURE 9.2: Flowers of *Melaleuca viridiflora*. Image: www.lullfitz.com.au/melaleuca-viridiflora with permission.

press. *Horticultural News* reported the loss of specimens on the *Canastota* the year before, but little was said, as nothing could be done, or undone. The disaster received scant publicity in the USA. Knowing the history of botanical collecting as he did, Wilson must have been philosophical about this terrible disappointment and the loss of two large consignments of specimens, photos, and (it seems) diary material. Knowledge of the Expedition's material loss had itself been lost over time, with an ensuing incorrect deduction that Wilson had collected little and taken few images while in Australia.[9]

The major remains of the Australian collection are of Western Australia. Wilson noted to Sargent, in a letter dated November 1920, that:

> I have seen all the different belts of vegetation in this part of the country & collected about five hundred specimens. I have seen all the important forest trees & taken photographs of them. In all I have taken eight dozen photographs. The Conservator of Forests has presented me with nearly fifty photographs & specimens of about 450 species of woody plants. These include a complete set of the Eucalypts of Western Australia all of which have been authentically named by Maiden. The Government Botanist is making up a set of duplicates for us also.[10]

Wilson's trip had added 'scores of books and hundreds of pamphlets' to the Arnold Arboretum, mainly sent by his Australian contacts. The Boston reporters now stated that 'the purpose of Mr Wilson's travels … was to establish more intimate connections between the Arboretum and other great gardens and arboretums of the world as well as to add to the library of the institution and to increase its great collection of photographs.'[11]

It is not known how much the Expedition cost. In a letter to Mr A. L. Endicott in Cambridge, Massachusetts, of August 1922, Sargent notes an enclosed cheque of $3,437.84 'being the unexpended balance of the credits issued to E. H. Wilson to cover the expenses of his recent journey'.[12] Wilson had travelled steerage to Australia, and had been given accommodation, assistance, and transport while in Australia, such was his fame.

With the rise in plant pathology in the 1920s and greater knowledge of disease epidemiology and transmission, new quarantine regulations had come into force in 1922 in the USA regarding the importation of plant material from overseas, and there was a complete ban on bringing in new tree specimens or seeds for horticulture in the USA. Sargent was very annoyed. He complained that 'the treatment of seeds in Washington by the employees of the Federal Horticultural Board means death.' The change in quarantine rules meant that seeds collected by Wilson on the Expedition to India, Africa, and Australasia were sent by him to England, not the USA. Sargent pointed out to the authorities, without effect, that 'by this ruling of the Federal Horticultural Board the Arboretum has lost an opportunity to improve the plantations and gardens of California and possibly of other states'.[13] Books about Wilson that focus on his successful plant introductions into the USA make no mention of the Australian expedition, because no new plants were brought in, nor could they be. Thus Stephen Spongberg's *A Reunion of Trees: The Discovery of Exotic Plants and Their Introduction into North American and European Landscapes*[14] does not mention the Expedition to Australasia, even though it was Wilson's last outside of the USA and took a great deal of his time and energy.

Reading Sargent's letter one hundred years later, I was struck by the time and commitment taken with botanical collecting by Wilson, and what was demanded of him from the Arnold Arboretum. Wilson himself acknowledged the great debt to plant collectors who had gone before him, and he noted the high calibre of the botanists who collected in Australia. There is a succinct sketch of plant exploration in Australia in *Smoke that Thunders*, in a chapter respectfully entitled 'Those who paid the Price'–many had died young through disease, the privations of long voyages, and misadventure. I had seen many of their names as I had gone through Harvard's collection of Australian plant specimens–names as collectors, as taxonomic authors of the species, or whom the species name honours. They included the seventeenth-century English buccaneer and privateer William Dampier (*Willdampia formosa*), Joseph Banks (the genus *Banksia*), who went with Captain James Cook, the Frenchman Jean-Baptiste Leschenault de la Tour (the genus *Leschenaultia*), Archibald Menzies (*Banksia menziesii*), who was the naval surgeon with Commander George Vancouver on HMS *Discovery* in 1791, the Kew gardener Peter Good (the family Goodeniaceae; *Banksia goodii*), who was assistant to the English botanist Robert Brown,[15] who named many Australian genera, including *Grevillea* and *Eremophila*; both Good and Brown sailed with the great botanical artist Ferdinand Bauer (*Banksia baueri*) on HMS *Investigator* with Matthew Flinders in 1801–3. Other names seen in the Harvard Herbaria collection were George Caley (*Banksia caleyi*), who explored and collected in eastern Australia from 1800 with his constant Aboriginal guide Daniel Moowattin. Kew's Allen Cunningham collected from 1816 inland in eastern Australia (the hoop pine, *Araucaria cunninghamii*), who went with Captain James Cook and Georgiana Molloy and James Drummond collected in Western Australia in the 1830s (*Acacia drummondii*), and these were all seen in the Harvard collection.

Yet more than plant collecting was here in Wilson's last great journey overseas. When I had begun, I had expected a relatively simple story of plants and collecting in the manner of his eastern Asian exploits, as they are now mainly remembered, though they include extensive landscape description. I had not expected the long passages of environmental concern recorded by the newspapers. With his anger and anguish clearly on show, and his frank comments to journalists, I felt that I had found my man, who had been in every way an elusive spirit in his scant writings in this Australian trip. This is a new story of Wilson, only touched upon by Wilson's biographer Roy Briggs in a few pages about this Australian sojourn.[16] It seemed that free from the strictures of tough collecting in remote sites, and now knowledgeably guided, with professional help beside him wherever he went, he was able to take on the different role that was expected of him in Australia – that of the well-travelled international expert and advisor. He was at the height of his powers, lionised in America, and enormously experienced, and it seems that he relished the new opportunity for his voice to be heard about his great passion – trees and their preservation – and to speak freely of the 'arboricidal mania' he saw in Australia. This was a completely different type of dialogue. From what I had found, it was an expedition that spoke of different things, and of the environmental movement that was alive in the 1920s.

I wondered what he did when he finally got home to Harvard in 1922. What had been expected of him on his return? *Smoke that Thunders*, written in 1927, gives no indication of the topics that Wilson talked about in scores of newspaper reports during his Australian trip. The most he says in *Smoke* is a criticism of the derogatory and inappropriate term 'scrub' for forests by settlers in Australia,[17] and that in New Zealand 'axe and fire have in less than a century played sad havoc' with the heritage of trees.[18] No more is said. On poor attitudes to trees or ruthless clearing and firing of forest, or clear-felling, he is silent. Instead, he returned to take up his role as President of the Horticultural Club of Boston, and we hear of apple varieties discussed while in Tasmania, and that he visited a chrysanthemum grower near Melbourne.[19] We read that South African proteas were very interesting, and that he found the baobab tree the ugliest tree he knew.[20] During the rest of the 1920s, Wilson wrote purely horticultural monographs on azaleas and the lilies of eastern Asia. His large book *China, Mother of Gardens*, published in 1929, is horticultural, and descriptive of landscape and Chinese society. In *Aristocrats of the trees*, published in 1930, he says nothing about arboricide and clear-felling of old aristocratic trees. The angry man in Australia is now silent.

Did Australia liberate him from some stricture of the horticultural man and enable him to speak his mind on a deeper subject, to take on a different mantle? It appears that he never did so again once he returned to Harvard. Was he protecting his readers and funders? If so, why? I was interested in these questions, and during my time in the Arboretum, I attended an excellent lecture on ecology, biodiversity, and ethics, where a member of the audience noted with indignant horror that 'Australia is killing its cats!' I went to respond, but the lecturer, who knew where I was from, indicated for me to remain silent, and I did so. Yet in a very different part of the planet from puss on the sofa in Boston, feral cats in Australia act as master hunters, and kill about one billion native animals per year; cats have brought an ecological and extinction crisis to Australia's unique mammals and marsupials. Perhaps Wilson felt that he needed to keep to his expected topic, and so as to not offend, stayed silent. I wondered what he said privately to Sargent about the loss and lack of protection of trees in Australia, given that one of the ambitions of the Expedition had been to do a census of the worlds' trees because they thought them to be in danger.

Wilson's silence about forest clearing in New Zealand is similar. In a letter to Sargent in 1921, he had written that 'New Zealand is forging a frightful toll for the indiscriminate and thoughtless introduction of … plants, animals & birds.'[21] But in *Smoke that Thunders*, despite some criticism of 'destructive intervention', he described New Zealand as 'Queen of the Southern Seas',[22] and described the country as 'Scenic isles mantled in green'.[23]

New Zealand today still puts out an image of '100 per cent Pure New Zealand', as its tourism industry has extolled since 1999. Yet 'over much of the wild New Zealand landscape, introduced plants, birds, and mammals are dominant.'[24] Biodiversity loss due to human activity in New Zealand has been ongoing since the arrival of the Māori in the early fourteenth century, and accelerated after European settlement in the early decades of the nineteenth century. Famously, the giant bird the moa was lost centuries ago, by about 1500, but today the influence of introduced plant species in New Zealand has meant that even the indigenous soil mycoflora has declined;[25] this is a startling thought. There are now efforts to reduce the spread and dominance of exotic species in New Zealand.[26] While Wilson would have heard stories about the loss of native species and the rapid takeover of introduced species from botanists and other scientists there, he saw with his own sharp eye that things were amiss in 1921, and visited sites where forests were clear-felled (Figure 9.3). Although I did not follow his New Zealand travels, there he also made very great criticisms to the press. Wellington's *Evening Post* ran the headline 'Wanton Waste. New Zealand's Burnt Forests. American Expert's Castigation'; Wilson suggested that New Zealand should 'set up and follow a new motto: 'use, not abuse, the bush'.[27] Despite these observations and concerns, he kept those thoughts from the reading public in his own publications. It is a curious insight into Wilson's professional world, and what he felt that he could say, and where. His reputation was not founded on bad-news stories, but on beautiful plants and practical plant-hunting in interesting but sometimes-challenging landscapes.

The absence of environmental critique and exposition on his return raised the question as to whether being away from a centre of fame in Harvard and now in a periphery region – the remote, new democracy of Australia, with emerging institutions and men trying to figure out how to farm and work with forest land that was proving so different to other parts of the world – provided Wilson with some liberation of thought and speech. He was free to be more than the plant collector, the importer of plants, and the horticultural guru that people in Harvard expected and lauded, and perhaps even took for granted.

On his return to Harvard, Wilson appeared to go back to the persona that he had established at Harvard, even though it might have been even confining. In Australia, he was the environmentalist and anything that he wanted to be; he was not constrained by any narrow expectations, because often the periphery is a freer place to be than the centre. Perhaps he felt more at home in a Commonwealth country, and was at ease enough to speak his mind. Further, he was asked his opinion about wider environmental issues wherever he went in Australia; perhaps he never was in America. This greater freedom is an aspect of Wilson that has never been explored.

There is likely another aspect of the lack of dissection of the Australian trip. Wilson himself might have felt it a failure, and Sargent also. Although I had uncovered about 600 plant specimens in the Harvard Herbaria, far more than expected, and 224 in the Western Australian Herbarium in Perth, collected by Wilson with Desmond Herbert or Charles Gardner, these numbers pale in comparison with the plant specimens collected by Wilson in China.[28] He had lost the entire collection from eastern Australia, and all of his images; he could no longer bring in seed or cuttings to the USA due to the changed quarantine regulations, and nothing collected in Australia could grow in the Arnold Arboretum with Boston's difficult combination of winter snow and hot, humid summers. It might be also that Wilson did not want to tarnish the international reputation that he had gained in China[29] by a fuller revelation of the disappointments of the

FIGURE 9.3: In New Zealand, showing clear-felling of a kauri forest. The partly scratched-out caption reads: NZ Forest Service & Prof. Wilson, [site unreadable], 1921. Image: Wilson Y-458 AA, but it was likely sent to Wilson later.

Australian expedition and its physical losses.[30] It might well have been that the less said about the Australasian Expedition the better. The 1927 book that describes the Expedition, *Plant Hunting*, is in keeping with the adventure stories of exploration common in Wilson's age, though that age was ending. In *Plant Hunting*, Wilson often writes in this romantic spirit. In Kenya, he saw distant volcanoes, and wrote that he was 'glad that I have lived and seen these monarchs of Africa's distant peaks' and that their 'lure tugged violently at the heart strings'.[31]

I was also struck by another significant omission in *Smoke that Thunders*. In *A Naturalist in Western China*, Wilson acknowledged his Chinese porters in some way – 'Peasants who served me faithfully throughout my journeys, and we parted with genuine regrets'.[32] But he did not acknowledge by name those who gave him tremendous assistance in Australia. Rather, he seems a man alone in *Smoke that Thunders*, travelling solo. While he noted to members of the press the help given to him in Australia, no one is mentioned by name in his books. He mentions Lane-Poole and Herbert only in private correspondence and in the journal article 'Notes from Australasia, No. 1'.[33] In the more widely available *Smoke that Thunders*, we can read of 'genial Forester Munro' in Kenya and 'big-hearted H.G. Deakin and his charming, capable wife' at the same forestry station,[34] but there is no Charles Lane-Poole, Desmond Herbert, Charles Gardner, Edward Swain, or Cyril White as a permanent record of those who assisted him in Australia. That is despite the obvious fact that without them he could not have seen what he did or known what he was looking at in this very different part of the planet. I hope to have amended their absence.

Wilson began his journey in Australia by noting the splendid diversity of the flora. All about him, he saw the extraordinary differences between the flora of the Northern Hemisphere, on which he was an expert, and the remarkable endemism of the flora of Australia, which was entirely new to him. It was then, and remains, another part of the planet. But, as he travelled, he was struck less and less by the botanical diversity, and became more concerned with the extensive clearing, continued clearing, and planned clearing that reflected a widespread disregard and disdain for trees that was supported by governments in Australia. Across the world today, woody vegetation is still most often cleared for agriculture, and often good biodiversity is lost if the land is seen as having productive value, while less productive land has been protected.[35] Wilson saw this skew in the 1920s, and thus decried the loss of magnificent timber trees for low-value dairy lands, a problem that reflected the dominance of the importance of agricultural production in government thinking and planning. With his enormous personal experience and knowledge, Wilson saw that trees were in trouble – that trees were suffering death by a thousand cuts. It continues today in rural and urban areas, with many local government councils trying to increase canopy cover, while others still have the negative views of trees that Wilson saw on his travels so long ago.

In the more than one hundred years since Wilson's Australian expedition, the world has slowly been turning to see the destruction of trees as Wilson and others of his time did, though much needs to be done. At our best today, we extoll the value of trees, and the special role of large, old trees, yet these are in decline globally. We act – sometimes – for the preservation of old growth forests, but many forest giants are left standing forlornly as solo trees that have lost their forest community, and are thus at risk of loss by wind, though we ourselves remain their greatest threat. Potential large, old trees of the future are destroyed before achieving their maturity and greatness as the biggest and grandest organisms on the Earth. Wilson saw all of this; he could see the dangers ahead for tree loss, and in his mind, these were no better revealed than what he saw freshly done in Australia over the entire continent. Australia has now been labelled a 'deforestation hotspot',[36] and Australians might well look back in dismay that calls for tree preservation and protection are not new positions in this country, or even of the past fifty years, but old ones that reach back into the nineteenth century. Importantly, these older pleas were made by persons in authority with leading positions in government organisations, but they had to fight their own governments, and mostly lost, or were replaced by others who did not share their vision of saving trees for future generations or the more careful use of forests.

'I am interested', Wilson said, 'in trees for trees' sake ... what one does not like to see, what hurts, is this thoughtless destruction of things so valuable.'[37]

He asked, 'How long is this sort of thing going to go on?' We, the future generations, do wonder what went wrong.

NOTES

1   Letter to Wilson from Sargent, 18 April 1921, catalogue no. 599, AA Archives.
2   Wilson's obituary, *Nature*, 1 November 1930, 126(3183):693.
3   JE Chamberlin 'The Arboretum's Ambassador', *Boston Evening Transcript Magazine section*, 30 September 1922, part 2. Other information about Wilson's travels in India and Africa are given in this long article.
4   Wilson's trip to Java and Malaya and his view of rubber plantations was noted by Briggs in *'Chinese' Wilson: A Life of Ernest H Wilson 1876–1930*, HMSO, London, 1993, p. 101.
5   Brisbane Courier, Tuesday 24 May 1921, p. 6, gives this date and ship.
6   Boston Evening Transcript, 30 September 1922.
7   'A cutting from the Journal of the Botanical Society of South Africa', from RH Compton, Kirstenbosch National Botanical Garden. The Compton Herbarium in Kirstenbosch is named in his honour. At the time of Wilson's visit, Compton was also the Professor of Botany at the University of Cape Town. Wilson was delighted to see that Kirstenbosch was 'devoted to the preservation of local flora and the cultivation of

indigenous plants.' AA Archives.

8. The *Boston Evening Transcript* gives the date of arrival in Boston as 24 August in their article of 30 September 1922.

9. RAEH Howard, 'Wilson as a botanist', *Arnoldia*, 1980, 40(4):154–93.

10. Wilson letter to Sargent, from Perth, 17 November 1920, AA Archives, W.XIV:A, box 21, folder 11. He also told Sargent that these would all be sent to Boston by the Conservator, Lane-Poole.

11. AA Archives, W.XV, box 26, folder 12.

12. Letter from Sargent to Mr Endicott, 25 August 1922, AA Archives.

13. Letter from Sargent to Mr McFarland, 10 January 1922, AA Archives, 634.

14. SA Spongberg, *A Reunion of Trees: The Discovery of Exotic Plants and their Introduction into North American and European Landscapes*, Harvard University Press, Cambridge, Mass., 1990.

15. Brown also described the phenomenon known as 'Brownian motion'.

16. Briggs, '*Chinese*' *Wilson*, 1993, pp. 98–101.

17. EH Wilson, *Smoke that Thunders*, Waterstone, London, 1985 edn, p. 109.

18. Wilson, *Smoke that Thunders*, 1985 edn, p. 223. Yet Wilson was scathing of the 'appalling waste' of timber he saw all over New Zealand, as reported, for example, in *New Zealand Times*, 11 March 1921.

19. Article in *Horticulture*, 1922, p. 131, AA Archives, W.XV, box 26, folder 12, titled 'Clippings on lectures by E. H. Wilson on his expedition to the Southern Hemisphere – Gardens of the World, 1922–1924'.

20. *The Florists' Exchange*, 17 November 1923 p. 1396, AA Archives, W.XV, box 26, folder 12. Report from the New York Florist's Club.

21. Letter to Sargent, from Melbourne, 25 April 1921, AA Archives.

22. Destructive intervention on p. 223, Chapter XXVII; Queen is the previous chapter.

23. Wilson, *Smoke that Thunders*, Waterstone, London, 1985 edn, part II, p. 97.

24. MS McGlone, K McNutt, SJ Richardson, PR Bellingham, and EF Wright, 'Biodiversity monitoring, ecological integrity, and the design of the New Zealand Biodiversity Assessment Framework', *New Zealand Journal of Ecology*, 2020, 44(2):3411.

25. A Williams, DA Norton, and HJ Ridgway, 'Different arbuscular mycorrhizal inoculants affect the growth and survival of *Podocarpus cunninghamii* restoration plantings in the Mackenzie Basin, New Zealand', *New Zealand Journal of Botany*, 2012, 50:473–9.

26. Ibid.

27. *Evening Post*, 11 March 1921.

28. His Chinese collection of specimens number about 65,000, as noted in Briggs, '*Chinese*' *Wilson*, 1993, p. 7. More recent numbers show fewer, as in Y Zou, K Shi, S Liao, Z Xiang, J Luo, X Nan, H Yan, Z Bao, W Nie, and R Wu, 'A survey and analysis of the history of Ernest Henry Wilson's specimen collections in China', *Forests*, 2024, 15(3):475.

29. See TM Holway, 'History or romance? Ernest H Wilson and plant collecting in China', *Garden History*, 2018, 46(1):3–26.

30. Note the work that went into sustaining Wilson's fame. This is summarised in XS Chacko, 'When life gives you lemons: Frank Meyer, authority, and credit in early twentieth-century plant hunting', *History of Science*, 2018, 56(4):432–69.

31. EH Wilson, *Plant Hunting*, volume 1, 1927, p. 69.

32. EH Wilson, *A Naturalist in Western China with Vasculum, Camera, and Gun*, 1913, preface.

33. On page 60, both Lane-Poole and Herbert are noted.

34. Wilson, *Smoke that Thunders*, Waterstone, London, 1985, p. 72.

35. LN Joppa and A Pfaff, 'High and far: biases in the location of protected areas', *PLOS One*, 2009, 4(12):e8273.

36. R Taylor, N Dudley, S Stolton, and A Shapiro, 'Deforestation fronts: 11 places where most forest loss is projected between 2010 and 2030', *Proceedings of the XIV World Forestry Congress, Durban, South Africa*, September 2015, pp. 7–11.

37. *Western Mail*, 25 November 1920.

EPILOGUE

My time in the United States was running out. In the Harvard Herbaria, Anthony Brach had imaged more than six hundred herbarium specimens collected in Australia on Wilson's Expedition to Australia or later sent by others for the Expedition's collection, giving me more help than I could ever thank him enough for. There was more for me to do, but no time to do it.

As the Arboretum closed for holidays, I caught the train to Connecticut to spend Christmas week with my friend Kerri, a writer and book critic I had met at Breadloaf Environmental Writers' conference in Vermont two years previously. Kerri was working on a book about environmental pollution in the Maine papermill town where she grew up.[1] Her house was unexpected – a beautiful clapboard home built by a General just after the Revolutionary War in 1784, when the Noongar lands that would later introduce Wilson to Australia's flora, and where Wilson would spend two months, were solely Noongar Country, and the European settlement of Australia and the immense changes it would bring to forests, landscapes, and Indigenous societies had not begun.

Christmas dinner was with Kerri and her husband, who had been a US coastguard, and a long-retired and eternally glamorous Hollywood actress, an artist related to the dramatist Arthur Miller, and a businesswoman. It seemed pure American. Intrigued by my hot-weather Christmases, my American friends seemed unsure about what to cook for Christmas; perhaps, I thought, it is too close to Thanksgiving, which Australia does not have. When I asked about their hopes and concerns for their country and what were the two main things that worried them most, they replied 'health and education' in unison and looked around at each other, somewhat disconcerted. They seemed relieved to be asked. We sat by the fire and talked as the December snow gathered outside.

The next day, the beeches bare and stark, the artist invited us to her house down the street. Kerri had never been, nor had another neighbour who joined us for our little expedition. Georgette's house was another wonderful eighteenth-century home, seemingly unchanged since its construction, and the dark and leafless garden was streaked with white and crusty ice. We stayed for hours talking as we roamed from room to room, absorbed in the colours of her extraordinary bird paintings, winter sun coming in the windows and catching all their yellows. A few days before, Kerri had wondered if I would want to go to New York on this day, since I had never been there, but I had declined, and the day with those nearby was for me the unique and richer choice. Perhaps she was surprised by my preference to stay in a small village rather than the famous city, but it is often the visitor from elsewhere – who sees the landscape with different eyes and hears the place with different ears – who helps us appreciate what we have where we are.

Wilson wrote in *Smoke that Thunders* that

> There are no happier folk than plant-lovers and none more generous than those who garden. There is a delightful free-masonry among them, they mingle on a common plane, share freely their knowledge and with advice help one another over the stepping stones that lead to success. It is truthfully said that a congenial companion doubles the pleasure and halves the discomforts of travel and so it is with the brotherhood who love plants.[2]

Throughout his trip to Australia, Wilson had been full of praise to the press for the assistance and good cheer he met everywhere, and his long obituary in the journal *Nature* described him as 'a very likeable man'.[3] Many of the men he met on his trip later became very well-known in Australia, and left their mark on Australian forestry or botany, such as Lane-Poole, Herbert, Gardner, and Swain. People had gone out of their way to help and to guide him through this other planet's 'new world'[4] of flora.

And well they might. The east coast of the United States is on the other side of the world from Australia, and even today, it is a tiring journey to make. In the 1920s and the days before widespread and easier international travel, most who came to Australia had travelled far and long and at great expense. Only the year before Wilson's trip, in 1919, the first pioneering and daring flight was made from London to Darwin in Australia's north; the trip took twenty-eight days. When Wilson was in Kalgoorlie in November 1920 and about to take the train to South Australia, Qantas – the Queensland and Northern Territory Airline Service – was founded in outback Queensland. Even from the mid-1930s, a Brisbane to London trip by air was by Qantas to Singapore, and then a connection with Imperial Airways (later British Airways) to get a flying boat service, with up to forty-three stops before arriving in London, two weeks after departing; it was prohibitively expensive.[5] I can recall as a child in the 1960s the respect shown to those who came to visit from afar, particularly those who, like Wilson four decades earlier, came to ask and find out what Australia was like, without prejudice or preconceived ideas. The respect and guidance that Wilson was shown in the early 1920s does not surprise me. Physical isolation, that great driver of the splendid and wondrous plant biodiversity and tree speciation that delighted Wilson, would have been a player in the treatment of him as well.

In the New Year, after the long haul back across the width of North America and the Pacific and into the Southern Hemisphere, I was home in Australia, blinking in bright summer sunshine. Some species of those archetypal Australian trees, the eucalypts, were pregnant with soon-to-emerge stamens in the garden – a legacy of their ancient Gondwanan source and Australia's slow evolutionary pathway generating unique species as she drifted northward in complete isolation for more than one hundred million years. Large prickly paperbarks (*Melaleuca linariifolia*), planted as street-trees, had flowered all late November and December, and had dropped their creamy-white blossoms with summer's heat in accord with their common name of 'Snow in Summer'; spent flowers lay upswept in heaps upon the footpaths and in the gutters.

NOTES

1   K Arsenault, *Mill Town: Reckoning with what remains*, St Martin's Press, New York, 2020.
2   Wilson, in the preface to *Smoke that Thunders*, 15 May 1927.
3   *Nature*, 126(3183):693.
4   EH Wilson, *Smoke that Thunders*, Waterstone, London, 1927 (1985 edn), p. 106.
5   M Llewellyn, 'What was flying like in the 1950s and 1960s compared to now?', 2017, accessed 30 January 2021 https://www.news.com.au/travel/travel-advice/flights/what-flying-was-like-in-the-1950s-and-1960s-compared-to-now/news-story/7f8a-6666f844a3c504baa36a111f60d6

APPENDIX 1

SELECTED IMAGES OF WILSON'S HERBARIUM SPECIMENS

The images presented here were selected from those found in the Harvard Herbaria in 2018. Selection was based on showing a range of species collected by Wilson, showing their beauty and variety of leaf forms, and catching examples of the nature of sclerophylly.

The Harvard Herbaria in Divinity Avenue consists of six integrated herbaria – the Herbarium of the Arnold Arboretum (about 1.5 million specimens), the Economic Herbarium of Oakes Ames (about 40,000 specimens), the Oakes Ames Orchid Collection (2,500 species), the Farlow Herbarium (about 1.4 million specimens), the Gray Herbarium (nearly 2 million specimens), and the locally focused New England Botanical Club Herbarium (350,000 specimens).

The search for Wilson's images was carried out in the Gray Herbarium, where the Australian specimens are held. Asa Gray (1810–88), Professor of Natural History at Harvard, was appointed head of the Harvard Botanic Garden in 1842, and founded a herbarium to help determine what was required in the botanic garden. He received exchanges of specimens from many of the most important collectors in the nineteenth century, and evidence of this could be found in the species files. His significant collection was bequeathed to Harvard University in 1864.

The images selected in this appendix represent a sampling across major Australian genera from Wilson's Expedition. Material is entirely confined to specimens collected by Wilson in Australia, as evidenced by the label and the writing in his hand. Many of the specimens observed and photographed appeared to have been examined in the 1960s by David Moresby Moore (1933–2013), a British botanist with expertise in taxonomy and systematics who worked at Harvard during that period. Moore reclassified many specimens from the designation given in 1920–1. His changes can be seen on some of the labels, with author abbreviation D. M. Moore. A full list of specimens collected is given in Appendix 3.

During Wilson's visit to Western Australia, he collected with Charles Lane-Poole in Perth's Darling Scarp, the tuart forest, and in the southern forests, as well as at Albany and the smaller town of Denmark. He travelled with Desmond Herbert into the Wheatbelt, and then he and Herbert were joined by Charles Gardner in the Goldfields. Both Herbert and Gardner sent copy specimens of those collected to the West Australian Herbarium, where two hundred and twenty-five specimens are now housed that have Wilson as the collector, usually with Herbert or Gardner. A map of the sites is given in Appendix 4 and a list of specimens collected.

Other collectors sent material towards the Expedition for E. H. Wilson. Many of these appeared to come from the eastern seaboard of Australia, and were likely sent in response to the loss of this part of the collection on the *Canastota*. These collectors included Edward Swain, Cyril Tenison White, Sethrick F. Kajewski[1], and also Captain A. L. Merrotsy (a surveyor who also collected fossils) in 1923, all in Queensland, Gardner in Western Australia, J. F. Bailey in Adelaide in 1923, from Victoria by A. D. Hardy in 1923 and C. E. F. Allen in the 1920s, and large numbers of previously collected specimens from Schock (mainly from 1916–18) in Western Australia. Sargent wrote to Cyril White in July 1921, with a confirmation of a request via Wilson to send specimens:

I am writing at Mr Wilson's suggestion to confirm the arrangement which he has made with you to pay up to the amount of thirty pounds for the service of an assistant to get together a selection of specimens of the woody plants of Queensland for this herbarium … I take this opportunity to thank you in the name of the Arboretum for the great assistance you have rendered Mr. Wilson during his short stay in your country. Please believe me, Faithfully yours, C. S. Sargent.[2]

When I left Harvard in early January 2019, I had captured the bulk of the collection in my search, but I had missed a handful of genera. The missed genera were housed in a section of the Grey Herbarium, where all specimens were wrapped in plastic due to an earlier insect infestation, the bane of herbaria. While still accessible, they would have been a slower job, and my time had run out. I had intended to return in late 2019 to complete the examination of material, but in August 2019, I sustained a 9-cm skull fracture and was not permitted to fly, and by February 2020, both Australia and the United States had closed their borders to international travel due to COVID-19. In 2021, Anthony Brach kindly searched out more specimens from the collection that belonged to the Expedition.

These specimens are shown in the following figures: Figures 3.1 to 3.12 (Chapter 3, botanic garden); Figures 4.1 to 4.21 (Chapter 4, the Wheatbelt); Figures 5.1 to 5.17 (Chapter 5, forests); Figures 6.1 to 6.10 (Chapter 6, arid); and Figures 7.1 to 7.12 (Chapter 7, eastern Australia).

NOTES

1. More were collected for the Arnold Arboretum by Kajewski. See CT White and SF Kajewski, *Ligneous plants collected for the Arnold Arboretum in North Queensland by S. F. Kajewski in 1929*, Contributions from the Arnold Arboretum of Harvard University series, no. 4, Arnold Arboretum, Jamaica Plain, Mass., 1933.
2. Letter from Sargent to CT White in Queensland, dated 12 July 1921, AA Archives.

APPENDIX 2

WILSON'S TRANSCRIBED DIARIES 1920-1, AUSTRALIA

The extant diaries that Wilson kept while in Australia are an incomplete record of his trip. The two diaries transcribed here are only those available and held in the Arnold Arboretum Archives. They report on one trip in Western Australia, from Perth, east via Toodyay and Merredin as far as Westonia, returning via Bruce Rock and York, and then one incompletely recorded trip from Sydney to the Dubbo region in mid New South Wales. Though there are no diary records of other trips, Wilson's images and specimens from the Harvard Herbaria show where he collected in other places in Australia. For example, trips not covered in his diaries include a trip in Western Australia to Albany, Denmark, Pemberton, Nannup, and Nornalup Inlet, and a trip with Charles Lane-Poole to the tuart forests (*Eucalyptus gomphocephala*) south of Perth near Busselton and Capel; evidence of these trips is from photos of Lane-Poole and Wilson, notes about that trip in press cuttings, and specimens collected.

Other records might appear at some stage, but it is likely that any other diaries recording Wilson's time in the eastern states of Australia in 1921 were lost at sea with the main collection, or that a diary was never made during those trips. For his 1927 publication *Plant Hunting, Volume 1* (Stratford, Boston, Mass), reprinted as *Smoke that Thunders* (Waterstone, London) in 1985, Wilson might have drawn on material not found today, or used his formidable memory, or perhaps letters to his wife, which were all destroyed following his death. *Plant Hunting* covers the entire trip, including India and Africa. Part II deals with Australia. Chapters concerning Australia in *Plant Hunting* are titled: A fortunate accident', 'Land of prehistoric plants' (Western Australia), 'Trees that cast no shadows' (Western Australia), 'Fairyland of flowers' (Western Australia), 'Botany Bay' (Sydney; though not mentioned in diaries, and is concerned with history), 'Treasures of the brush' (eastern Australia), 'A marvellous dowry' (general horticulture), 'My lady *Acacia*' (on that genus), and 'Those who paid the price' (history of plant collecting in Australia). Tasmania is the subject of three chapters—'Isle of enchantment', 'Here and there', and 'Midst blue-gums and tree-ferns'. His travels in eastern Australia recorded in *Plant Hunting* give details of plant species and where they were observed, suggesting that he did not rely on memory alone.

A professional historical transcriber made the first foray into Wilson' diaries, which were handwritten in pencil and are often very difficult to read, and then I did further work to decipher more of the text. However, quite a few words have not been determined.

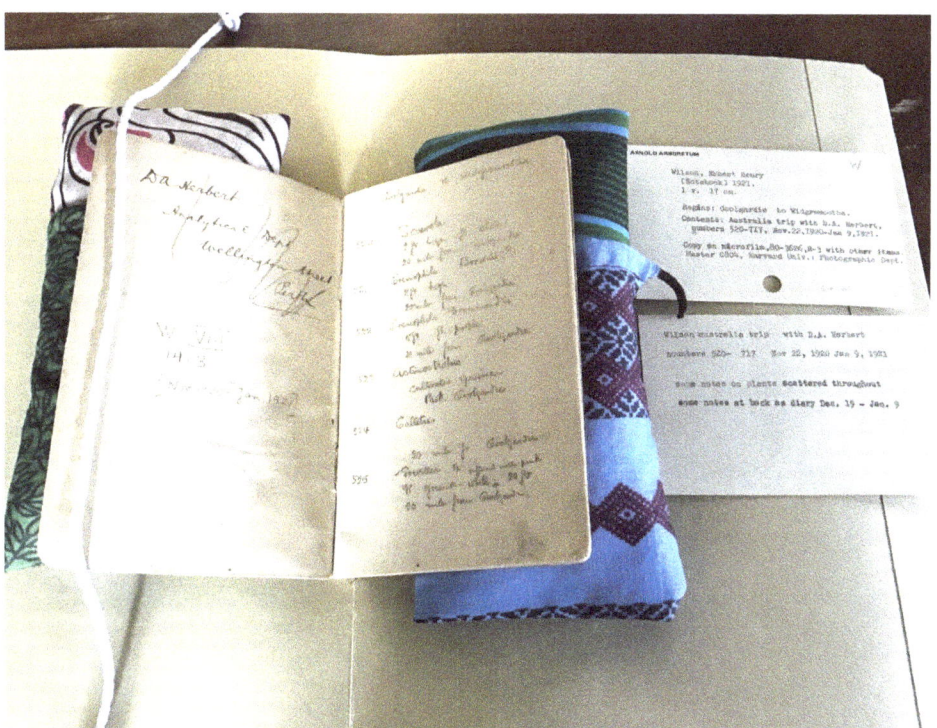

FIGURE A2.1: First page of Wilson's diary 'Coolgardie to Widgiemootha (*sic*)', photographed at the Arnold Arboretum, Harvard. Image: M Grose.

WILSON'S AUSTRALIAN DIARY ENTRIES, AS TRANSCRIBED

NOTES

1. Underlined words duplicate Wilson's underlining in his diary.
2. [ ] A blank space in brackets indicates a word that could not be transcribed.
3. [species] A word enclosed in a brackets is a guess at a word based on Wilson's handwriting; if a comment is also made by the author, it is in parentheses.
4. I have clarified the dates and added the year.
5. Capitals, including capitals of species names, are as Wilson wrote them, and dates are as given in his diaries.
6. Place names are generally as written by Wilson; some have changed, such as Hynes Hill, which is now Hines Hill.
7. Question marks in the text or tables are Wilson's.

DIARY ONE: III EHW, SERIES VIII, BOX 14, FOLDER 2

Perth to Westonia

[p. 1] 21 October 1920. Leave Perth about 9.45am in motor & follow a good road for 54 miles to Toodyay. The road bends through the Darling range the highest [we] reached being less than [1100ft]. Through the range the flora is wonderful. Flowers of bright colours [on] every [side]. Especially Proteaceae & Myrtaceae, with Xanthorrhoea Preissii in great abundance on all sides. The tallest flower would be 10ft with two to three [crowns] & 4ft high spikes of white flowers Another small [species] X. was also common & in one place we saw Kingia with its ring of short [brown] stick-like inflorescences. The Zamia was abundant. [p. 2] Banksia Grandis with 8 inch high cones of pale lemon-yellow flowers [ ] a [ ] low branching [ ] undergrowth. Marri, Jarrah & Wandoo were the chief trees. About 10 mile from Toodyay the flora changed abruptly. Jam –tree (*Acacia acuminata*) – a low twiggy covered tree with narrow leaves – fairly dominant with a little York

Gum here & there & Flooded Gum alongside the watercourses. We saw one Sandal-wood tree in fruit In swamps a Melaleuca was abundant. From Toodyay to Northam where we stayed the night is 20 miles through [p. 3] Jam country much of which is under cultivation. Wheat in one place was ripe. [ ] [crop] is [sown] [over] Marri & this is the eastern limit of the species. Northam is on the Avon which is [wider] than Swan River & an important tributary. Anyway at Northam we saw a black swan on the river. The [feature] of the day was the abrupt & astonishing change from a rich & varied flora of brilliant coloured flowers to the rolling farm Jam tree country very poor in species. Round Toodyay Rhodanthe Manglesii & the yellow flower Helichrysum apiculatum were [p. 4] abundant carpeting the ground for many miles. The houses mostly of one storey with verandah & roof of corrugated iron painted dull red (iron – oxide) are fenced & have neat small garden Roses [ ] & the Boursault [flower] also [carpet]. Lady Banks' Rose, the Cherokee & new garden roses are favourites. In one garden I saw Jap. Wisteria in full bloom. Geranium especially [big] leaf are wonderful. Trained up the side of houses & over the fence they are just one mass of pink & scarlet. Melia Azedarach is a favourite street tree & by the side of the Avon at Northam a Tamarisk has been planted.

[pp. 5–7] October 21st–22nd 1920

#	Plant and name as given by Wilson	Comment	Place collected
1	Eucalyptus rudis Endl.	'Flooded gum or Swamp gum'	Toodyay
2	Melaleuca sp. (?Preissii)		Banks of Avon River, Toodyay
3	Jacksonia Sternbergiana	'Stinkwood'	Toodyay
4	Gastrolobium spinosum	'Prickly Poison'	Toodyay
5	Loranthus liniphyllus	from 'Jam' (Acacia acuminata)	Toodyay
6	Acacia cyanophylla	'Black Wattle' extra large leaves	
7	Dryandra floribunda	'Parrot Bush'	Toodyay
8	Fusanus acuminatus	'Quandong' or 'Native Peach', 'Austral Peach'	Toodyay
9	Eucalyptus loxophleba	'York Gum'	Toodyay
10	Boursault rose (Naturalised)		Toodyay
11	Casuarina (?distyla)	'Sheoke'	Toodyay
12	Pimelea argentea		Toodyay
13	Tamarisk naturalised		On banks of Avon, Northam

#	Plant and name as given by Wilson	Comment	Place collected
14	Eucalyptus calophylla	'Red Gum' an isolated tree in the Jam formation	Toodyay
15	Eucalyptus pulchella	[ ]	Northam
16	Santalaceae. Exocarpus spartea R.Br.	12'	Northam
17	Eucalyptus longicornis F.v. M syn. E. oleosa var. longicornis	a 'Mallee'	Cunderdin
18	Calythrix sp. (Myrtaceae)	18' alt.	Cunderdin
19	Verbenaceae sect. Lachnostachydae	? Newcastlia	Cunderdin
20	Myrtaceae ?Baeckea or Leptospermum	shrub 4' high	Cunderdin
21	? Burtonia	2'	Cunderdin
22	? Eremala	2' Myrt.	Cunderdin
23	? Melaleuca	3'	Cunderdin
24	Baekea crassifolia	(prostrate)	Cunderdin
25	? Melaleuca	5' 'Tea Tree'	Cunderdin
26	? Leguminosae	Bush 4'	Cunderdin
27	Grevillea	18'	Tammin
28	Euc loxophleba, Benth	probably loxophleba but of the habit of salubris ('Gimlet')	Tammin
29	Grevillea	10'	Tammin
30	Grevillea	10'	Tammin
31	Casuarina ?distyla	'Sheoke' 12' (see 11) probably the same	Tammin
32	Daviesia euphorbioides	4' 'Centipede Bush'	Hyne's Hill
33	Callitris robusta ?	'Pine' 18'	Hyne's Hill
34	Salicicornia australia	6'' to 1' 'Samphire' from banks of salt lake	Hyne's Hill

[p. 8] 22nd October 1920. Leave Northam at 9 o'clock, & travel due east 105 miles to Merredin. First we traversed Jam country & then saw plains of no [size] Through strips of [Pindan] country. The latter is covered with a low 2-4 ft scrub growth of xerophytes. Near Northam Xanth. reflexa was common among Jam. We met with Salmon Gum & Gimlet gums 25 miles east of Northam & they continue with us. Mallee ([Euc] low [pictured] as a bush or slender low tree was common from 50 m east of Northam. Near Merredin (12 miles east) saw the first Callitris – a pyramidal tree 9 ft tall.

[p. 9] The Salmon Gum is a handsome tree with smooth white to pinkish trunk dividing 20 ft up into several ascending [ ] spreading stems which [ ] form in [ ]

[ ] [ ]. The twigs are reddish, the [pendants] yellowish & the leaves lustrous dark green & the effect with sunlight shining through is striking. The gimlet is much the same in appearance save that it is a smaller tree with duller trunk which is fluted & twisted like a screw hence the name gimlet. Many of the churches here showing flowers & a Grevillea with several spikes of orange flowers rich in honey was especially attractive.

[p. 10] The country was largely rolling plains & much of it under crop of which wheat in many areas was good. It was ripe & ripening. The sand [Prot .] & the [Pindan] are not so [ ] made [salt]. We found one small saltlake & a few shallow ponds but no streams Water is all that is needed to make this land teem with farms. The Jam country is good wheat land but supports a very meagre flora. The wood is not attacked by white ants and is [used] for fence posts.

[pp. 11–14] 23rd October 1920

(An asterisk * indicates that the fuller name was added to the diary by another hand. Wilson notes: 's.p. = sand plain'; sp. = species)

#	Plant and name as given by Wilson	Comment	Place collected
35	Oxylobium parviflorum	(Legum.) 'Box Poison' 3' high	Burracoppin
36		1' high	sand plain Merredin
37	Grevillea	full size	s.p. Merredin
38	Pimelea sp.	1' -1'6"	s.p. Merredin
39	Leguminaceae		sandplain Merredin
40	Leguminaceae		s.p. Merredin
41	Scaevola		s.p. Merredin
42	Scaevola		s.p. Merredin
43	Verticordia	2' high	s.p. Merredin
44	Dampiera		
45	? Dodonea aurea		s.p. Merredin
46	Proteaceae		s.p. Merredin
47	? Jacksonia hemisericia Herbert *		s.p. Merredin
48	Pimelea		s.p. Merredin
49	Psammarrya chorichoides		s.p. Merredin
50	Triumfetta		s.p. Merredin
51	Eriostemon (Proteaceae)	2' high	s.p. Merredin

#	Plant and name as given by Wilson	Comment	Place collected
52	Gastrolobiium crassifolium	'Narrow-leaf Poison'	s.p. Merredin
53	Balaustion sp. (Myrt)	creeping	s.p. Merredin
54	Grevillea sp.		s.p. Merredin
55	Baekea robusta F.v. M		s.p. Merredin
56	Diploleana	1'	Westonia
57	Sterculiaceae (? Guichenotia)	3' high	Westonia
58	Acacia		Westonia
59	Duboisia hopwoodii	'Pituri' 10'	s.p. Westonia
60	Eremophila [prysporeneae]	9'	Westonia
61	Santalaceae (Leptomeria?) Preissiana R.Br.		sand plain Westonia
62	Acacia		sand plain Westonia
63	Pityrodia ? racemosa		sand plain Westonia
64	Verticordia		s.p. Westonia
65	Melaleuca		s.p. Westonia
66	Eriostemon ? (Rutaceae)		s.p Westonia
67	?		s.p. Westonia
68	a, b. Acacia spp.		s.p. Westonia
69	Sterculiaceae		s.p. Westonia
70	Grevillea		s.p. Westonia
71	Hakea		s.p. Westonia
72	Baekea (?) Myrtaceae		s.p. Westonia
73	Persoonia sp.		s.p. Westonia
74	Myrtaceae		s.p. Westonia
75	Cassia		s.p. Westonia
76	Baekea		s.p. Westonia
77	Acacia	'Dead Finish' (double underlined by Wilson)	Westonia
78	Grevillea		Westonia
79	Myrtaceae		Westonia
80	?		Westonia
81	Persoonia (?same as 73)		Westonia
82	Grevillea		Westonia
83	Grevillea		Westonia
84	Hakea		Westonia
85	Grevillea (?)		Westonia
86	Melaleuca		Westonia
87	Eucalyptus Salubris	'Gimlet Gum' suckers leaves	Merredin

#	Plant and name as given by Wilson	Comment	Place collected
88	Cassia	5'	Merredin
89	Daviesia	3'	Merredin
90	Acacia	5'	Merredin
91	Fusanus acuminata	'Quandong'	Merredin
92	Acacia sp.	5'	Merredin
93	Daviesia (?)	3'	Merredin
94	Atriplex sp.	4'	Merredin
95	Persoonia	(see 81&72)	Burracoppin
96	Cassytha		Burracoppin
97	Eucalyptus pyriformis (?var. Rhameliana)		Burracoppin
98	Casuarina sp.		s.p. Westonia
99	Casuarina	9' (also of Westonia)	s.p. Merredin
100	Proteaceae (?Adenanthos)		Burracoppin
101	Hakea		s.p. Westonia
102	Gyrostemon ramulosus		s.p. Westonia
103	Acacia		s.p. Westonia
104	Melaleuca cordata	3'	s.p. Westonia
105	Hibbertia	2'	s.p. Westonia
106	Melaleuca	4'	s.p. Westonia
107	Myrtaceae (?Kunzea)	4'	s.p. Westonia
108	Hemiandra sp.		s.p. Westonia
109	Hemigenia humilis Benth.		s.p. Westonia
110	Darwinia thryptomeneoides Herbert		s.p. Westonia
111	Microcybe parviflora Herbert		s.p. Westonia
112	Leguminosae	2'	s.p. Westonia
113	Callistemon	4'	Bruce Rock
114	Lysinema	3'	Bruce Rock
115	Isopogon	1' 6"	Bruce Rock
116	Baekea ?	3'	Bruce Rock
117	Kunzea ?	6'	Totadgin
118	Pultenaea astipulia Herbert	2'	Totadgin
119	Daviesia	2'	Totadgin
120	Sterculiaceae		Merredin
121	Eremophila	(creeping)	Merredin
122	Duboisia ?	2'	Merredin
123	Jacksonia ?	2'	Merredin
124	Gyrostemon ?	2-3'	Merredin

#	Plant and name as given by Wilson	Comment	Place collected
125	Dodonea	2' 6"	Merredin
126	Callistemon	3'	Merredin
127	Thryptomeria (?)	6'	Merredin
128	Pityrodia	sp to 3'	s.p. Merredin
129	Calythrix	1' 6"	Merredin
130	Dampiera	2'	Merredin
131	Alyxia buxifolia R.Br.		Merredin
132	Fusanus spicatus	'Sandalwood' 12'	Burracoppin
133	Santalaceae Exocarpus aphylla R.Br	9'	Burracoppin
134	Eucalyptus salubris	'Gimlet'	Curabbin
135	Eucalyptus erythronema	a 'Mallee' 6'	Totadgin
136	Melaleuca	7'	Totadgin
137	Daviesia	2'	Totadgin
138	Acacia (aff heteroclita ?)	2'	Totadgin
139	Kunzea ?	1'	Totadgin
140	Sterculiacene	9'	Bruce Rock
141	Conospermum	3'	Bruce Rock
142	Xanthorrdosa nana [D.A.H]	1'	Bruce Rock
143	Tetratheca	9'	Bruce Rock
144	Casuarina	4'	Bruce Rock
145	Eucalyptus foecunda Schon.		Curabbin
146	Isopogon	1' 6"	Bruce Rock

[p. 15] 22nd October 1920 (*sic*) (Wilson's error; it was 23 October). Merredin to Westonia 54 miles across flat or slightly undulating country covered with low scrub with Salmon Gum & Gimlet everywhere. Here & there a few Wandoo but no other trees of size. The Jam tree was scattered here & there & in one place was [covered] with Salmon Gum. We saw a few small trees of Sandal wood but all the large trees have been felled. The variety of flowers was astonishing Superficially they all look alike but in blossom the variety is manifest. They are mostly prickly in character & many of them especially so. The soil was like yesterday [p. 16] attenuating strips of sand & laterite & it is on the sand that the flowers are most abundant. Near Burracoppin where we lunched we [peered] through the 1st rabbit fence - wire netting 4 ft which [dictates] [ ] [ ] from coast to coast. Westonia is a gold-mining camp – a scattered neat village of corrugated iron houses & shops including two Hotels. We stay in the Edna-May which is named for the goldmine. Things are no longer brisk here & the mine seems to be [panning] out. Saturday night in a mining camp is not a quiet place. Singing & chat extended far into the night. Hotel people very civil & obliging.

[p. 20] 24th October 1920. Westonia [trek] to Merredin and then to Bruce Rock 70 miles in all. From Merredin we journeyed south-west through flat scrub-clad country with here & there agricultural clearings. At Bruce Rock where is much farms & some fields were [intended] to yield 18 bushels to the acre. The rainfall is from 12-14 inches so such crops are remarkably good. With water everything is possible here. At Bruce Rock staying in State Hotel tariff 10/- for day £2.15.0 per week. A very excellent hotel. On sandy flat gathered a new Xanthorrhoea – a [ ] small [ ] [leaves] & a short inflorescence of white flower.

[p. 21] The extraordinary thing about the sand plains is the quantity of [supporting] plants normally topped in [Nature] [ ] 9 – 18 inch high & [ ] [ ]. They are of all colours & [ ] are indigo blue Dampiera was a wondrous sight – so too was an 18 inch high Datonea with clear primrose yellow flo. Curious plants are the Proteaceous [Isopogon] with cane like fruits, much divided spring foliage & [ ] [ ] white & pink flo. Among the taller [ ] Grevilleas Hakea & Dereos & Melaleucas abound & Leptospermum are [ ]. Green Parrots are common & so too is the Pied Magpie. [p. 22] Indeed we saw many birds & the old nest of a Mallee Hen. Rabbits unfortunately are not rare here we have seen but one dead snake & the common & racehorse Goanna (Iguana) are the only reptiles. The country is slightly undulating & from the crest of a slight rise the view of sand & [farmed] patches is very expansive. There are no good trees left [ ] – [cropping] there even were many of size [ ] fire by man in his effort to 'improve the country' has destroyed even where these were. Agriculture is necessary but its initiation is expensive - extravagantly so. & I fear that someday the price will be exacted.

[p. 23] Among the low [caec…] plants none are prettier than the Stylidiums with white, pink orange & yellow flos, very floriferous & attractive. The Melaleucas are

[numerous] & variable with white, pink to red flos. Verticordias, pink & yellow, from a ft to 1½ ft high are among the most common and attractive of W. A. [ ].

[pp. 24–6] Oct 25

#	Plant and name as given by Wilson	Comment	Place collected
148	Thomasia sp.	in mallee 3'	Bruce Rock
149	Stylidium		Bruce Rock
150	Grevillea		s.p Quairading
151	Grevillea	6'	Yoting
152	Melaleuca coronicarpa Herbert*	4'	Bruce Rock
153	Melaleuca	5'	Bruce Rock
154	Dodanea	5'	Kwolyin
155	Melaleuca	3'	Bruce Rock
156	Isopogon	2'	Bruce Rock
157	Rhamnaceae	2'	Kwolyin
158	Hemiandra(?)	1'	Kwolyin
159	Melaleuca		Yoting
160	Chorizema (?)	1'	Yoting
161	Polygalea (?)	1'	Yoting
162	Logannia [ ] Herbert *		Yoting
163	Solanum	4'	Bruce Rock
164	Pimelia		Bruce Rock
165	Thomasia	4'	Bruce Rock
166	Pimelea		Bruce Rock
167	Leguminosae (?)	9'	Bruce Rock
168	Pimelea		Bruce Rock
169	Eremophila	5'	Quairading
170	Hemiandra (?)	4'	Yoting
171	Callistemon	(host plant for (?) Leptomeria) 2'	Yoting
172	Gastrolobium parvifolium	'Berry Poison' 2'	s.p. Yoting
173	Rhamnaceae (?)	2'	Yoting
174	Daviesia uniflora Herbert * (see note A below)	2'	s.p. Yoting
175	Calythrip	1'	s.p Yoting
176	Grevillea	2'	s.p Yoting
177	Epacridaceae (Leucopogon or [Ly...]	1'	s.p Yoting
178	Leptospermum (?)		s.p Yoting
179	Acacia	2'	s.p Yoting

#	Plant and name as given by Wilson	Comment	Place collected
180	Chenopodeaceae ?Rhagodia	3'	s.p Yoting
181	Leschenaultia		s.p Yoting
182	Santalaceae	Parasitic 2' 6" Leptomeria spinosa [ ] #171	s.p Yoting
183	Daviesia	3'	s.p Yoting
184	Hakea	1' (young fruits glutinous)	s.p Yoting
185	Myrtaceae	1' 6"	s.p. Quairading
186	Isopogon	2'	s.p. Quairading
187	Atriplex [ ]	4'	Yoting
188	Eucalyptus macrocarpa	8'	Quairading
189	Banksia attenuata	10'	Quairading
190	Xylomelum	12'	s.p Yoting
191	Banksia	10'	s.p Quairading
192	Hakea sp.	3'	s.p Yoting
193	Cypress	12'	s.p Yoting
194	Eucalyptus redunca var elata	Wandoo	Quairading
195		'Westonia sand plain perhaps already numbered but number missing'	
196	Actinostrobus acuminatus	12'	Yoting
197	Banksia grandis		

[p. 27] October 25th. Bruce Rock to Quairading via Kwolyin 67 miles over undulating country most of which is farmed. Scenery much the same as yesterday. At Yoting saw a Green Pine & near Quairading it was common growing from 6 to 10 ft high. It is densely pyramidal in habit & this & its green colour make it conspicuous from a distance. The bark is rough & fibrous. A Banksia with its erect pale yellow flos was common near Quairading. The vegetation in general is similar to that of yesterday. Prickly plants abound & gay coloured [ ] [ ] [ ]

[p. 28] The farming is not seriously [ ]– the ground is little more than scratched & the yield in good years 24 or even thirty bushels of wheat per acre. Wheat [ ] [are] the work though but evidently the land [ ] [ ] good [ ]. In dry seasons on the sandy soil (which is [des…] [ ] the crops fail. The clearing of the ground for farm land is [stubbornly] done & I can't help but think that hills of [ ] left as windbreaks would be advantageous. Winds are drying if nothing else & [that] they blow[

across] here is proved by the last two days when a strong westerly gale has blown the whole time.

[p. 29] Rabbits are plentiful near No 2 fence & evidently a bad pest.

[p. 30]. **October 26th**: Quairading to York via Green Hills 57 [or 67] miles through agricultural land very little [fauna] or scrub left. York is as old as Perth & famously was the starting point for the trek to the goldfields. It is a nice little town on the banks of the Avon which is really the Upper Swan. The Hotel -Palace kept by T.C [S…] is a model of what such places should be. Clean well-kept with good bed & sanitary arrangements & with courteous & obliging help it is all a man can wish for. In York there is a local botanist named [p. 31] O H Sargent - a chemist. He is a little bearded man & evidently well acquainted with the local flora. We spent an hour with him in the evening. He is at work on a handbook of the flora of Western Australia & has been for the last 5 years. He has a private herbarium & has promised to duplicate specimens. [Pretty] plants. On muddy flats over the waters of the Avon the [Pop] [ ] Melaleuca (M.(Preisiei ? or raphiophylla?) is abundant as a picturesque tree 20 ft [a sp w] white flos. Flooded gum is [p. 32] common in the [ ] places. [Caperina?] glauca [is rare], has relatively [smooth] bark, & [confined]to fairly moist places. [Another][common] species is Callitris Heugelii with black, furrowed fibrous bark, & a neat rounded crown. It appears indifferent to situation. York gum is common here & Salmon & Gimlet have reappeared. Much [Jam] wood [grown here] [indication] of good wheat & [cheap] land. In good years 25 – 30 & even 35 bushels of wheat are yielded from here & that of [Oats] is [heavier].

Chauffer's name William Finlay

[p. 33] **October 27th**: York to Perth 60 miles over a good road through the Darling ranges. Much of the road was through open park like forests. Jarrah, Red gum & Wandoo. Near York, we saw the Powder Gum [matched] with Wandoo tho whiter bark. Banksia Grandis was a fine sight with 10' high pale yellow cones. Flowers were less numerous but the forest scenery was much better than else-where.

The road trip was about 465 miles & across Hills.[Great] flowers & good [p. 34] agricultural lands & has given a good idea of the vegetation of W.A. in the latitude of Perth. The features of the trip have been the outstanding variety of xerophilric [sic] plants with brilliantly coloured fl. & mostly with [berries] & prickly leaves. The abundance of [ants] & the plague of flies in the bush. The yield of wheat from a 10-12 inch rainfall & this mostly by scratching the soil.

We had no rain the whole journey which has been one of the most profitable & enjoyable in my experience. For much of this I am indebted to my extraordinary [companion] D.A. Herbert.

[pp. 35–6]

#	Plant and name as given by Wilson	Comment	Place collected
198	Gastrolobium calycinum	'York Road Poison' 3'	The Lakes
199	Grevillea Wilsonii	2'	The Lakes
200	? Rutaceae		The Lakes
201	Kingia australis	15 '	The Lakes
202	Xanthorrhoea gracilis		The Lakes
203	Dryandra	6'	The Lakes
204	Isopogon	2'	The Lakes
205	Gastrolobium villosum	'Crimp-Leaved Poison' 2 '	The Lakes
206	Pittosporaceae (Billardiera?)	3'	The Lakes
#	Eucalyptus calophylla	seedling	York
207	Casuarina glauca	65' by Avon River	York
208	Hakea aemulata	25'	York
209	Chamelaucium uncinatum	'Geraldton Wax Plant' nat of Geraldton WA cult. at York 15'	York
210	Boronia?	10'	York
211	Acacia	6'	York
212	Daviesia	3'	York
213	Scaevola	3'	York
214	Gastrolobium oxylobioides	'Champion Bay Poison' 4'	York
215	Burtonia ?	2'	York
216	Grevillea	4'	York
217	Casuarina sp.	4'	s.p. York

[p. 37–43] 'South-West. Wilson & Lane-Poole Coll.' (Two pages of plant names only; the handwriting is in pencil and is exceptionally small, even for Wilson.)

#	Plant and name as given by Wilson	Comment	Place collected
218	Casuarina thymoides (?)	2-5'	N. Dandalup
219	Casuarina humilis	2'	N. Dandalup
220	Callistemon	<1' 'early [ ]'	N. Dandalup
221	Euc Lane-Poolei	25-40' 8' girth	N. Dandalup
222	Oxylobium cuneatum	3'	N. Dandalup
223	Kunzea acuminatum	1-2'	N. Dandalup
224	Dasypogon		N. Dandalup
225	Hemigenia		N. Dandalup
226	Lambertia	5'	N. Dandalup
227	Isopogon		N. Dandalup
228	Euc marginata	'Jarrah'	N. Dandalup
229	Casuarina fraseriana		N. Dandalup
230	Viminaria denudata		Cannington
231	Xanthorrhoea Preisii	'Blackboy'	nr Hamel
232	Boronia lanuginosa	1'	nr Hamel
233	Banksia littoralis		Harvey
234	Hakea linearis (?)		Waterloo
235	Calothamnus ?	3'	Waterloo
236	Hakea sp.	low spreading shrub up to 3'	Waterloo
237	Isopogon		
238	Brachysema	(Ed note: Swan River Pea, another poison)	Big Brook
239	Leucopogon		Big Brook
240	Hibbertia		
241	Leguminoseae		
242	Euc. megacarpa		Big Brook
243	Boronia	rose red 2'	Big Brook
244	Leptospermum	8'	Big Brook
245	Eriostemon	2'	Big Brook
246	Casuarina decussata	5' × 7 or 8'	Big Brook
247	Acacia armata		
248	Thomasia	1' 6"	Karri forest, Big Brook
249	Diplolaena?	8'	Big Brook
250	Acacia	6'	Big Brook
251			
252	Thomasia	5'	Big Brook
253	Hovea elliptica (?)		Big Brook

#	Plant and name as given by Wilson	Comment	Place collected
254	Brachyscome sp?		Big Brook
255	Pimelea obovata	'Banjine'	Big Brook
256	Pimelea sp.		Big Brook
257	Leguminaceae	in thickets	Big Brook
258	Daviesia ?		Big Brook
259	Isopogon		Big Brook
260	Polygalaceae	4'	Big Brook
261	Tetratheca	2'	Big Brook
262	Chorizema ?		
263	Podocarpus Drouynianus		Big Brook
264	Dampiera		Big Brook
265	Oxylobium		Big Brook
266	Hypocalymma robustum	(Ed: 'Swan River myrtle')	Big Brook
267	Kennedya		Big Brook
268	Callitris robusta	(planted at Donnelly River)	Rottnest Island
269	Legum.		Donnelly River
270	Tetratheca (?)		Donnelly River
271	Spyridium or Trymalium [ ] Billardierii (?)	'Hazel'	Donnelly River
272	Oxylobium callistachys		Donnelly River
273	E. haematoxylon	30'	N. Jarrahdale
274	Dasypogon Hookeri		N. Jarrahdale
275	Santalaceae		N. Jarrahdale
276	Daviesia cordata		Greenbushes
277	Legum		s.p. nr Busselton
278	Sphaerolobium ?		s.p. nr Busselton
279	Legum		Busselton
280	Burtonia (?)		Busselton
281	Labichea landeolata		Busselton
282	[ ] ?		Busselton
283	Adenanthos barbigera (?)		Busselton
284	Adenanthos		Busselton
285	Proteaceae		Busselton
286	Melaleuca		Busselton
287	Verticordia		Busselton
288	Eriostemon		Busselton
289	Thomasia		Busselton
290	Pimelea		Sabina River

#	Plant and name as given by Wilson	Comment	Place collected
291	Grevillea		Sabina River
292	Boronia		Sabina River
293	Eriostemon		Sabina River
294	Pimelea		Sabina River
295	Isopogon		Capel
296	Acacia		Capel
297	Candollea	6′	Capel
298	Leucopogon australis ?		Capel
299	Acacia		Busselton
300	Acanthocarpus Preisii		Busselton
301	Scaevola crassifolia		Busselton
302	Melaleuca	30′ paperbark	Busselton
303	Euc cornuta	Yate	Sabina River
304	Kennedya prostrata		Sabina River
305	Agonis flexuosa		Sabina River
306	Hardenbergia comptoniana		Capel
307	Acacia		
308	Jacksonia furcellata		Capel
309	Pimelea		Capel
310	Tetratheca		Capel
311	Thomasia		Capel
312	Kunzea ?		Bunbury
313	Daviesia		Bunbury
314	Proteaceae		Bunbury
315	Pimelea		Bunbury
316	Melaleuca		nr Bunbury
317	Euc.		(cult) Hamel
318	E. tetraptera		(cult) Hamel
319	Legum.		Pinjarrah
320	Mirbelia ?		Pinjarrah
321	Tetratheca		Jarrahdale
322	Boronia		Jarrahdale
323	Petrophila		nr Busselton

[pp. 44–53] Albany trip 'Wilson Herbert 6/11/20'

#	Plant and name as given by Wilson	Comment	Place collected
324	Legu.	prostate 6'	Torbay
325	Pultanea ?	prostrate	Torbay
326	Plumbaginaceae	1' 6"	Albany
327	Boronia	(swamp)	Albany
328	Baekea	3-4'	Albany
329	Andersonia	3'	Albany
330	Olax phyllanthi	2'	Albany
331	Phyllota (?)	1'	Albany
332	Boronia	4'	Albany
333	Thomasia	2'	Albany
334	Persoonia elliptica	10'	Albany
335	Hypocalymna cordifolium*		Albany
336	Legum.	(prostrate)	Albany
337	Myrt (?)		Albany
338	Acacia	4'	Albany
339	Pimelea	2'6"	Albany
340	Leucopogon australis		Albany
341	Melaleuca	4'	Albany
342	Boronia	2'6"	Albany
343	Lysinema	2'	Albany
344	Leucopogon		Albany
345	Conospermum	3'6"	Albany
346	[Sypaphea]	2'6"	Albany
347	Casuarina Fraseriana	up to 50'	Albany
348	Banksia coccinea	up to 12'	Albany
349	Melaleuca	2'6"	Albany
350	?	4'	Albany
351	Euc marginat	Curly Jarrah 50'	Albany
352	Andersonia	1'	Albany
353	Legum	2'	Albany
354	Hovea chorizemifolia		Albany
355	Stylidium scandens		Albany
356	Epacrid?	2'	Albany
357	Jacksonia	2'	Albany
358	Melaleuca	3'	Albany

#	Plant and name as given by Wilson	Comment	Place collected
359	Legum.	2′	Albany
360	Leptospermum	5′	Albany
361	Persoonia	2′	Albany
362	Callistemon	6′	Albany
363	Leptospermum	4′	Albany
364	Xanthosia 'Southern Cross'	2-3′	Albany
365	Leucopogon	2′	Albany
366	Sphaerolobium (?)	2′ 6″	Albany
367	Legume	1-2′	Albany
368	Pimelea	2′	Albany
369	Acacia cyclops	8′	King River, Albany
370	Isopogon ?	7′	Albany
371	Dryandra	8′	Albany
372	Legume	1′ 6″	Albany
373	Legume	1′	Albany
374	Isopogon occidentalis (?)	3′	Albany
375	Isopogon or Petrophile?	2′	Albany
376	Isopogon or '	2′	Albany
377	Callistemon	5′	Albany
378	Proteaceae	3′	Albany
379	Pultanea?	3′	Albany
380	Legume [...]	2′	Albany
381	Marianthus ?	3′	King River, Albany
382	Santalaceae	2′	Albany
383	Santalaceae	2′	Albany

West Denmark and Frankland River

#	Plant and name as given by Wilson	Comment	Place collected
384	Narrow leaf P[...]	8′	W Denmark
385	Legume	8′	"
386	Leucopogon		
387	Polygala	2-4′	W. Denmark
388	Yellow Tingle	up to 100′	Frankland R.
389	Euc. ficifolia	α 40′	"
390	Boronia	8′	"

#	Plant and name as given by Wilson	Comment	Place collected
391	Euc. megacarpa	80'	30 miles W. Denmark
392	Hibbertia?	5-8'	W. Denmark
393	Agonis	12'	Frankland River
394	Leucopogon	3'	Frankland R
395	Red Tingle	α180'	"
396	Thomasia	8'	Frankland River
397	Boronia lanuginosa	8'	" "
398	Legume	6'	"
399	Chorizema	2'	" .... "
400	Tetragona	5'	" "
401	Paper Bark		" "
402	Euc. diversicolor Karri		" "
403	Choretrum lateriflorum	3'	W Denmark
404	Legume	8'	"
405	Callistemon	8'	"
406	Adenanthos	8'	"
407	Banksia littoralis	50'	Denmark
408	Euc. ficifolia		Cultivated

Mundaring/Perth 14/XI

#	Plant and name as given by Wilson	Comment	Place collected
409	Trymalium Billardieri (*sic*)		
410			
411	Synaphea polymorpha		
412	Gompholobium tomentosum		Applecross
413	Anthocercis littorea		
414	Eucalyptus rudis		Applecross
415	Eucalyptus rudis		Applecross
416	Eucalyptus rudis		Applecross
417	Pimelea suaveolens		
418	Styphelia verticillata		
419			
420	Leptomeria Cunninghamii R.Br.		
421			
422	Sphaerolobium medium		

#	Plant and name as given by Wilson	Comment	Place collected
423	Grevillea Endlicheriana		
424	Adenanthos barbiger		
425	Diplolaena Dampieri		
426	Stypandra glauca		Nodding blue lily
427	Acacia saligna		
428	Leschenaultia biloba		
429			
430			
431	Petrophile linearis		Pixie mops
432			
433	Dasypogon bromeliifolius		Pineapple bush
434	Gompholobium tomentosum		
435	Casuarina dystyla		
436	Hypocalymma robustum		Swan River myrtle
437	Calectasia cyanea		
438	Laxmannia grandiflora		
439	Conostephium pendulum		Pearl flower
440	Synaphea polymorpha		
441	Hakea glabella		
442	Banksia ilicifolia		
443	Hypocalymma angustifolium		
444	Phyllanthus calycinus		
445–451			
452	Dampiera linearis		
453	Hibbertia hypericoides		
454	Hibbertia hypericoides		
455			
456			
457	Kennedia prostrata		
458	Grevillea synapheae		
459	Acacia pulcherrima		
460	Acacia alata		
461	Acacia Baileana		
462	Acacia cyanophlla		
463			
464	Agonis		

#	Plant and name as given by Wilson	Comment	Place collected
465	Chorizema ilicifolia		G
466			
467	Bossiaea ornata F v. M.		
468	Darwinia citriodora		
469	Grevillea bipinnatifida		
470	Astroloma pallidum R. Br.		
471	Dryandra nivea		
472	Bossiaea ornata Fv M		
473	Clematis aristata		
474	Kennedia coccinea		
475			
476	Hibbertia montana, Stend.		
477	Chorizema Dicksonii Graham.		
478	Leucopogon australis		
479			
480	Persoonia elliptica		

Zoological Gardens

#	Plant and name as given by Wilson	Comment	Place collected
481	Callitris robusta	Cult. [...]	Rottnest
482	Callitris		Cultivated, said to come from near Perth
483	Eucalyptus torquata		Cult.
484	Eucalyptus cornuta	Yate.	Cult
485	Euc. pyriformis		Cult.
486	Euc. Lehmannii		Cult.
487	Hemiandra incana		Wild
488	Dodonea viscosa		Cult.
489	Euc. megacarpa		Cult.
490	Euc. oleosa?		Mr Robinson's garden
491	Beaufortia		Mr Robinson's garden
492	Calothamnus homalophyllus		Mr Robinson's garden
493	Euc. Lehmannii		Mr Robinson's garden
494	Callitris		Cult. Said to come from Murray R.
495	Red flowered gum		Cult. Perth

#	Plant and name as given by Wilson	Comment	Place collected
496	Callitris		Mr Robinson's garden
497	Calltris robusta		Near naval base, Fremantle
498	Callitris robusta		Peppermint Grove
499	Banksia Menziesii		Near Applecross
500	Macrozamia Fraserii		Near Fremantle
501	Nuytsia floribunda		Near Canning Bridge
502	Scaevola crassifolia		Fremantle nr Naval Base
503	Hibiscus Huegelii		Nr Naval Base
504	Personia [sic] saccata		Nr Naval Base
505	Hibiscus Huegelii		Nr Naval Base
506	Spyridium		Nr Naval Base
507	Gastrolobium callistachys		'loc (?)WA'

DIARY TWO: III EHW, SERIES VIII, BOX 14, FOLDER 3

Inside cover: DA Herbert's address crossed out:
D A Herbert Analytical Dept Wellington Street Perth
W VIII
14: 3

[Nov 1920 – Jan 1921]
[p. 1–4] Coolgardie to Widgiemooltha

#	Plant and name as given by Wilson	Comment	Place collected
520	Scaevola	2ft high fl white	20 miles from Coolgardie
521	Eremophila Brownii (?)	2ft high	20 miles from Coolgardie
522	Eremophila Drummondii	5ft fl purple	20 miles from Coolgardie
523	Actinostrobus	Cultivated specimen	Park, Coolgardie
524	Callitris		20 miles from Coolgardie
525	Grevillea sp.	Fl greenish white suffused with pink	25 miles from Coolgardie
526	Myrtaceae	Fl. White—shrub of 6 feet erect	20 miles from Coolgardie
527	Shrub	Fl pink, shrub 18 ins. high	Widgiemooltha
528	Dodonea aciphylla ?	Erect shrub 5-6 feet	Widgiemooltha

#	Plant and name as given by Wilson	Comment	Place collected
529	Trichinium obovatum	Fl pink—diffuse 2 ft high	Widgiemooltha
530	Eucalyptus calycogona var. gracilis	Snap & rattle 15 feet	20 miles from Coolgardie
531	Compositae	3 ft high	20 miles from Coolgardie
532	Verbenaceae ?		20 from Coolgardie
533	Melaleuca sp.	Small tree, 15 feet. Fl. White	20 miles from Coolgardie
534	Cassia eremophila A. Gunni	Sp in seed	Widgiemooltha
535	Eucalyptus Griffithsii	30 to 60 feet fl white filaments slightly pink	Widgiemooltha
536	Eucalyptus Le Souefii	Goldfields blackbutt 30 to 45 feet	Widgiemooltha
537	Eucalyptus Flocktoniae	'White Gum' filaments gellow 50 to 60 feet	Widgiemooltha
538	Euc. salubris F.v.M	'Gimlet' 50 feet	Widgiemooltha
539	Eucalyptus campaspe	'Gimlet' 35-45 ft	Widgiemooltha
540	Eucalyptus Stricklandi	30 feet fl greenish-yellow	Widgiemooltha
541	Eremophila Oldfieldii?	Fl red 4 to 10 feet erect & spreading 'Camel Poison'	Widgiemooltha
542	Eremophila sp.	Fl purple 3 ft spreading	Widgiemooltha

[p. 5–8] 'Collected between Widgiemootha 22nd Nov. 1920'

#	Plant and name as given by Wilson	Comment	Place collected
543	Eucalyptus torquata	Fl red 20-30 feet	Widgiemooltha
544	Fusanus spicatus R. Br.	'Sandalwood'	Widgiemooltha
545	Euc. Ewartina	15 ft	Widgie
546	Casuarina sp	12 ft--- cones	20 miles from Coolgardie
547	Melaleuca sp	6 ft	20 miles S of Coolgardie
548	Dodonaea sp	6ft	20 m S. of Coolgardie
549	549 to 560 incl missing		
561	Eucalyptus oleosa	'Morrell' 60 ft	Widgiemooltha
562	Eucalyptus torquata	Fl. White – tree of 25 ft	Widgiemooltha
563	Eucal. torquata	Ft. red—tree of 25 ft	Widgiemooltha
564	Eucalyptus transcontinentalis	'White Gum'	20m S. of Coolgardie
565	Acacia	6 ft	Widgiemooltha
566	Eremophila	6ft fl blue	20 miles south Coolgardie
567	Saltbush	[abundant covered]	Widiemooltha
568	Marsdenia Leichhardtiana	fl. greenish white twinning [ ] [around]	Widgiemooltha

#	Plant and name as given by Wilson	Comment	Place collected
569	Acacia	6-12 ft around W-Moo Bush to small tree	(Widgiemooltha)
481		[with][st...] [Transcontinental Euc]	
570	Acacia	With—similar to No.569	
571	Callitris		12 miles WNW Kalgooorlie. on photographs

[p. 9–15] Round Adelaide Hills S.A. 10-XII-20.

#	Plant and name as given by Wilson	Comment	Place collected
572	Callitris propinqua	Tree 20-30 ft &2-5 ft common in Hills	
573	Hakea	3-5 ft Hills, common	Millpark
574	Banksia marginata	3-10 ft flo yellow in Hills common	
575	Acacia	5-10 ft	Nr Gawler
576	Muhlenbergia	Thickets	Nr Gawler
577	Baeckia	5-7 ft common on Hills	
578	to next missing		
588	Dodonaea	5-7 ft common	
589	Euc. [ ]?	80-100 ft &10-15 ft flo pink common	
590		2-5 ft flo yellow	Mt Crawford
591		1-3 ft flo yellow & [ ] common	
592	[ ]	flo yellow	Mt Crawford
593		Flo pale yellow common	Mt Crawford
594	Daviesia	3-5 ft	Mt Crawford
595		2-3 ft flo white common on Hills	
596	Leptospermum	3-6 ft – flo white common	
597		Flo [ ] common on Hills	
598	Pimelia	2-3 ft flo white	Mt Crawford
599	Hibbertia	2 ft	Mt Crawford
600		2-3 ft fo white	Mt Lofty
601	Grevillea	Prostrate or nearly so flo [cone] fruit	Mt Crawford
602	Euc. cosmophyllla	6-25 ft [swamp] Kuitpo [ ]	[ ]
603	Exocarpis cupressiformis	15-30 ft × 1-5 ft fruit [ ] common in Hills	
604	Euc.	Tree 40 ft × 8 ft	Mt Crawford

#	Plant and name as given by Wilson	Comment	Place collected
605	Callistemon	12 ft flo scarlet [ ] Kuitpo	
606	Euc [ ]	Flo white tree up to 100 ft & 5-8 ft common	
607	Pimelia	[rare]	Mt Crawford
608	Acacia pycnantha	Up to 18 ft × 1 ft Tanning Wattle [ ]	
609	Eucl obliqua	Up to 90 ft & 20 ft [Pink] [ ] [ ]	
610	Euc. fores	50 ft × 6 ft pink form	Miponga [sic]
611	Melaleuca	2-4 ft flo pink common	Miponga
612	Casuarina	5-7 ft common	Miponga
613	Epacris	2 ft flo red	Miponga
614	Epacris	2 ft flo white swamps	Miponga

[p. 16–19] New South Wales
Trip to Castlecrag State Forest December 19th 1920

#	Plant and name as given by Wilson	Comment	Place collected
615	Euc	Tr to 80 ft × 15 ft flo white	
616		[ ] 4 ft flo yellow	
617	Acacia [ ]	[ ] Wattle 6-12 ft flo pale yellow, bark [ ]	
618	Hakea	3-5 ft	
619	Leptospermum	3-5 ft flo white	
620		Flo yellow about 1-2 ft	
621		Flo yellow about 1-2 ft	
622		C 5 ft flo red purple [ ] white	
623		1 ft flo pale yellow	
624	Pimelia	1-2 ft flo white	
625		Flo white	
626	Euc siderophloia	Up to 100 ft × 15 ft flo white	
627	Isopogon	2-3 ft	
628	Melaleuca	4 ft	
629		[ ] 4 ft flo yellow + brown	
630	Melaleuca	3-4 ft flo pink	
631	Xanthorrhoea [ ]		
632		6-8 ft flo white	
633	Euc.	50 ft × 6 ft flo white	

#	Plant and name as given by Wilson	Comment	Place collected
634	Casuarina torulosa	Up to 60 ft x 4 ft forest near Sydney	
635		4- 6 ft flo white	

The above plants collected on a trip to the Castlereagh Forest reserve some 40 miles west of Sydney visited on Saturday Dec. 19. The Chairman of the Forestry **[p. 20]** Commission W. Dalrymple Hay took us by motor. It was a pleasant ride over poor roads & we had many fine views of Sydney & its suburbs. The forest was acquired 4 years ago & is nearly all pole-growth of Iron bark (Euc. Siderophloia) with a little white [fern] mixed **[p. 21]** where a fine view of the valley of the Hawkesbury River is obtainable. The old town is very English. It is a splendid example of what can be obtained in a few years by keeping out fire. On some [other] [farm] [wooded] Pinus insignis & P. muricate have been experimentally planted. We lunched at Windsor the 2nd oldest town in N.S.W

[p. 22]. December 19th 1920: Sydney B.G

(Followed by two pages of plant names with mainly exotics such as poplar, copper beech, oak).

[p. 23] American trees. Followed by plant names. (not given here)
[p. 24] S. African tree. Followed by two plant names (not given here)
[p. 25] Kaffir plum 15 ft though missing
[p. 26]. Australian & N. Zealand. Followed by plant names (not given here).
[p. 27 blank]
[p. 28]. The gardens occupy a narrow valley & sloping banks & are laid out in terraces running parallel with the Naval Anchorage on Sydney Harbour. Spacious lawns, green at the time of my visit. Made of Buffalo grass [ ] mown and springy to walk on. Trees dotted all over borders innumerable crowded with shrubs & herbs. The show of colour due to familiar plants like Cannas, Phlox & [ornamentals], Verbenas Salvias, Roses, Hibiscus, Rosa sinensis, Oleander Agapanthus, Spartium Junceum, Veronicas Russelia Evening Primrose. **[p. 29]** Tree labels wood painted white, black lettering bold, & perhaps too large though the size varied with that of the tree trunk [straw] latch zinc bracket white, black, lettering.

Coreopsis, Penstemons, Geraniums, California Poppy Scabiosa Anthiriums Hydrangea Flax. Centranthus ruber Moraea Robinsoniana from Lord Howe Island with iris-like leaf & [p] white flos on [ ] [good yard] high inflorescence. Flws white with yellow crescent [ ] of outer segment. A lovely herb Pinks, Marigolds Fuchsias, Delphiniums.

[p. 30–1] Remarkable as very interesting trees

Keteleeria Fortunei - soft flat topped, green cones. Not [bark] like that of the Davidiana, suggests a [Hendrik].

Macrozamia spiralis N.S.W. [? Burrawang] 8 ft girth of [good] high trunk glorious cones erect, more than a foot high, black green erect [ ] [ ] smooth berries

Ficus macrophylla [ ] 'Moreton Bay Fig' with a spread of 120 ft × 100 ft in epulis trunk more than 20 ft tall & massive spreading trunk with [ ] thick aerial root masses.

Brachychiton rupestris Queensland Bottle tree short fat bottle-like trunk

Pinus Ternifolia Guatemala 6ft × 30 ft flat- topped

[pp. 32–7] Collected Gosford & Narara 28/12/20

#	Plant and name as given by Wilson	Comment	Place collected
636	Leptospermum	1-2 ft flo white	The Heads Sydney Harbor [sic]
637	[ ]	6 ft fruits	Narara
638	[Eugenia] [?]	5 ft × 2 ft flo white	Narara
639		Bush to 12 ft flo white	Narara
640	[ ]	5-6 ft woody	Gosford
641		Tree 40 ft × 5 ft	Roadside Narara
642	Tristania conferta	40-100 ft × 5-15 ft flo white	Narara
643		Tree 35 ft flo purple [ ]	Gosford Nursery
644	Callistemom lanceolata	6 ft flo [ ] [ ]	Gosford Nursery
645	[Eugenia?] [?]	Tree 50 ft × 5 ft fruits	Narara
646	[ ]		Narara
647	[ ]	6-8 ft flo white	Narara
648		6-8 ft fruit [orange]	Narara
649	Acacia mollissima	50 ft flo pale yellow	Narara
650	[ ]	Flo pink	Narara
651	Myrtus [ ]	12- [ ] ft flo white	Narara

#	Plant and name as given by Wilson	Comment	Place collected
652		climber	Narara
653	Trema ?	12-50 ft	Narara
654	Melaleuca	10-12 ft flo white	Narara
655		Bush 6-12 ft flo rich yellow	Towards Narara
656	Angophora linnifolia ?	20 ft flo yellow	Narara
657	Prostanthera	6 ft flo ruby purple fruits	Narara
658	[ ]	12 ft flo white	Forests Narara
659	Callitris cupressis cult.	20 ft [ ]	Gosford Nursery
660	Callitris [ ]	40 ft × 5 ft cult	Gosford Nursery

[p. 38–9] **Trips to forests N of Sydney N.S.W**
Dec 27th: Leave Sydney 6.50pm for Taree 200 miles north by train arrive 4 am. Taree is situated on the Manning River with a rich alluvial plain. The raising of cattle & dairying on the

[p. 39] **December 28th**: Visited Coopernook state forest some 25 miles from Taree. Chiefly & in many places pure Blackbutt (E. pilularis) pole growth up in 150 ft like ships mast under-growth chiefly Casuarina torulosa very little [ ] which is kept down by fire. En route saw fine stands Water Oak, Casuarina glauca, a tall species. One some [ ] the Stag's Horn fern. [South] of Manning river grows Sinceu officinalis – often tree is 20 to 26 ft dry oppressively hot with thunderstorms in evening.

[p. 40] **December 29th**: Visited forested country round the upper waters of the Manning River. Mostly pole growth of Gray Gum (Euc. Punctata) Iron bark (Euc. Paniculata) spotted gum (E. maculata) & Blue gum (E. saligna) & White mahogany (e. acmenoides). The spotted gum with its spotted bark and white [ground (short for background?) / frond] is the 'Lady of the Forest' of this part of N.S.W. In open pasture I saw a [fine] tree of Red Cedar (Cedrela australis) and was told that formerly much good cedar grew here. By side of river saw large [tree] of Casuarina cunninghamiana. River Oak the largest of the brush [ ]

[p. 41] **December 30th**: Visited the Rain forest on the Bulgar [sic] plateau some 40 miles N.W. of Taree [2600 ft tall] Here wonderful forest of [ ] hardwood including Brushbox Tristania conferta with a grey scaly bark, [between] [ ] -red & dense canopy of [luscious] foliage. Tallow-wood (E. microcorys) & E. saligna was also [dominant] here, harder growth dense jungle with many lianas. Platycerium grande was occasionally seen also an Alsophila 10 – 11 ft tall! A Black boy & many other interesting plants. The plateau is volcanic with basalt [ ] with rich humus. Saw the Ellenborough Falls 507 ft tall over basalt cliff.

[p. 42] December 31st: Visited Tuncurry Prison Camp – a Repository for good conduct prisoners where they are found employment in afforestation work. Here some 900 acres are under young Pinus [ ] & pinaster, soil sandy & Eucalyptus – [(word hidden by Archive stamp)] of no value to prisoners looks promising. Euc scrub, so much Eucalypt forest & by side of stream forest in which Livistona australis & Araucaria [ ] is common.

[p 43] January 4th 1921: Left Sydney 9 o'clock & Crossed the ferry to Milsons Point. There join car with M.W. Pope & M.... Judy his wife & small daughter. [Motor] [ ] lunching at Camden an old town & then on through Picton & other places to Moss Vale where we [stayed] the night. Remained most of the day. Journey through dairying land to sheep land. Forest destroyed through [work] of small [stuff] in [common] En route gathered the following species of Eucalypts.

[p. 44] Picton to Mossvale 4-1-21 (Author note: this list is not in Wilson's hand)

#	Plant and name as given by Wilson	Comment	Place collected
661	E. Crebra	Narrow-leafed Ironbark	
662	E. maculosa (check)	White Gum	
663	E Piperita	Sydney Peppermint	
664	E. maculosa (check)	White Gum	
665	E dives	Broad leaf Messmate	
666	E radiata	Narrow leaf messmate	
667	E punctata	Grey Gum	
668	E longifolia	[ ] Woolly butt	
669	E paniculata	Grey Ironbark	
670	Xylomelum pyriforme	Pear	
671	Euc sieberiana	Coast Ash	
672	Euc haemastoma	Scribbly Gum	

#	Plant and name as given by Wilson	Comment	Place collected
673	Euc leptospermum		
674	Euc tereticornus	Red Gum	
675	Euc eugenioides	White Stringbark	
676	Euc mannifera grandiflora	Red Mahogany	

[p. 46] Moss Vale to Gundagai 5 Jan 1921

#	Plant and name as given by Wilson	Comment	Place collected
677	Juniperis communis	[ ]	[ ] (locality)
678	Euc polyanthemos		
679	Euc elaeophora	Apple Box [ ]	

#	Plant and name as given by Wilson	Comment	Place collected
680	Euc viminalis	(Manna Gum or Ribbon Gum)	
681	Euc macrorhyncha	(Brown Stringybark)	
682	Euc tereticornis	(Broadleafed) Red Gum	
683	Euc melliodora	(Yellow Box)	
684	Euc Macarthurii		
685	Euc cineria [sic]	(Argyle apple)	
686	Euc coriacea	(Snow Gum)	
687	Euc. Smithii (check)	(Gully Ash)	

[p. 47–9] Tumut — Rules Point

#	Plant and name as given by Wilson	Comment	Place collected
688	Euc rostrata		
689	Callitris calcarata	Black cypress pine	
690	Euc Cambageii		
691		Shrub 3 ft	
692		Bush 6 ft	
693		Shrub 2 ft	
694	Euc Camphora	(White Sally)	
695	Euc Stellulata	Black Sally	
696	Euc Coriacea	Pyriform fruit	
697	Euc Coriacea	[ ] (Sessile)	
698	Euc rubida	[ ]	

#	Plant and name as given by Wilson	Comment	Place collected
699	Baeckea		
700	Legume		
701	Euc Dalrympleana (1)		
702	" " (2)		
703	Liliaceae		
704	Myrtacea		
705	Daviesia lattifolia [sic]		
706	Acacia decurrens (dealbata)		
707	Proteaceae		
708	Epacris		
709	Tic bush		
710	Euc gigantea syn Delegatensis	(Alpine Ash)	
711	Dodonaea		
712	Oxylobium?	1 ft flo yellow	
713	Acacia [ ]	6 ft	
714	Melaleuca	6-8 ft flo white	
715	E globulus		
716		Messmate	Rules Point
717	Callitris robusta		

[p. 50] Jan 5th: Moss Vale to Gundagai 160 miles via Goulburn over a good road & through rich pastoral land – one of the best for sheep in N.S.W. Much being plateau land with a low – average rainfall. Saw & explored many different species of Eucalyptus none of them [ ]. Euc. Macarthurii in open pastures looks very like an English Elm. A flowering little species is E. cinerea with purple furrowed bark, & glabrous opposite leaves. At Goulburn are many planted exotic trees & a P. canariensis. 90 ft – was mostly of note. At Moss Vale saw a fair Pinus distus & on the road a good Pinus ponderosa.

[p. 51] Jan 6th: From Gundagai To Tumut mostly along banks of River of that name. Delayed Tumut for car repairs. Visit river course with fine avenue of English Elm. Tumut old town with many fine Lombardy poplars. Eng. Oak, Planes etc. From Tumut commence ascent of main divide & after 47 miles reach a hotel on summit --4000 ft. Hotel on edge of low forest with long open fern-clad valley in front. Running N. of evidently old lake. 12 miles N. from here rises the Murrumbidgee River. Ascent over fair road very [bumpy] in places. Gundagai to Rules Point 70 miles.

[p. 52] **January 7th**: Had easy day round Rules Point. Collected Euc. rubida (Red Gum) E. stellulata (Black Sally) & E. Dalrympleana & E. coriacea (Snow Gum) all four common trees round here. The Snow Gum & Black Sally reach the altitudinal limits of the forest in the district & evidently withstand considerable ice & snow. The day was perfect. Clear sunshine with cool wind & not the least bit hot.

[p. 53] **January 8th**: Visited Jounama state forest where the planting of exotic conifers is contemplated. The Eucalypts are poor consisting of E. Dalrympleana Messmate & snow Gum. I think the experiment has every chance of success. Saw E. Gigantea (Mountain Ash) which yields the best Timber in these forests. Its peculiarity is that unlike nearly all Eucalypts it does not [shrink]. E. Dalrympleana…

[p. 54] **January 9th**: visited Jillabenan cave

(No more entries except 5 pages along, on the inside back cover, he writes 'Julius. Orange to Dubbo')

Note A: '34. Daviesia uniflora Herbert (1922: 37), Crisp (1995: 1244). Type: 'Locality-Yoting, in sand plain. Collectors—Herbert & Wilson No. 174. Date—November, 1920.' Lectotype (Crisp 1995: 1244): PERTH; isolectotype: MEL80533'. From: MD Crisp, L, Cayzer, GT Chandler, and LG Cook, 'A monograph of *Daviesia* (Mirbelieae, Faboideae, Fabaceae)', *Phytotaxa*, 2017, 300(1):87 cited.

NOTES

1. However, *Templetonia sulcata*, another leafless shrub is known as 'centipede bush', while *Daviesia eu*phorbioides is known as 'Wongan Cactus'. This entry as it appears in Wilson's diary was likely incorrectly identified in 1920.
2. Wilson is noting the gleam and glitter found in many eucalypts when the sun strikes their leaves.
3. Gawler is approximately 42 km (26 miles) north of Adelaide.
4. Narara is a suburb just north of Gosford NSW.
5. This is likely Kunzea ambigua
6. This is the 'Main Divide' of the Great Dividing Range.

APPENDIX 3

SPECIMENS FROM WILSON'S AUSTRALASIAN EXPEDITION IN THE HARVARD HERBARIA

Table 3a gives a list of families and genera in the Harvard Herbaria that were collected by Wilson on his Australasian Expedition. It also gives an indication of which of these families I examined. Some of the families not examined were in an annex that had been under quarantine.

TABLE 3A: Families and genera from Wilson's Australasian Expedition in the Harvard Herbaria, with an indication of those families examined.

Family	Genera included	Examined?
Myrtaceae	*Eucalyptus, Melaleuca, Agonis, Verticordia, Callistemon, Kunzea, Darwinia, Calothamnus,*	Yes
Proteaceae	*Banksia, Dryandra, Grevillea, Conospermum, Lambertia, Isopogon, Adenanthus, Petrophile*	Yes
Cupressaceae	*Callitris*	Yes
Fabaceae	*Acacia, Bossiaea, Hovea, Hardenbergia, Jacksonia, Phyllota, Aotus, Oxylobium, Cassia, Chorizema, Daviesia, Kennedia,*	Yes
Rutaceae	*Boronia, Correa, Chorilaena, Eriostemon,*	Yes
Goodeniaceae	*Dampiera, Scaevola,* Leschenaultia, *Goodenia*	Yes
Casuarinaceae	*Casuarina*	Yes
Santalaceae	*Santalum*	Yes
Stylidiaceae	*Stylidium*	Yes
Scrophulariaceae	*Eremophila*	Yes
Loranthaceae	*Nuytsia, Amyema*	Yes
Thymeleaceae	*Pimelea*	Yes
Dasypogonaceae	*Kingia*	Yes
Asphodelaceae	*Xanthorrhoea*	Yes
Olacaceae	*Olax*	Yes
Dilleniaceae	*Hibbertia*	No
Malvaceae	*Thomasia*	No
Apiaceae	*Xanthosia*	No
Asparagaceae	*Acanthocarpus*	No
Ericaceae	*Leucopogon*	No

Table 3b is a list of the species Wilson collected in Australia on the expedition that I imaged in 2018; others remain in another part of the Harvard Herbaria where access was difficult in 2018. In total, 524 plant specimens were imaged. As noted in Appendix 1, many additional specimens were collected by others on behalf of the Wilson Expedition. These come from various people who used the Expedition label 'Arnold Arboretum. Expedition to Australasia, India and Africa 1920-22'; most are from the eastern States of Australia.

Note: An interesting Australian item not belonging to the Wilson Expedition but related to the United States was *Angophora lanceolata* var. *angustifolia*, collected at Wollongong and near Sydney by the 'U.S. South Pacific Exploring Expedition under the command of Capt. Wilkes, U. S. N., 1838-42.' and distributed to Harvard by the Smithsonian Institution from that expedition.

TABLE 3B: Specimens found at Harvard Herbaria collected by E. H. Wilson in Australia. These are listed as found under their families. A number in parentheses (e.g. 2) indicates the number of sheets made of that plant or species. A catalogue number was given by Wilson on most sheets. A dash (-) indicates no number was written, or it was unreadable; many had Wilson's writing on the tag on the sheet. Wilson often wrote the genus but not species, so many of the species names listed below were not given by Wilson but by later taxonomists, notably Moore in the 1960s. Blanks against a number indicate that no information given, but the number was in his diaries.

No. given by Wilson	Date collected	Species name	Place collected and comments
		Myrtaceae	
490	?	Eucalyptus oleosa	
561	?	Eucalyptus oleosa	
388		Eucalyptus Guilfoylei	[Yellow tingle] Frankland River
242		Euc megacarpa	
391		Euc megacarpa	
489		Euc megacarpa	
493 (2)		Euc megacarpa	
486		Euc megacarpa	
496		Euc macrocarpa	
188 (2)		Euc macrocarpa	
-	Oct 23 1920	Euc erthyronema	Totadgin Rock
484		Euc. cornuta	[yate]
303		Euc. cornuta	
602		Euc cosmophylla	S Aust
402		Euc cosmophylla	

No. given by Wilson	Date collected	Species name	Place collected and comments
535		Euc ficifolia	
539		E campaspe	
530		E campaspe	
28		E. loxophlobia	
145		E. foecunda	
228		E. marginata	
357		E. marginata	
604		E. obliqua	Tasmania
97 (3)		E. pyriformis	Burracoppin
194		E. redunca	
418		E. rudis	
415		E. rudis	
414		E. rudis	
15 (3)		[no comments made]	
87		[no comments made]	
538		[no comments made]	
134		[no comments made]	
485		E. pyriformis	Zoological Gardens, Perth
540 (2)		E. stricklandii	Widgimooltha
317		Eucalyptus indet.	
724		Eucalyptus indet.	Tumut NSW
722		Eucalyptus indet.	Bribie Is
564 (2)		E. transcontinentalis	S Coolgardie
562		E. torquata	Widgimooltha
543		E. torquata	
563 (2)		E. torquata	
483		E. torquata	
14 (several)		Eucalyptus calophylla	
458		Eucalyptus ficifolia	
495		Eucalyptus ficifolia	
389		Eucalyptus ficifolia	Frankland River
740		Eucalyptus trachypholia	Bribie Is, Q.
117		Melaleuca	Totadgin
349	Nov 6 1920	Melaleuca thymoides	Albany
316	Nov 6 1920	Melaleuca thymoides	Albany

No. given by Wilson	Date collected	Species name	Place collected and comments
286	Nov 6 1920	*Melaleuca thymoides*	Albany
25		*Melaleuca uncinata*	Cunderdin
65		*Melaleuca uncinata*	Cunderdin
86 (2)		*Melaleuca uncinata*	Westonia
156		*Melaleuca indet. pungens?*	Westonia
126		*Melaleuca indet.*	'Callistemon' Merredin
302		*Melaleuca hamulosa*	
341		*Melaleuca indet.*	
533		*Melaleuca sheathiana*	Coolgardie
475		*Melaleuca scabra*	Mundaring
139		*Melaleuca scabra*	'Kunzea' Mundaring
107		*Melaleuca scabra*	'Kunzea' Mundaring
74		*Melaleuca scabra*	'Kunzea' Mundaring
775		*Melaleuca quinquenervia*	
718		*Melaleuca quinquenervia*	Bribie Is Q
23		*Melaleuca pentagona*	Cunderdin
-		*Melaleuca minor*	
235(2)		*Melaleuca lateritia*	'Calothamnus'
547		*Melaleuca acuminata*	Coolgardie
154		*Melaleuca cordata*	
401		*Melaleuca cuticularis*	Frankland River
358		*Melaleuca densa*	Albany
153		*Melaleuca glaberrima*	Bruce Rock
152		*Melaleuca*	Westonia
111		*Melaleuca*	Westonia
55		*Baeckea robusta*	
53		*Balaustion pulcherrimum*	
159	Oct 25 1920	*Beaufortia bracteosa*	'Mel' Toting
405		*Beaufortia decussata*	'Callistemon' W. Denmark
377		*Beaufortia decussata*	'Callistemon' W. Denmark
220		*Beaufortia macrostemon*	'Callistemon'
605		*Callistemon*	
362		*Callistemon speciossus*	Albany
734 (2)		*Callistemon viminalis*	Bribie Is Q; 'Mel' Fraser Island
780		*?Callistemon*	'Callistemon' Gympie, Q

No. given by Wilson	Date collected	Species name	Place collected and comments
492		*Calothamnos quadriform?/ homolophulus?*	Bruce Rock WA
113		*Calothamnos quadriform?/ homolophulus?*	Bruce Rock WA
18	Oct 1 1920	*Calytrix brachyphylla*	Yoting WA
175	Oct 1 1920	*Calytrix strigosa*	Yoting
129	Oct 22 1920	*Lhotskya violaceae*	'Calythina?' Merredin
209		*Chamelaucium uncinatum*	'Geraldton Wax' York; home garden.
76	Oct 23 1920	*Darwinia thryptomenoides*	'Baectia'
56		*Dawinia pupurea*	'Dip [...]'
468		*Darwinia citriodora*	
–	May 3-9 1921	*Corymbia (Indet.)*	'indefined'
171		*Eremaea pauciflora*	'Callistemon'
421	Nov 14 1920	*Verticordia acerosa*	Mundaring
43	Oct 23 1920	*Verticordia chrysantha*	Merredin
287		*Verticordia labrantha*	
64	Oct 25 1920	*Verticordia picta*	'Sand Plains Westonia'
464		*Agonis linearifolia*	
393		*Agonis linearifolia*	Frankland River
244		*Agonis parviceps*	'Leptospermum' 'Karri Forest'
–		*Thyptomene ericaea*	S. Aust
116	Oct 24 1920	*Thyptomene urceolaris*	'Tho?' 'Bruce Rock'
127	Oct 22 1920	*Thyptomene indet.*	[great flowers]
758	May 1921	*Syzygium coolminianum*	'Nr Gympie, Q'
778 (4)	1921	*Syncarpia hillii*	Q
185	Oct 25 1920	*Pilanthus peduncularis*	Quariding, WA
79		*Indet.*	Micromyrtus sp. Westonia
712		*Leptospermum polygalifolium*	
223 (2)		*Kunzea recurve ? Kunzea micromera*	'Kunzea recrva' Dandalup
312		*Kunzea micrantha*	Bunbury
360	Nov 6 1920	*Leptospermum*	'Leptosperm' Albany
443		*Hypocalymma angustifolium*	Mundaring
266		*Hypocalymma angustifolium*	Big Brook
355		*Agonis*	
741		*Angophora*	

No. given by Wilson	Date collected	Species name	Place collected and comments
328	Nov 6	*Baeckea asterioides*	Albany
24		*Baeckea crassifolia*	
178		*Baeckea crispiflora*	
72		*Baeckea grandis*	
		Proteaceae	
233 (2)	Nov 1920	*Banksia littoralis*	Harvey WA
407	Nov 1920	*Banksia littoralis*	Denmark WA
711 (3)	June 3 1921	*Banksia integrifolia*	Stradbroke Island
442	Oct 14 1920	*Banksia ilicifolia*	Not stated but WA
197	Oct 24 1920	*Banksia grandis fruits*	The Lakes, Darling Range, near Perth; many sample of fruits with no label
720	May 25 1921	*Banksia serratifolia far aemulata*	'Banksia aemula'
721 (2)	May 25 1921	*Banksia serrata*	Bribie Is, Q.
781 (2)	Jun 3 1921	*Banksia latifolia*	Stradbroke Island
311-(2)	May 2 1921	*Banksia spinulosa*	'Banksia (sp?)' Winton, NSW
782	May 25 1921	*Banksia aemula*	Brobie Is, Q
189 (3)	Oct 25 1920	*Banksia attenuata*	'See fruit collection' Quairading
348 (2)	Nov 6 1920	*Banksia coccinea*	Albany
574 (5)	Dec 10 1920 to Mar 7 Apr 1921	*Banksia marginat*	Adelaide, near Hobart, and Stanley, Tasmania
100		*Persoonia angustiflora*	'Adenanthos'
480	Oct 14 1920	*Persoonia elliptica*	Mundaring
334	Nov 6 1920	*Persoonia elliptica*	Albany
95 (2)		*Persoonia saundersiana*	
81		*Persoonia saundersiana*	
210		*Persoonia striata?*	York
713	1921	*Persoonia cornifolia*	[eastern Australia]
147	Oct 24 1920	*Petrophile circinata*	Bruce Rock
370 (2)	Nov 5 1920	*Petrophile diversifolia*	'Isopogon' Albany
156	Oct 25 1920	*Petrophile ericifolia*	Bruce Rock
431		*Petrophile linearis*	Mundaring
186	Oct 25 1920	*Petrophile media*	'Isopogon' Quariding
376	Nov 6 1920	*Petrophile media var juncifolia*	'Isopogon' Albany
227		*Petrophile media var juncifolia*	

No. given by Wilson	Date collected	Species name	Place collected and comments
204	Oct 27 1920	*Petrophile serruriae*	'Isopogon' The Lakes nr York
375	Nov 6 1920	*Petrophile rigida*	'Isopogon' Albany
447	Nov 14 1920	*Synaphea petiolaris*	Mundaring
346	Nov 6 1920	*Synaphea polymorpha*	Albany
411	Oct 14 1920	*Synaphea polymorpha*	Mundaring
226	Nov 1920	*Lambertia multiflora*	Dandalup
259	Nov 1920	*Isopogon sphaerocephalus*	Big Brook
237	Nov 1920	*Isopogon sp*	'Sandplains' W […]
115	Oct 24 1920	*Isopogon divergens*	Bruce Rock
295	Nov 1920	*Isopogon formosus*	Capel
374	Nov 6 1920	*Isopogon formosus*	Albany
234	Nov 1920	*Hakea varia*	'Hakea linearis?' Watheroo [?]
550	Nov 22 1920	*Hakea multilineata*	Coolgardie
-		*Hakea platysperma*	
573	Dec 10 1920	*Hakea rostrata*	Millbrook SA
-	Apr 13 1921	*Hakea sericea*	'Hake acicularis' near lake south Tasmania
479	Oct 1920	*Hakea ambigua*	'Acacia' Mundaring
208	Oct 26 1920	*Hakea commuta*	
236	Nov 1920	*Hakea ceratophylla*	Westonia
84	Oct 23 1920	*Hakea meissneriana*	'Hakea' Westonia
71	Oct 23 1920	*Hakea meissneriana*	'Hakea' Westonia
184	Nov 25 1920	*Hakea incrassata*	Quairiding
441	Oct 14 1920	*Hakea glabella*	Mundaring
458		*Grevillea glabrata*	'Grevillea syncphae' Mundaring
525	Nov 22 1920	*Grevillea nematophylla*	S. Coolgardie
26	Oct 21 1920	*Grevillea paniculata*	Cunderdin
54	Oct 23 1920	*Grevillea paradoxa*	Merredin [MG note very prickly good image]
451		*Grevillea pilulifera*	Mundaring
101	Oct 21 1920	*Grevillea pterosperma*	'Hakea' Westonia
291	Nov 1920	*Grevillea [ ]*	Sabine River near Busselton
78	Oct 23 1920	*Grevillea teretifolia*	Westonia
216	Oct 6 1920	*Grevillea tridentifera*	York
176	Oct 24 1920	*Grevillea uncinulaata*	Yoting
199 (2)	Oct 27 1920	*Grevillea wilsonii*	The Lakes
601	Dec 10 1920	*Grevillea lavandulacea*	Mt Crawford, SA
82 (2)	Oct 23 1920	*Grevillea apiciloba*	Westonia MG: 'Black toothbrushes'

No. given by Wilson	Date collected	Species name	Place collected and comments
70	Oct 23 1920	*Grevillea apiciloba*	Westonia
30		*Grevillea pritzelii*	
29	Oct 21 1920	*Grevillea integrifolia*	Tammin
46	Oct 23 1920	*Grevillea hakeoides*	'Proteaceae' Merredin
?	Oct 14 1920	*Grevillea endlicheriana*	Mundaring
37	Oct 23 1920	*Grevillea eryngioides*	Merredin
150	Oct 25 1920	*Grevillea eryngioides*	Quairiding
27 (2)	Oct 23 1920	*Grevillea exelsior*	Tammin
151	Oct 25 1920	*Grevillea eriostachya/ excelsior?*	Yoting
432	Oct 1920	*Grevillea crithmifolia*	Mundaring
85	Oct 23 1920	*Grevillea didymobotrya*	Westonia
469	Oct 14 1920	*Grevillea bipinnatifida*	Mundaring
411	Oct 14 1920	*Dryandra nivea*	Mundaring
7	Oct 21 1920	*Dryandra sessilis*	'D. floribunda. Parrot bush' Toodyay'
345	Nov 6 1920	*Conospermum caeruleum*	Albany
141	Oct 22 1920	*Conospermum stoechadis*	Bruce Rock
203	Oct 27 1920	*Dryandra bipinnatifida*	
371 (2)	Nov 6 1920	*Dryandra formosa*	Albany
361	Nov 1920	*Persoonia microcarpa*	Albany
284	Nov 1920	*Adenanthos barbigera*	Busselton
424	Oct 14 1920	*Adenanthos barbigera*	Mundaring
314	Nov 1920	*Adenanthos meissneri*	'Grevillea' Bunbury
285	Nov 1920	*Adenanthos melis*	Busselton
283	Nov 1920	*Adenanthos obovata*	'Adenanthos [...]'
406	Nov 1920	*Adenanthos obovata*	West Denmark
		Casuarinaceae	
31	Oct 21 1920	*Casuarina campestris*	12' tall Tammin
99 (2)	Nov 23 1920	*Casuarina corniculata*	Shrub 9', sandplain, Merredin
733 (2)	May 26	*Casuarina cunninghamiana*	Bribie Island, Queensland
701	June 3 1921	*Casuarina glauca*	Stradbroke Island, Queensland
546	Nov 22 1920	*Casuarina hemsii*	12'; 20 mils S of Coolgardie
11	Oct 21 1920	*Casuarina huegelliana*	Small tree 10–15'; Toodyay; 'took 2 photos'
219	Oct 26 1920	'*Casuarina humilis*'	York
144	Oct 24	'*Casuarina humilis*'	Bruce Rock

No. given by Wilson	Date collected	Species name	Place collected and comments
435	Oct 14 1920	'Casuarina humilis'	Mundaring
719	25 April 1921	Casuarina littoralis	Small tree 15–30'; Bribie Island
-[missing] (2)	9 April 1921	Casuarina monolifera	(1) 'the country around Strahan' (2) 'small tree. Open country near Strahan' Western Tasmania
552 (2)	Nov 1920	Casuarina obesa Miq. Ssp obesa	Tree 20'; Lake Lefroy, Western Australia
207 (2)	Oct 26 1920	C. obesa / C. glauca	Tree 65' {...} Avon River, York
246 (2)	Nov 1920	Casuarina decusssata	Big Brook, WA
702	June 3 1921	Casuarina equise .... var subsp. ....inana	Stradbroke Island
218 (2)	Nov 1920	Casuarina thuyoides	2–3'; near Dandanup
		Dasypogonaceae	
224	Nov 1920	Dasypogon bromelifolium R.Br	Dandanup
274	Nov 1920	Dasypogon hookeri	Near Jarrahwood
433	Oct 14 1920	Dasypogon bromelicifolius	Mundaring
437	Oct 14 1920	Calectassa cyanea	Mundaring
201	Oct 27 1920	Kingia australia	15' grass tree; The Lakes near York
		Asphodelaceae	
202 (2)	Oct 27 1920	Xanthorrhoea gracilis	(1) The Lakes near York (2) Flower spike
231 (3)	Nov 1920	Xanthorrhoea preisii	Flowers white; abundant sand plain; Hammel WA (2) flowers (3) 'Blackboy'
		Rutaceae	
329	Nov 6 1920	Boronia elatior	'Boronia 4 ft swamp sand Albany'
243 (2)	Nov 1920	Boronia crenulata	'Boronia 2 ft flowers red' (2) Big Brook
292	Nov 1921/20	Boronia crenulata	'Boronia 4 ft flowers rose purple' Sabine River
342	Nov 6 1920	Boronia crenulata	'Boronia 2 ½ ft'; Albany
332 (2)	Nov 6 1920	Boronia denticulata Sm.	'Boronia 4 ft' Albany (both)
232 (3)	Nov 1920	Boronia elatior	'Boronia lanuginosa' Near Hamel, W. Aust
397 (2)	Nov 1920	Boronia elatior	'Boronia lanuginosa 5 ft' and 8 ft' Frankland River
327	Nov 6 1920	Boronia elatior	'Boronia' 4-6 ft; 'swamp near Albany'
390	Nov 10 1920	Boronia gracilipes	'Boronia shrub 8 ft' Frankland River
249 (2)	Nov 1921	Chorilaena quercifolia	'Diplolemma bush 8 ft' Big Brook

No. given by Wilson	Date collected	Species name	Place collected and comments
-	Mar 24 1921	*Correa reflexa*	Tasmania
245	Nov 1921 (but actually 1920)	*Correa angustifolia*	'Eriostemon Bush 2 ft fls pink' Big Brook
391 (2)	Nov 4 & 6 1920	*Correa angustifolia*	As per 245. W. Denmark.
	Mar 24 1921	*Correa reflexa*	Eaglehawk Neck, Tasmania '_____'
245	Nov 1920	*Correa angustifolia*	'Eriostemon' bush 2 ft fls pink; Big Brook
51 (2)	Oct 23 1920	*Eriostemon brucei var. aphylla*	'Eriostemon' 2 ft Merredin
293 (2)	Nov 1920	*Eriostemon spicatus*	'Eriostemon 2 ft' Sabine River
288	Nov 1920	*Eriostemon spicatus*	'Eriostemon' Busselton
66 (2)	Oct 21/25 1920	*Eriostemon thryptomenoides*	'Eriostemon' sand plains Westonia
425	Oct 14 1920	*Diplolaema drummondii*	'Diplolaema Dampierii' Mundaring
		Fabaceae	
77	Oct 23 1920	*Acacia colletiodes*	'Dead Finish'/ site not given
569 & 570	Nov 20 1920	*Acacia aneura*	Mulga; collected 'Transcontinental Railway 481 mile station'
247		*Acacia pulchella*	Big Brook WA
296		*Acacia pulchella*	Capel WA
459	Oct 24 1920	*Acacia pulchella*	Mundaring WA
unclear	Dec 10 1920	*Acacia pycantha*	Adelaide
-	June 1 1921	*Acacia cincinnata*	'Slender tree' near Brisbane (1);
	May 25 1921	*Acacia cincinnata*	Bribie Island
-	Dec 1920	*Acacia calamifolia*	
250	Nov 1920	*Acacia biflora*	Big Brook, WA
68	Oct 25 1920	*Acacia bauverdiana*	Westonia WA GOOD SPECIMEN
461	Oct 14 1920	*Acacia baileyana*	Mundaring 'planted' specimen
-		*Acacia cunninghamii*	Thursday Island
6	Oct 21 1920	*Acacia cyanophylla*	'Black Wattle' Toodjay WA
462	Oct 14 1920	*Acacia cyanophylla*	Mundaring WA
369 (2)	Nov 1920	*Acacia cyclops*	Albany WA
427	Oct 14 1920	*Acacia dentiferra* (?)	'Acacia saligna' Mundaring
463	Oct 1920		'Acacia saligna'; 'dam full' Mundaring
68	Oct 25 1920	*Acacia desertorum*	Westonia
103	Oct 21 920	*Acacia close to desertorum* (Moore 1967)	
575	Dec 10 1920	*Acacia euthycarpa*	Near Lawler, SA; has a big packet of seeds
565	Nov 22 1920	*Acacia erinacea*	Widgimooltha; harsh, prickly

No. given by Wilson	Date collected	Species name	Place collected and comments
92	Nov 921	Acacia erinacea	Merredin
-(2)	June 7 1921	Acacia falcata	Townsville
755	May 29 1921	Acacia falcata	Bennakin, Q
763	May 19 1921	Acacia falcata	Fraser Island
62	Oct 22 1920	Acacia fragilis	'sandplains Westonia'
611 (2)	June 6 1921	Acacia helosericea	Port Darwin
179	Oct 24 1920	Acacia lasiocarpa var sedifolia	Yoting. Very harsh, very prickly
-/(2)	June 1921	Acacia leptocarpa?	'Acacia' Thursday Island
-	Dec 1920	Acacia longifolia	S.A
211	Oct 26 1920	Acacia meisneri	'Blue bush' York
58	Oct 23 1920	Acacia merrallii	Westonia
338	Nov 6 1920	Acaca nigricans R.Br	Albany
785	May 25 1921	Acaci pugioniformis Wendl.	
247 (2)	Nov 1920	Acacia pulchella R.Br.	Karri forest, Big Brook
459	Oct 24 1920	Acacia pulchella	Mundaring
608	Dec 10 1920	Acacia pycantha Benth	Near Adelaide
90	Oct 23 1920	Acacia sericocarpa WV Fitzgerald	Merredin
556	Nov 27 1920	Acacia signata F. Muell	Coolgardie
789	May 30 1921	Acacia suareolens	'Acacia' Fraser Island, Q.
703	29-4-1921	Acacia suaveolens	Woods, Bennackin?
788	May 30 1921	Acacia ulieifolia	'Acacia' fraser Island
714 (2)	May 25 1921	Acacia ulieifolia	Bribie Island, Q
138	Oct 23 1920	Acacia sp.	Totadgin
307	Nov 1921	Acacia (indet)	'Tuart forest on limestone' Capel
75	Nov 1920	Cassia pleurocarpa	'Cassia' Westonia
?	Nov 21 1920	Cassia sp.	Widgiemooltha
88	Oct 23 1920	Cassia hemophila? spe	Merredin
-	June 1921	Cassia alata	Thursday Island
404 (2)	Nov 1920	Aotus ericoides	W. Denmark
353	Nov 6 1920	Aotus ericoides	Albany (note: not identified by Wilson)
263	Nov 1920	Chorizema rhomeuum	
160	Oct 25 1920	Chorizema aciculane	Yoting
-	Oct 1920	Oxylobium sarinatum	Near Busselton
449	Oct 1920	Chorizema dicksonii	Mundaring
399	Nov 1920	Chorizema ilicifolium	Frankland River

No. given by Wilson	Date collected	Species name	Place collected and comments
465 (2)	Oct 14 1920	*Chorizema ilicifolia*	'Chorizema ilicifiia' Mundaring
269 (2)	Nov 1920	*Bossiaea laidlawiana* ??	Donnelly River
241 (2)	Nov 1920	*Bossiaea linophylla*	Big Brook
257	Nov 1920	*Bossiaea linophylla*	Karri forest Big Brook
254	Nov 1920	*Bossiaea ornata*	'Brachysema' Big Brook
320	Nov 1920	*Bossiaea ornata*	'Mirbelia' Pinjarrah
472	Oct 1920	*Bossiaea ornata*	Mundaring
467	Oct 1920	*Bossiaea ornata*	Mundaring
336	Nov 6 1920	*Brachysema praemorsum*	'Prostrate......sandplains' Albany
-	1921	*Dillwynia seicea*	'D.... floribunda.' Tasmania (?)
93	Oct 23 1920	*Daviesia acanthocluna*	Merredin
-	Oct 25 1920	*Daviesia acanthocluna*	Yoting, sand plain
331	Nov 6 1920	*Daviesia alternifolia*	Phylota ....' Albany
119	Oct 22 1920	*Daviesia costata*	Totadgin
296	Nov 1921	*Daviesia cordata*	Greenbushes
32	Oct 22 1920	*Daviesia euphorbioides*	'Centipede bush' Hynes Hill WOW Known now as Wongan Cactus
313	Nov 1920	*Daviesia incrassata*	Bunbury
448	Oct 1920	*Daviesia incrassata*	Mundaring
212	Oct 26 1920	*Daviesia microphylla*	York
137	Oct 23 1920	*Daviesia nematophylla*	Totadgin
89	Oct 23 1920	*Daviesia nematophylla*	Merredin
429	Oct 1920	*Daviesia pectinata*	Mundaring
174	Oct 25 1920	*Daviesia uniflora*	Yoting
7**	-	*Kennedia rubicunda*	Fraser Island
304	Nov 1920	*Kennedia prostrata*	Sabine River
457	1920	*Kennedia prostrata*	Mundaring
474	1920	*Kennedia coccinea*	Mundaring
267	Nov 1920	*Kennedia carinata*	Karri forest, Big Brook
-	Oct 21 1920	*Jacksonia sternbergiana*	Toodjay
774	1921	*Jacksonia scoparia*	Fraser Island
357	Nov 6 1920	*Jacksonia spinosa*	Albany
123	Oct 22 1920	*Jacksonia hakeoides*	Merredin
308	Nov 1920	*Jacksonia furcellata*	Capel
354 (3)	Nov 6 1920	*Hovea chorizemifolia*	Albany & Big Brook
21	Oct 21 1920	*Gompholobium aristatum*	'Buitonia' Cunderdin
379	Nov 6 1920	*Gompholobium capitatum*	'Pultunaea' Albany

No. given by Wilson	Date collected	Species name	Place collected and comments
373	Nov 6 1920	*Gompholobium knightianum*	' ———'
372	Nov 6 1920	*Gompholobium polymorphum*	' ———'; 'sandplain' Albany
319	Nov 1920	*Gompholobium polymorphum*	' ———'; Pinjarra
112	Oct 23 1920	*Gompholobiuim pachyphllum*	' ———'; 'sandplains' Westonia
238	Nov 1920	*Gompholobium ovatum*	'Bracheama' Big Brook
215	Oct 26 1920	*Gompholobium shultleworthii*	'Burtonia' York
434 (2)	Oct 14 1920	*Gompholobium tomentosum*	Mundaring & Applecross, Perth
359	Nov 6 1920	*Gompholobium venustum*	' ———'; Albany
-	Nov 1920	*Hardenbergia componiana*	Capel
507	Nov 20 1920	*Gastrolobium callistachys*	Perth
100?	Oct 27 1920	*Gastrolobium calycinum*	'York Rd Poison' The Lakes York
52	Oct 23 1920	*Gastrolobium floribundum*	'G. crassifolium' 'Narrow leaf poison' Merredin
214	Oct 26 1920	*Gastrolobium oxylobioides*	'Champion Bay Poison' York
172	Oct 24 1920	*Gastrolobiium parvifolium*	'P.... poison' Yoting
4	Oct 21 1920	*Gastrolobium spinosum*	Toodjay
205	Oct 27 1920	*Gastrolobium villosum*	??'..... Poison' The Lakes
-	Nov 1920		'Narrow-leaf Poison 8' West Denmark'
-	March 24 1921	*Genista monospessulana*	Near Hobart
47 (2)		*Jacksonia*	Westonia?
Scrophulariaceae			
541	Nov 22 1920	*Eremophila angustifolia*	'Erempholia oldfieldii' Widgimooltha
521	Nov 22 1920	*Eremophila clavata*	'20 miles S Coolgardie'
60	Oct 23 1920	*Eremophila decipiens*	'Eremophila 5' ' Westonia
169	Oct 25 1920	*Eremophila drummondii*	'Eremophila' Quairading
542	Nov 22 1920	*Eremophila glabra*	'Emu Bush' Widgimooltha
-	Nov 22 1920	*Eremophila ionantha*	'Eremophila' near Coolgardie
522	Nov 22 1920	*Eremophila ionantha*	'20 miles form Coolgardie'
566	'22 XI 1920'	*Eremophila scoparia*	'Eremophila 20 miles S Coolgardie'
Goodeniaceae			
264	Nov 1920	*Dampiera cuneata*	'D' Big Brook
44	Oct 23 1920	*Dampiera dielsii*	'D' Merredin

No. given by Wilson	Date collected	Species name	Place collected and comments
130	Oct 22 1920	*Dampiera stewardii*?	Merredin
-	Oct 23 1920	*Goodenia indet.*	'Scaevola' Merredin
428 (2)	Oct 14 1920	*Leschenaultia biloba*	Mundaring
181	Oct 25 1920	*Leschenaultia brevifolia*	'L.....' Yoting
301	Nov 1920	*Scaevola crassifolia*	Busselton
213	Oct 26 1920	*Scaevola helmsii*	York
41	Oct 23 1920	*Scaevola helmsii*	Merredin
520	Nov 22 1920	*Scaevola spinescens*	'20 miles s Coolgardie'
Stylidiaceae			
419	Nov 14 1920	*Stylidium brunonianum*	Mundaring
149	Oct 24 1920	*Stylidium dichotomum*	'Stylidium' Bruce Rock
355	Nov 6 1920	*Stylidium scandens*	Albany
Cupressaceae			
- (5)	Oct 25 1920	*Actinostrobus acuminatus*	Yoting
521	Nov 21 1920	*Actinostrobus* (*indet.*)	Coolgardie
796	May 25 1921	*Callitris columellaris*	'C. mimosa' Bribie Island
700	May 19 1921	*Callitris columellaris*	'Callitris mimosa' Bribie Island
-	May 24 1921	*Callitris columellaris*	'C. avenosa' Fraser Island
-	May 19 1921	*Callitris columellaris*	'C. avenosa' Fraser Island
-	June 1 1921	*Callitris rhomboides*	Stradbroke Island, Q.
-(2)	June 1 1921	*Callitris cupressiformis*	'C rhomboides' Stradbroke Island
481 (2)	Nov 1920	*Callitris robusta*	Rottnest Island, off Perth, and Zoological Gardens, Perth
268 (2)	Nov 1920	*Callitris robusta*	'Planted at Donnelly River said to have come from Rottnest'
571?(2)	Nov 22 1920	*Callitris preissii*	'Callitris ———' '20 miles from Coolgardie'. Second given as indet.
572	Dec 10 1920	*Callitris preissii*	'C. proproinquies' 'Hills near Adelaide; 'tree 20-35 ft x 2-5 ft girth'
-	April 24 1921	*Callitris preissii ssp murrayensis*	'Callitris' 'Murray River Pine' Victoria
571 (3)	Nov 23 1920	*Callitris hugelii*	'Callitris ———' '12 miles WNW of Kalgoorlie'; 'on photo'
-	Apr 29 1921	*Callitris* (*indet.*)	Callitris ———'; Macedon, Victoria
494	Nov 1920	*Callitris* (*indet.*)	'Said to have..... ? Murray River'; Perth
-	Apr 24 1921	*Callitris* (*indet.*)	'Start [...]'; Pimble near Melbourne
-	22-X-1920	*Callitris* (*indet.*)	'Tree to 9-20' 12 miles from Merredin'
33tag (2)	22 Oct 1920	*Callitris* (*indet.*)	'Edge of salt pan. Hynes Hill WA' and 'Tree 9-20' of pyramidal habit. Hynes Hill near Merredin

No. given by Wilson	Date collected	Species name	Place collected and comments
-	June 1 1921	Callitris (indet.)	'Tree 25-40'/ 2-8'. Stradbroke Is. near Brisbane, Q.
-	Apr 29 1921	Callitris (indet.)	'State Nursery Macedon'
-	Apr 24 1921	Callitris (indet.)	Bulla near Melbourne
482	Nov 1920	Callitris (indet.)	'Is said to grow wild near Perth'
524	Nov 22 1920	Callitris (indet.)	'20 miles from Coolgardie
		Casuarinaceae	
98 (2)	Oct 23 1920	Casuarina	'6' sandplain.' Westonia. Wilson has date as 1921 but must be 1920
551	Nov 22 1920	Casuarina acutivalis	'Casarina' tree 20' 20 miles S of Coolgardie'
		Santalaceae	
603		Exocarpus cupressiformis	
786		Leptomeria acida	Bribie Is, Q.
182	Oct 25 1920	Leptomeria cunninghamii	'Santalum'/ Parasitic. Mundaring/ Yoting?
382	Nov 6 1920	Leptomeria squarrulosa	
132 (2)		Santalum spicatum	'Sandalwood' Burracoppin
544		Santalum spicatum	Widgiemooltha
8		Santalum spicatum	Toodjay
91	Oct 23 1920	Santalum acuminatum	Merredin
729		Santalum candolii	Fraser Is, Q
403 (2)	Nov 1920	Choretrum lateriflorum	W Denmark
		Loranthaceae	
5		Amyema preisii	'Loranthus' Toodjay (a mistletoe)
769		Amyema cambagei	
		Thymelaeaceae	
294	Nov 1920	Pimelia	Pimelia Sabine River
256	Nov 1920	Pimelia lehmannii var ligustrinoides	'Pimelia 3-4 ft forests, Big Brook
166	Oct 25 1920	Pimelia	'Pimelia' Bruce Rock
48	Oct 23 1920	Pimelia imbricata	'Pimelia 1-1 ½ ft tall' Merredin
455	Oct 1920	Pimelia hispida	'Pimelia 2 ft fls white Mundaring near Perth'
339	Nov 6 1920	Pimelia hispida	'Pimelia 2 ½ ft Albany'
290	Nov 1920	Pimelia hispida	Pimelia Sabine River
-		Pimelia hispida	Pimelia Mundaring, nr Perth

No. given by Wilson	Date collected	Species name	Place collected and comments
		Olacaceae	
330		*Olax phyllanthi*	Albany

APPENDIX 4

A NOTE ON WILSON'S SPECIMENS IN AUSTRALIAN HERBARIA

The Western Australian Herbarium is the only Australian herbarium that houses specimens collected by Wilson during the 1920–2 Expedition, and they are limited to the collections he made in Western Australia. The herbarium houses 220 of his specimens, many collected by Wilson with DA Herbert, the Government Botanist, and some with Charles Gardner. Julia Percy-Bowers, the curator, noted (2019) that these do not have the Expedition label, but have either Charles Gardner's personal herbarium label or the label of the Western Australian Herbarium written by Gardner; clearly, Gardner must have submitted this collection. The Western Australian Herbarium shows that Wilson collected from the sites shown on the map in Figure 4a. Access to this collection is available.

FIGURE 4A: Map showing sites where Wilson collected specimens, according to information in the Western Australian Herbarium.

While no other Australian herbaria house plant specimens collected by Wilson, some specimens are shown in the Australasian Virtual Herbarium.

Sheet Number	Family	Genus	Species	Rank	Infraspecies	Author	Manuscript	Determiner Name
2630273	Boraginaceae	Halgania	anagalloides			Endl.		
2630397	Boraginaceae	Halgania	anagalloides			Endl.		
1939637	Casuarinaceae	Allocasuarina	acutivalvis	subsp.	acutivalvis	(F.Muell.) L.A.S. Johnson		L.A.S. Johnson
2145499	Casuarinaceae	Allocasuarina	campestris			(Diels) L.A.S.Johnson		S. Curry
1554107	Casuarinaceae	Allocasuarina	corniculata			(F.Muell.) L.A.S. Johnson		J.A. Wege
2213281	Casuarinaceae	Allocasuarina	humilis			(Otto & A.Dietr.) L.A.S.Johnson		S. Curry
2793180	Celastraceae	Psammomoya	choretroides			(F.Muell.) Diels & Loes.		
2459906	Chenopodiaceae	Atriplex	paludosa	subsp.	baudinii	(Moq.) Aellen		Paul G. Wilson
2615312	Chenopodiaceae	Rhagodia	preissii	subsp.	preissii	Moq.		
1990047	Dasypogonaceae	Dasypogon	bromeliifolius			R.Br.		
3041891	Dilleniaceae	Hibbertia	amplexicaulis			Steud.		
3068390	Dilleniaceae	Hibbertia	commutata			Steud.		
3071979	Dilleniaceae	Hibbertia	furfuracea			(DC.) Benth.		K.R. Thiele
3035913	Dilleniaceae	Hibbertia	hypericoides	subsp.	hypericoides	(DC.) Benth.		K.R. Thiele
3108236	Dilleniaceae	Hibbertia	stellaris			Endl.		
2962632	Elaeocarpaceae	Tetratheca	efoliata			F.Muell.		J. Thompson
2963884	Elaeocarpaceae	Tetratheca	hirsuta	subsp.	hirsuta	Lindl.		E.M. Joyce
2862751	Elaeocarpaceae	Tetratheca	hirsuta	subsp.	viminea	(Lindl.) Joyce		E.M. Joyce
2862638	Elaeocarpaceae	Tetratheca	hirsuta	subsp.	viminea	(Lindl.) Joyce		E.M. Joyce
2862409	Elaeocarpaceae	Tetratheca	hirsuta	subsp.	viminea	(Lindl.) Joyce		E.M. Joyce
2120038	Ericaceae	Andersonia	caerulea			R.Br.		C.A. Gardner
2155028	Ericaceae	Andersonia	caerulea			R.Br.		
3000060	Ericaceae	Leucopogon	australis			R.Br.		M. Hislop
5701023	Ericaceae	Leucopogon	glabellus			R.Br.		C.A. Gardner
3009173	Ericaceae	Leucopogon	parviflorus			(Andrews) Lindl.		J.M. Powell
8532788	Ericaceae	Leucopogon	polystachyus			R.Br.		M. Hislop
5904137	Ericaceae	Lysinema	pentapet-alum			R.Br.		K.R. Thiele
4386892	Ericaceae	Lysinema	pentapet-alum			R.Br.		K.R. Thiele
5170265	Euphorbiaceae	Adriana	quadripartita			(Labill.) Müll.Arg.		C.L. Gross & M.A. Whalen
661953	Fabaceae	Acacia	cyclops			G.Don		R.S. Cowan
663506	Fabaceae	Acacia	cyclops			G.Don		R.S. Cowan

Name Comment	Plant Description	Site Description	Vegetation	Locality	Collector	Collector's Number	Collection Date	Type Status
	9 inches high.			Bruce Rock	E.H. Wilson & D.A. Herbert	s.n.	/11/1920	
				Merredin	Wilson and Herbert	120	/11/1920	
			Sandplain.	Westonia	Wilson & Herbert	s.n.	/11/1920	
				Tammin	Wilson & Herbert	s.n.	/11/1920	
Cited by J.A. Wege in Nuytsia 17:409(2007). L. Biggs, 04/02/2008.	Shrub.	Sand.	Sand heath.	Burracoppin	Wilson & Herbert	EW 99	/11/1920	HOLO
				Bruce Rock	Wilson & Herbert	s.n.	/12/1920	
			Sandplain.	Merredin	Herbert & Wilson	49	/10/1920	
				Yoting	E.H. Wilson & D.A. Herbert	s.n.	/11/1920	
				Merredin	Wilson & Herbert	94	/11/1920	
				North Dandalup	Wilson & Herbert	s.n.	/11/1920	
				Mundaring	E.H. Wilson & D.A. Herbert	s.n.	/11/1920	
				Mundaring	E.H. Wilson	476	/11/1920	
				W Denmark	Wilson & Herbert	392	/11/1920	
				Mundaring	E.H. Wilson	453	/11/1920	
				Big Brook	Wilson and Herbert	240	/11/1920	
				Bruce Rock	Wilson & Herbert	s.n.	/11/1920	
				Mundaring	Wilson & Herbert	s.n.	/11/1920	
				Capel	Wilson & Herbert	s.n.	/11/1920	
				Jarrahdale	Wilson & Herbert	321	/11/1920	
				Capel	Wilson & Herbert	s.n.	/11/1920	
				Albany, King George's Sound	E.H. Wilson & D.A. Herbert	352	/11/1920	
				Albany	E.H. Wilson & D.A. Herbert	329	/11/1920	
				Albany	Wilson & Herbert	s.n.	/11/1920	
Agrees with type (R.Br. 2348) but the spikes are looser: it thus agrees much closer with Drumm. 319.				Albany,	E.H. Wilson & D.A. Herbert	s.n.	/11/1920	
				Capel	Wilson & Herbert	s.n.	/11/1920	
	Shrub 3 feet high.			Frankland River	E.H. Wilson & D.A. Herbert	394	/11/1920	
				Bruce Rock,	E.H. Wilson & D.A. Herbert	s.n.	/11/1920	
				Albany	Wilson and Herbert	s.n.	/11/1920	
				Naval Base, Coogee	E.H. Wilson	sn.	11/20/1920	
				Albany	E.H. Wilson & D.A. Herbert	s.n.	/11/1920	
				Albany	Wilson & Herbert	s.n.	/11/1920	

Sheet Number	Family	Genus	Species	Rank	Infraspecies	Author	Manuscript	Determiner Name
129097	Fabaceae	Acacia	divergens			Benth.		B.R. Maslin
119261	Fabaceae	Acacia	erinacea			Benth.		B.R. Maslin
644021	Fabaceae	Acacia	heteroneura	var.	heteroneura	Benth.		R.S. Cowan
310042	Fabaceae	Acacia	lioderma			Maslin		Maslin B.R.
100854	Fabaceae	Acacia	meisneri			Meisn.		B.R. Maslin
135550	Fabaceae	Acacia	merrallii			F.Muell.		B.R. Maslin
455989	Fabaceae	Acacia	pulchella	var.	pulchella	R.Br.		B.R. Maslin
177717	Fabaceae	Acacia	pulchella	var.	pulchella	R.Br.		B.R. Maslin
177709	Fabaceae	Acacia	pulchella	var.	pulchella	R.Br.		Maslin B.R.
678392	Fabaceae	Aotus	intermedia			Meisn.		
677957	Fabaceae	Aotus	intermedia			Meisn.		
2112132	Fabaceae	Bossiaea	aquifolium	subsp.	laidlawiana	(Tovey & P.Morris) J.H.Ross		T.R. Lally
2720337	Fabaceae	Bossiaea	linophylla			R.Br.		
2720345	Fabaceae	Bossiaea	linophylla			R.Br.		
2720957	Fabaceae	Bossiaea	ornata			(Lindl.) Benth.		
2906236	Fabaceae	Callistachys	lanceolata			Vent.		J.R. Wheeler
3552500	Fabaceae	Chorizema	dicksonii			Graham		J.M. Taylor
3548570	Fabaceae	Chorizema	rhynchotropis			Meisn.		J.M. Taylor
3355136	Fabaceae	Chorizema	spathulatum			(Meisn.) J.M.Taylor & Crisp		J.M. Taylor
5211247	Fabaceae	Daviesia	angulata			Benth.		M.D. Crisp
5197821	Fabaceae	Daviesia	benthamii			Meisn.		M.D. Crisp
5147689	Fabaceae	Daviesia	cordata			Sm.		M.D. Crisp
5189721	Fabaceae	Daviesia	decurrens	subsp.	decurrens	Meisn.		M.D. Crisp
2727358	Fabaceae	Daviesia	grahamii			Ewart & Jean White		M.D. Crisp
5210755	Fabaceae	Daviesia	nematophylla			Benth.		M.D. Crisp
5202000	Fabaceae	Daviesia	physodes			G.Don		M.D. Crisp & G.T. Chandler
5200261	Fabaceae	Daviesia	uniflora			D.A.Herb.		M.D. Crisp
1010999	Fabaceae	Daviesia	uniflora			D.A.Herb.		M.D. Crisp
1042734	Fabaceae	Gastrolobium	brownii			Meisn.		M.D. Crisp
2800446	Fabaceae	Gastrolobium	calycinum			Benth.		G.T. Chandler
2801264	Fabaceae	Gastrolobium	floribundum			S.Moore		G.T. Chandler

Name Comment	Plant Description	Site Description	Vegetation	Locality	Collector	Collector's Number	Collection Date	Type Status
				Big Brook, Pemberton	Wilson & Herbert	s.n.	/11/1920	
				Merredin	E.H. Wilson & D.A. Herbert	s.n.	/11/1920	
			sandheath	Westonia	E.H. Wilson & D.A. Herbert	103	/11/1920	
				Albany	E.H. Wilson & D.A. Herbert	s.n.	/11/1920	
				York	E.H. Wilson & D.A. Herbert	211	/11/1920	
				Westonia	E.H. Wilson & D.A. Herbert	57	/11/1920	
				Capel	E.H. Wilson & D.A. Herbert	s.n.	/11/1920	
				Big Brook, Pemberton	E.H. Wilson & D.A. Herbert	s.n.	/11/1920	
				Big Brook, Pemberton	E.H. Wilson & D.A. Herbert	s.n.	/11/1920	
				Albany	E.H. Wilson & D.A. Herbert	353	/11/1920	
				Near Denmark	Wilson & Herbert	s.n.	/11/1920	
				Donnelly River	Wilson & Herbert	s.n.	/11/1920	
				Big Brook	Wilson & Herbert	s.n.	/11/1920	
				Big Brook	Wilson & Herbert	s.n.	/11/1920	
				Mundaring	E.H. Wilson	467	/11/1920	
				Donnelly River	Wilson & Herbert	s.n.	/11/1920	
				Mundaring Weir	E.H. Wilson	477	/11/1920	
				Yoting	Wilson and Herbert	s.n.	/11/1920	
			Sandheath.	Near Busselton	E.H. Wilson & D.A. Herbert	s.n.	/11/1920	
				Mundaring	Wilson and Herbert	s.n.	/11/1920	
			Sandplain.	Yoting	Wilson & Herbert	s.n.	/11/1920	
				Greenbushes,	Wilson and Herbert	s.n.	/11/1920	
				Mundaring	Wilson & Herbert	s.n.	/11/1920	
				Totadgin	Wilson & Herbert	s.n.	/11/1920	
				Merredin	Wilson and Herbert	s.n.	/11/1920	
				Bunbury	Wilson & Herbert	s.n.	/11/1920	
Flowers yellow, keel purple.				Yoting	Wilson & Herbert	192	/11/1920	
			Sandplain.	Yoting	Wilson & Herbert	174	/11/1920	LECTO
				W Denmark	E.H. Wilson & D.A. Herbert	s.n.	/11/1920	
				York	Wilson & Herbert	s.n.	/11/1920	
		Sand.	Heath.	Merredin	E.H. Wilson & D.A. Herbert	s.n.	/11/1920	

Sheet Number	Family	Genus	Species	Rank	Infraspecies	Author	Manuscript	Determiner Name
2827034	Fabaceae	Gastrolobium	parvifolium			Benth.		G.T. Chandler
2864207	Fabaceae	Gompholobium	laxum			(Benth.) Chappill		J.A. Chappill
2835363	Fabaceae	Gompholobium	shuttleworthii			Meisn.		
2877066	Fabaceae	Gompholobium	venustum			R.Br.		
2804204	Fabaceae	Hovea	elliptica			(Sm.) DC.		J.H. Ross
1911716	Fabaceae	Jacksonia	furcellata			(Bonpl.) DC.		
1024272	Fabaceae	Jacksonia	nematoclada			F.Muell.		M.D. Crisp
1023217	Fabaceae	Jacksonia	nematoclada			F.Muell.		C. Wilkins
7811640	Fabaceae	Jacksonia	nematoclada			F.Muell.		C. Wilkins
1919326	Fabaceae	Jacksonia	ramulosa			Chappill		J. Chappill
2870827	Fabaceae	Kennedia	carinata			(Benth.) Domin		M.R.F. Taylor
711233	Fabaceae	Latrobea	brunonis			(Benth.) Meisn.		
2881659	Fabaceae	Mirbelia	dilatata			R.Br.		
2881705	Fabaceae	Mirbelia	dilatata			R.Br.		
2886367	Fabaceae	Mirbelia	trichocalyx			Domin		M. Crisp & J. Taylor
1025287	Fabaceae	Phyllota	luehmannii			F.Muell.		M.D. Crisp
4510186	Fabaceae	Psoralea	pinnata			L.		C. Stirton
634352	Fabaceae	Pultenaea	reticulata			(Sm.) Benth.		
2919656	Fabaceae	Sphaerolobium	hygrophilum			R.Butcher		R. Butcher
2920352	Fabaceae	Sphaerolobium	macranthum			Meisn.		R. Butcher
2921898	Fabaceae	Sphaerolobium	medium			R.Br.		
2921960	Fabaceae	Sphaerolobium	scabriusculum			Meisn.		
2900130	Fabaceae	Templetonia	smithiana			J.H.Ross		I.R. Thompson
629391	Fabaceae	Urodon	dasyphyllus			Turcz.		T.D. Macfarlane
2851814	Fabaceae	Viminaria	juncea			(Schrad. & J.C.Wendl.) Hoffmanns.		
8419027	Frankeniaceae	Frankenia	setosa			W.Fitzg.		M.A. Whalen
2543699	Goodeniaceae	Dampiera	juncea			Benth.		S. Curry
2547295	Goodeniaceae	Dampiera	linearis			R.Br.		M.T.M. Rajput & R. Carolin
2606380	Goodeniaceae	Goodenia	helmsii			(E.Pritz.) Carolin		R. Carolin
2606305	Goodeniaceae	Goodenia	helmsii			(E.Pritz.) Carolin		
2650215	Goodeniaceae	Scaevola	crassifolia			Labill.		R. Carolin & S. Dyer

Name Comment	Plant Description	Site Description	Vegetation	Locality	Collector	Collector's Number	Collection Date	Type Status
				Yoting,	Wilson & Herbert	172	/11/1920	
				Cunderdin	Wilson & Herbert	s.n.	/11/1920	
				York	E.H. Wilson & D.A. Herbert	s.n.	/11/1920	
				Albany	E.H. Wilson & D.A. Herbert		/11/1920	
				Big Brook [near Pemberton]	Wilson & Herbert	s.n.	/11/1920	
				Capel	Wilson & Herbert	s.n.	/11/1920	
		Sandplain.		Merredin	Wilson & Herbert	47	/11/1920	HOLO
		Sandplain.		Merredin	Wilson & Herbert	47	/11/1920	HOLO
				Merredin	Wilson & Herbert	47	/11/1920	ISO
				Merredin	Wilson & Herbert	s.n.	/11/1920	
				Pemberton	E.H. Wilson & D.A. Herbert	s.n.	/11/1920	
				Torbay	Wilson & Herbert	s.n.	/11/1920	
				Busselton	Wilson & Herbert	s.n.	/11/1920	
				Frankland River	Wilson & Herbert	s.n.	/11/1920	
		Sandplain.		Merredin	E.H. Wilson & D.A. Herbert	s.n.	/11/1920	
				Totadgin	E.H. Wilson & D.A. Herbert	118	/11/1920	TYPE
				Albany	Wilson & Herbert	380	/11/1920	
				Big Brook	Wilson & Herbert	s.n.	/11/1920	
				Albany	Wilson & Herbert	s.n	/11/1920	
				Busselton	Wilson & Herbert	s.n.	/11/1920	
				Big Brook	Wilson & Herbert	s.n.	/11/1920	
		Sandplain.		Near Busselton	Wilson & Herbert	s.n.	/11/1920	
				Hines Hill	Wilson and Herbert	s.n.	/11/1920	
		Sand.	Heaths.	Merredin	Wilson & Herbert	s.n.	/11/1920	
				Cannington	Wilson & Herbert	s.n.	/11/1920	
	Flowers pink.			Kalgoorlie	E.H. Wilson & C.A. Gardner	578	11/23/1920	
				Near Merredin	E.H. Wilson & D.A. Herbert	s.n.	/11/1920	
				Pemberton	E.H. Wilson & D.A. Herbert	s.n.	/11/1920	
				York	E.H. Wilson & D.A. Herbert	s.n.	/11/1920	
				Merredin	E.H. Wilson & D.A. Herbert	s.n.	/11/1920	
				Busselton	E.H. Wilson & D.A. Herbert	301	/11/1920	

Sheet Number	Family	Genus	Species	Rank	Infraspecies	Author	Manuscript	Determiner Name
3292355	Gyrostemonaceae	Gyrostemon	racemiger			H.Walter		A.S. George
3293939	Gyrostemonaceae	Gyrostemon	subnudus			(Nees) Baill.		C.A. Gardner
3470660	Haloragaceae	Glischrocaryon	angustifolium			(Nees) M.L.Moody & Les		
3652521	Lamiaceae	Cyanostegia	angustifolia			Turcz.		
2013983	Lamiaceae	Hemiandra	pungens			R.Br.		C.A. Gardner
3733459	Lamiaceae	Westringia	rigida			R.Br.		B.J. Conn
1574140	Loganiaceae	Logania	tortuosa			D.A.Herb.		B.J. Conn
1599402	Loganiaceae	Logania	tortuosa			D.A.Herb.		B.J. Conn
1599399	Loganiaceae	Logania	tortuosa			D.A.Herb.		B.J. Conn
2280787	Malvaceae	Brachychiton	gregorii			F.Muell.		G. Guymer
1046101	Malvaceae	Lasiopetalum	membranaceum			(Steud.) Benth.		K. Shepherd
2702142	Malvaceae	Seringia	velutina			(Steetz) F.Muell.		C.F. Wilkins
2706520	Malvaceae	Thomasia	grandiflora			Lindl.		K. Shepherd
2932628	Malvaceae	Thomasia	paniculata			Lindl.		K. Shepherd
2708019	Malvaceae	Thomasia	purpurea			(Aiton) Gay		C.A. Gardner
2708027	Malvaceae	Thomasia	purpurea			(Aiton) Gay		W.E. Blackall
2709384	Malvaceae	Thomasia	sp. Vasse (C. Wilkins & K. Shepherd CW 581)				PN	K. Shepherd
2709414	Malvaceae	Thomasia	sp. Vasse (C. Wilkins & K. Shepherd CW 581)				PN	K. Shepherd
3416526	Myrtaceae	Baeckea	sp. Koonadgin (B.L. Rye & M.E. Trudgen BLR 241137)				PN	
2284790	Myrtaceae	Beaufortia	bracteosa			Diels		
2285894	Myrtaceae	Beaufortia	decussata			R.Br.		
2288869	Myrtaceae	Beaufortia	macrostemon			Lindl.		
2328933	Myrtaceae	Calothamnus	quadrifidus	subsp.	asper	(Turcz.) A.S.George & N.Gibson		A.S. George
3570835	Myrtaceae	Calytrix	leschenaultii			(Schauer) Benth.		L.A. Craven
3521222	Myrtaceae	Calytrix	strigosa			A.Cunn.		L.A. Craven
1630229	Myrtaceae	Chamelaucium	pauciflorum	subsp.	Perenjori (B.J. Conn 2181)		PN	
1098861	Myrtaceae	Corymbia	ficifolia			(F.Muell.) K.D.Hill & L.A.S.Johnson		N. Robinson

Name Comment	Plant Description	Site Description	Vegetation	Locality	Collector	Collector's Number	Collection Date	Type Status
		Sandplain.		Westonia	E.H. Wilson & D.A. Herbert	102	/11/1920	
				Merredin	Wilson & Herbert	124	/11/1920	
				Merredin	Wilson and Herbert	s.n.	/11/1920	
				Westonia	Wilson & Herbert	57	/11/1920	
				North Dandalup	Wilson & Herbert	s.n.	/11/1920	
				Yoting	Wilson & Herbert	s.n.	/11/1920	
				Yoting	Herbert & Wilson	162	/11/1920	ISO-LECTO
				Yoting	Wilson & Herbert	s.n.	/11/1920	LECTO
Presumably not type material. B.J. Conn 01/1992.	Flowers bluish-white.			Yoting	Wilson & Herbert	596	/11/1920	
				8 miles S of Widgiemooltha	E.H. Wilson & C.A. Gardner	CAG 1055	11/22/1920	
				Capel	E.H. Wilson & D.A. Herbert	s.n.	/11/1920	
				Bruce Rock	Herbert & Wilson	145	/11/1920	
				Busselton	Wilson & Herbert	s.n.	/11/1920	
				Frankland River	E.H. Wilson & D.A. Herbert	s.n.	/11/1920	
				Albany	E.H. Wilson & D.A. Herbert	s.n.	/11/1920	
				Albany	E.H. Wilson & D.A. Herbert	s.n.	/11/1920	
				Big Brook, Pemberton	Wilson & Herbert	s.n.	/11/1920	
				Big Brook [Pemberton]	Wilson & Herbert	s.n.	/11/1920	
			Sandheaths.	Westonia	Wilson and Herbert	72	/11/1920	
				Yoting	E.H. Wilson & D.A. Herbert	s.n.	/11/1920	
				W of Denmark	Wilson & Herbert	s.n.	/11/1920	
				North Dandalup	E.H. Wilson	s.n.	/11/1920	
				Cultivated: Dr Robinson's garden	Wilson & Herbert	491	/11/1920	
				Cunderdin	Wilson & Herbert	s.n.	/11/1920	
				Yoting	Wilson & Herbert	s.n.	/11/1920	
		Sandplain.		Westonia	Wilson & Herbert	110	/11/1920	TYPE
				Frankland River	E.H. Wilson & D.A. Herbert	s.n.	/11/1920	

Sheet Number	Family	Genus	Species	Rank	Infraspecies	Author	Manuscript	Determiner Name
1375342	Myrtaceae	Corymbia	haematoxylon			(Maiden) K.D.Hill & L.A.S.Johnson		N. Robinson
3363457	Myrtaceae	Ericomyrtus	tenuior			(Ewart) Rye		B.L. Rye
1326236	Myrtaceae	Eucalyptus	campaspe			S.Moore		M.I.H. Brooker
1337467	Myrtaceae	Eucalyptus	diversicolor			F.Muell.		M.I.H. Brooker
1063537	Myrtaceae	Eucalyptus	guilfoylei			Maiden		M.E. French
1522949	Myrtaceae	Eucalyptus	rudis			Endl.		M.I.H. Brooker
1444239	Myrtaceae	Eucalyptus	sargentii	subsp.	sargentii	Maiden		
3563871	Myrtaceae	Euryomyrtus	leptospermoides			(C.A.Gardner) Trudgen		M. Trudgen
2151650	Myrtaceae	Homalospermum	firmum			Schauer		J. Thompson
2352095	Myrtaceae	Hypocalymma	robustum			(Endl.) Lindl.		C. Parker
2416433	Myrtaceae	Kunzea	micrantha	subsp.	micrantha	Schauer		N. Robinson
9057978	Myrtaceae	Kunzea	micrantha	subsp.	micrantha	Schauer		H.R. Toelken
1557238	Myrtaceae	Melaleuca	carrii			Craven		L.A. Craven
3528553	Myrtaceae	Melaleuca	cuticularis			Labill.		B.A. Barlow
3529134	Myrtaceae	Melaleuca	densa			R.Br.		B.A. Barlow
3656039	Myrtaceae	Melaleuca	hamata			Fielding & Gardner		L.A. Craven & B.J. Lepschi
3600033	Myrtaceae	Melaleuca	incana	subsp.	incana	R.Br.		B.A. Barlow & F.C. Quinn
3601943	Myrtaceae	Melaleuca	lateritia			A.Dietr.		K.J. Cowley
1638777	Myrtaceae	Melaleuca	marginata			(Sond.) Hislop, Lepschi & Craven		M. Hislop et al.
1572865	Myrtaceae	Melaleuca	parviceps			Lindl.		L.A. Craven
3605612	Myrtaceae	Melaleuca	platycalyx			Diels		K.J. Cowley
3608794	Myrtaceae	Melaleuca	rhaphiophylla			Schauer		B.J. Lepschi
3596966	Myrtaceae	Melaleuca	thymoides			Labill.		B.A. Barlow
3653714	Myrtaceae	Melaleuca	thymoides			Labill.		B.A. Barlow
3653765	Myrtaceae	Melaleuca	thymoides			Labill.		B.A. Barlow
4509692	Myrtaceae	Rinzia	schollerifolia			(Lehm.) Trudgen		M. Trudgen
1133896	Myrtaceae	Scholtzia	eatoniana			(Ewart & Jean White) C.A.Gardner		
1133918	Myrtaceae	Scholtzia	eatoniana			(Ewart & Jean White) C.A.Gardner		
3885941	Myrtaceae	Taxandria	linearifolia			(DC.) J.R.Wheeler & N.G.Marchant		J.R. Wheeler
1896083	Myrtaceae	Verticordia	acerosa	var.	preissii	(Schauer) A.S.George		E.A. George

Name Comment	Plant Description	Site Description	Vegetation	Locality	Collector	Collector's Number	Collection Date	Type Status
				Near Jarrahwood	Wilson & Herbert	s.n.	/11/1920	
				Yoting	E.H. Wilson & D.A. Herbert	178	/11/1920	
	Young tree			Not given	C.A. Gardner & E.H. Wilson	543	11/19/1920	
				Frankland River	E.H. Wilson & D.A. Herbert	s.n.	/11/1920	
	Up to 100 ft.			Frankland River	E.H. Wilson & D.A. Herbert	s.n.	/11/1920	
				3 miles E of Northam	Wilson & Herbert	s.n.	/11/1920	
				Cunderdin	E.H. Wilson & D.A. Herbert	s.n.	/11/1920	
				Merredin	Wilson & Herbert	s.n.	/11/1920	
				Albany	Wilson & Herbert	s.n.	/11/1920	
Cited by Strid & Keighery in Nordic J.Bot.22:549(2002). C. Parker 30/6/2004.				Big Brook, Pemberton	Wilson & Herbert	s.n.	/01/1920	
Cited as Kunzea micrantha Schauer subsp. micrantha by Toelken in J.Ad.Bot. Gard. 17:73(1996)				North Dandalup	E.H. Wilson & D.A. Herbert	s.n.	/11/1920	
				North Dandalup	Wilson & Herbert	s.n.	/11/1920	
				Cunderdin,	Wilson & Herbert	23	/11/1920	
				Frankland River	Wilson & Herbert	s.n.	/11/1920	
				Albany	Wilson and Herbert	s.n.	/11/1920	
				Cunderdin	Wilson & Herbert	s.n.	/11/1920	
				Albany	E.H. Wilson & D.A. Herbert	s.n.	/11/1920	
				Waterloo	Wilson & Herbert	s.n.	/11/1920	
Cited by M. Hislop et al. in Nuytsia 21: 153(2011)	Shrub 4 ft.			Bruce Rock	E.H. Wilson & D.A. Herbert	DAH 152	/11/1920	HOLO
				Mundaring,	Wilson & Herbert	475	/11/1920	
				Bruce Rock	Wilson & Herbert	s.n.	/11/1920	
		Banks of Avon River.		Toodyay	Wilson & Herbert	s.n.	/11/1920	
				Near Bunbury	Wilson and Herbert	s.n.	/11/1920	
				Busselton	Wilson and Herbert	s.n.	/11/1920	
				Albany	Wilson and Herbert	s.n.	/11/1920	
				Albany	Wilson & Herbert	337	/11/1920	
	Diffuse shrub; flowers pink.			Cunderdin.	E.H. Wilson & D.A. Herbert	24	/11/1920	
				Cunderdin.	E.H. Wilson & D.A. Herbert	s.n.	/11/1920	
				Mundaring	Wilson & Herbert	464	/11/1920	
				Mundaring	Wilson & Herbert	s.n.	/11/1920	

Sheet Number	Family	Genus	Species	Rank	Infraspecies	Author	Manuscript	Determiner Name
2063808	Myrtaceae	Verticordia	chrysantha			Endl.		A.S. George
2063794	Myrtaceae	Verticordia	chrysanthella			A.S.George		E.A. George
1638432	Pittosporaceae	Billardiera	fusiformis			Labill.		
3385760	Pittosporaceae	Cheiranthera	filifolia			Turcz.		
3382710	Pittosporaceae	Marianthus	erubescens			Putt.		
2432587	Podocarpaceae	Podocarpus	drouynianus			F.Muell.		
4040643	Polygalaceae	Comesperma	scoparium			J.Drumm.		
4040899	Polygalaceae	Comesperma	spinosum			F.Muell.		
4091957	Proteaceae	Acidonia	microcarpa			(R.Br.) L.A.S.Johnson & B.G.Briggs		T.R. Lally
1720368	Proteaceae	Adenanthos	meisneri			Lehm.		E. Charles Nelson
1720724	Proteaceae	Adenanthos	meisneri			Lehm.		E. Charles Nelson
1721720	Proteaceae	Adenanthos	obovatus			Labill.		P.G. Wilson
1715976	Proteaceae	Adenanthos	sp. Whicher Range (G.J. Keighery 9736)				PN	
1722638	Proteaceae	Banksia	attenuata			R.Br.		A.S. George
1801724	Proteaceae	Grevillea	excelsior			Diels		T.R. Lally
2357577	Proteaceae	Grevillea	hookeriana subsp. apiciloba / hookeriana subsp. digitata					
2357453	Proteaceae	Grevillea	hookeriana subsp. apiciloba / hookeriana subsp. digitata					
2072831	Proteaceae	Grevillea	integrifolia			(Endl.) Meisn.		T.R. Lally
1850881	Proteaceae	Grevillea	paniculata			Meisn.		D.J. McGillivray
2415380	Proteaceae	Grevillea	pilulifera			(Lindl.) Druce		D.J. McGillivray
1839136	Proteaceae	Grevillea	pterosperma			F.Muell.		D.J. McGillivray
6158951	Proteaceae	Hakea	ceratophylla			(Sm.) R.Br.		R.M. Barker
6159850	Proteaceae	Hakea	incrassata			R.Br.		R.M. Barker
1891278	Proteaceae	Hakea	meisneriana			Kippist		L. Haegi
6276210	Proteaceae	Hakea	meisneriana			Kippist		L. Haegi
3363627	Proteaceae	Isopogon	attenuatus			R.Br.		T.R. Lally

Name Comment	Plant Description	Site Description	Vegetation	Locality	Collector	Collector's Number	Collection Date	Type Status
		Sandplain.		Merredin	Wilson & Herbert	s.n.	/11/1920	
		Sandplain.		Merredin	Wilson & Herbert	s.n.	/11/1920	
				King River, Albany	Wilson & Herbert	s.n.	/11/1920	
		Sandplain.		Westonia	E.H. Wilson & D.A. Herbert	195	/11/1920	
				Merredin	Wilson & Herbert	s.n.	/11/1920	
				Fly Brook	E.H. Wilson & D.A. Herbert	263	/11/1920	
				Yoting	Wilson & Herbert	s.n.	/11/1920	
		Sandplain.		Merredin	Wilson & Herbert	s.n.	/11/1920	
				Albany	Wilson and Herbert	361	/11/1920	
				Busselton	E.H. Wilson & D.A. Herbert	s.n.	/11/1920	
				Bunbury	Wilson & Herbert	s.n.	/11/1920	
				Busselton	Wilson & Herbert	s.n.	/11/1920	
				Busselton	E.H. Wilson & D.A. Herbert	s.n.	/11/1920	
				Harvey	Wilson & Herbert	s.n.	/11/1920	
				Tammin	Wilson & Herbert	s.n.	/11/1920	
	Shrub 2 - 3 ft high.	Sand.	Heath.	Westonia	Wilson & Herbert	s.n.	/11/1920	
				Westonia	Wilson & Herbert	s.n.	/11/1920	
				Tammin	E.H. Wilson & D.A. Herbert	s.n.	/11/1920	
				Cunderdin	Wilson & Herbert	s.n.	/11/1920	
				Mundaring	E.H. Wilson & D.A. Herbert	s.n.	/11/1920	
				Westonia	Wilson & Herbert	101	/11/1920	
				Waterloo	Wilson & Herbert	s.n.	/03/1920	
		Sandplain.		Yoting	E.H. Wilson & D.A. Herbert	s.n.	/03/1920	
		Sandplain.		Westonia	Wilson & Herbert	s.n.	/11/1920	
				Westonia	Wilson and Herbert	s.n.	/11/1920	
				Waterloo, near Brunswick	E.H. Wilson & D.A. Herbert	237	/11/1920	

Sheet Number	Family	Genus	Species	Rank	Infraspecies	Author	Manuscript	Determiner Name
1901990	Proteaceae	Isopogon	divergens			R.Br.		
3419118	Proteaceae	Isopogon	sphaerocephalus			Lindl.		D.B. Foreman
1924761	Proteaceae	Lambertia	multiflora	var.	darlingensis	Hnatiuk		J.R. Wheeler
4092163	Proteaceae	Persoonia	angustiflora			Benth.		P. Weston
4091418	Proteaceae	Persoonia	angustiflora			Benth.		
4093038	Proteaceae	Persoonia	elliptica			R.Br.		P.H. Weston
4173570	Proteaceae	Persoonia	quinquenervis			Hook.		P. Weston
4173805	Proteaceae	Persoonia	saccata			R.Br.		P. Weston
4173929	Proteaceae	Persoonia	saundersiana			Meisn.		P. Weston
1920308	Proteaceae	Petrophile	brevifolia			Lindl.		D.B. Foreman
3431479	Proteaceae	Petrophile	juncifolia			Lindl.		B.L. Rye
1858386	Proteaceae	Petrophile	serruriae			R.Br.		D.B. Foreman
1515772	Rhamnaceae	Cryptandra	apetala	var.	anomala	Rye		F. Udovicic
1519573	Rhamnaceae	Trymalium	odoratissimum	subsp.	trifidum	(Rye) Kellermann, Rye & K.R.Thiele		
948551	Rutaceae	Boronia	crenulata			Sm.		
946400	Rutaceae	Boronia	crenulata	subsp.	pubescens	(Benth.) Paul G.Wilson		Paul G. Wilson
947091	Rutaceae	Boronia	gracilipes			F.Muell.		
948667	Rutaceae	Boronia	heterophylla			F.Muell.		
963984	Rutaceae	Boronia	molloyae			J.Drumm.		
950106	Rutaceae	Chorilaena	quercifolia			Endl.		
956511	Rutaceae	Crowea	angustifolia	var.	platyphylla	Benth.		
1616560	Rutaceae	Microcybe	ambigua			(C.A.Gardner) Paul G.Wilson		Paul G. Wilson
903728	Rutaceae	Philotheca	spicata			(A.Rich.) Paul G.Wilson		
902691	Rutaceae	Philotheca	spicata			(A.Rich.) Paul G.Wilson		
907375	Rutaceae	Philotheca	thryptomenoides			(S.Moore) Paul G.Wilson		
2441586	Santalaceae	Leptomeria	cunninghamii			Miq.		B.J. Lepschi
2441926	Santalaceae	Leptomeria	cunninghamii			Miq.		B.J. Lepschi
2444615	Santalaceae	Leptomeria	pauciflora			R.Br.		B.J. Lepschi

Name Comment	Plant Description	Site Description	Vegetation	Locality	Collector	Collector's Number	Collection Date	Type Status
				Bruce Rock	C.A. Gardner [Wilson & Herbert]	s.n.	/11/1920	
				Big Brook, Pemberton	E.H. Wilson & D.A. Herbert	259	/11/1920	
				North Dandalup	E.H. Wilson & D.A. Herbert	s.n.	/11/1920	
Possible Type of Persoonia angustiflora var. burracoppinensis D.A.Herb. See P.H. Weston in Fl. Australia 16: 112 (1995)				Burracoppin	E.H. Wilson & D.A. Herbert	s.n.	/11/1920	TYPE?
Possible Type of Persoonia angustiflora var. burracoppinensis D.A.Herb. See P.H. Weston in Fl. Australia 16: 112 (1995)	Flowers yellow.			Burracoppin	Wilson and Herbert	0.458	/11/1920	ISO
	Shrub 10 ft high.			Albany	D.A. Herbert & E.H. Wilson	334	/11/1920	
				York	E.H. Wilson & D.A. Herbert	210	/11/1920	
				Naval Base, Rockingham	E.H. Wilson & D.A. Herbert	504	/11/1920	
				Burracoppin	Wilson & Herbert	s.n.	/11/1920	
		Sanplain.		Quairading	E.H. Wilson & D.A. Herbert	s.n.	/11/1920	
				30 miles E of Perth at junction: The Lakes	Wilson and Herbert	s.n.	/11/1920	
				North Dandalup	Wilson & Herbert	s.n.	/11/1920	
				Kwolyin	E.H. Wilson & D.A. Herbert	157	/11/1920	
				Donnelly River	E.H. Wilson & D.A. Herbert	s.n.	/11/1920	
				Albany	Wilson & Herbert	s.n.	/11/1920	
	Shrub 2 ft.			Pemberton.	Wilson & Herbert	243	/11/1920	
				Frankland River.	Wilson & Herbert	s.n.	/11/1920	
				Albany.	Wilson & Herbert	s.n.	/11/1920	
				Frankland River.	Wilson & Herbert	377	/11/1920	
				Big Brook	Wilson & Herbert	249	/11/1920	
				Big Brook,	Wilson & Herbert	s.n.	/11/1920	
Cited by Paul G. Wilson in Fl. Australia 26:484 (2013)		Sandplain.		Westonia, between Merredin and Southern Cross	E.H. Wilson & D.A. Herbert	711	/11/1920	HOLO
				Busselton	Wilson & Herbert	288	/11/1920	
				Sabina River	Wilson & Herbert	s.n.	/11/1920	
		Sand.	Heath.	Westonia	Wilson & Herbert	66	/11/1920	
				Mundaring	E.H. Wilson	s.n.	/10/1920	
				Mundaring	E.H. Wilson	s.n.	/10/1920	
		Sand.	Sandheath.	Yoting	E.H. Wilson & D.A. Herbert	182	/11/1920	

Sheet Number	Family	Genus	Species	Rank	Infraspecies	Author	Manuscript	Determiner Name
2443597	Santalaceae	Leptomeria	preissiana			(Miq.) A.DC.		B.J. Lepschi
2750619	Sapindaceae	Dodonaea	adenophora			Miq.		J.G. West
2750651	Sapindaceae	Dodonaea	adenophora			Miq.		J.G. West
3796450	Scrophulariaceae	Eremophila	decipiens	subsp.	decipiens	Ostenf.		R.J. Chinnock
3162621	Stylidiaceae	Stylidium	scandens			R.Br.		
3404625	Thymelaeaceae	Pimelea	angustifolia			R.Br.		B.L. Rye
3404870	Thymelaeaceae	Pimelea	angustifolia			R.Br.		B.L. Rye
3406857	Thymelaeaceae	Pimelea	argentea			R.Br.		
3409449	Thymelaeaceae	Pimelea	ciliata	subsp.	ciliata	Rye		B.L. Rye
3409821	Thymelaeaceae	Pimelea	clavata			Labill.		
3409805	Thymelaeaceae	Pimelea	clavata			Labill.		
3444821	Thymelaeaceae	Pimelea	hispida			R.Br.		
3446352	Thymelaeaceae	Pimelea	imbricata	var.	piligera	(Benth.) Diels		B.L. Rye
3445720	Thymelaeaceae	Pimelea	imbricata	var.	piligera	(Benth.) Diels		B.L. Rye
3446964	Thymelaeaceae	Pimelea	imbricata	var.	piligera	(Benth.) Diels		B.L. Rye
3448940	Thymelaeaceae	Pimelea	longiflora			R.Br.		
3451623	Thymelaeaceae	Pimelea	rosea	subsp.	rosea	R.Br.		B.L. Rye
3473880	Thymelaeaceae	Pimelea	spectabilis			Lindl.		
2021080	Xanthorrhoeaceae	Xanthorrhoea	gracilis			Endl.		D.J. Bedford
2021374	Xanthorrhoeaceae	Xanthorrhoea	nana			D.A.Herb.		D.J. Bedford

Name Comment	Plant Description	Site Description	Vegetation	Locality	Collector	Collector's Number	Collection Date	Type Status
	Shrub 4 ft high. Parasite.	Sandplain.	On Acacia signata.	Westonia	Herbert & Wilson	61	10/23/1920	
				Merredin	Wilson and Herbert	s.n.	/11/1920	
				Merredin	Wilson and Herbert	125	/11/1920	
				Westonia	Wilson & Herbert	60	/11/1920	
				Albany	Wilson and Herbert	355	/11/1920	
				Bruce Rock	Wilson & Herbert	168	/11/1920	
		Sandplain.		Merredin	Wilson & Herbert	38	/11/1920	
				Toodyay	Wilson & Herbert	s.n.	/11/1920	
				Mundaring	Wilson & Herbert	455	/11/1920	
				Pemberton	Wilson & Herbert	255	/11/1920	
				Big Brook	Wilson & Herbert	s.n.	/11/1920	
				Albany	Wilson & Herbert	339	/11/1920	
	Female.	Sandplain.		Merredin	Wilson & Herbert	s.n.	/11/1920	
				Bruce Rock	Wilson & Herbert	166	/11/1920	
				Bunbury	Wilson & Herbert	315	/11/1920	
See C.S.P. Foster et al. in Austral.Syst. Bot. 29:194 (2016)				Albany	Wilson & Herbert	368	11/6/1920	
				Capel	Wilson & Herbert	37	/11/1920	
				Big Brook	Wilson & Herbert	s.n.	/11/1920	
				"The Lakes" York Road	E.H. Wilson & D.A. Herbert	s.n.	/11/1920	
				Bruce Rock	E.H. Wilson & D.A. Herbert	s.n.	/11/1920	

SELECTED BIBLIOGRAPHY

Detailed references are found in each chapter's endnotes. Included here are more general books, and lists of those most relevant to the study of Wilson and his trip to Australia that were examined; some are referred to in the main text.

ON WILSON

RW Briggs, *'Chinese' Wilson, A life of Ernest H. Wilson 1876–1930*, (The great plant collectors), London, HMSO (Her Majesty's Stationery Office), 1993.

KS Clausen and SY Hu, 'Mapping the collecting localities of E. H. Wilson in China', *Arnoldia*, 1980, 40(3):139–45.

S Dietrich and MR Dietrich, 'Ernest 'Chinese' Wilson's re-imagined legacy in Sichuan', *Circulation*, 2019, 9(2), http://hdl.handle.net/2027/spo.7977573.0009.209.

M Flanagan and T Kirkham, *Wilson's China, A century on*, Kew, London, 2009.

M Grose, 'Searching for Wilson's expedition to Australia', *Arnoldia*, 2019, 76(4):2–13.

RA Howard, 'E. H. Wilson as a botanist', part I, *Arnoldia*, 1980, 40(3):102–38.

RA Howard, E. H. Wilson as a botanist', part II, *Arnoldia*, 1980, 40(4):154–93.

R Wu, Y Zou, S Liao, K Shi, X Nan, H Yan, J Luo, Z Xiang, and Z Bao, 'Shall we promote natural history collection today?—Answered by reviewing Ernest Henry Wilson's plant collection process in China', *Science of the Total Environment*, 2024, 915:170179.

K Yin, *Tracing one hundred years of change: Illustrating the environmental changes in Western China*, Encyclopedia of China Publishing House, Beijing, 2010.

Y Zou, K Shi, S Liao, Z Xiang, J Luo, X Nan, H Yan, Z Bao, W Nie, and RA Wu, 'Survey and analysis of the history of Ernest Henry Wilson's specimen collections in China', *Forests*, 2024, 15(3):475.

BY WILSON (with date of first publication)

EH Wilson, *A naturalist in western China with vasculum, camera and gun: Being some account of eleven years' travel*, vol. 1, Cambridge University Press, 1913.

EH Wilson, *The romance of our trees*, Kessinger Publishing, Montana, 1920.

EH Wilson (with Alfred Rehder), *Monograph of azaleas:* Rhododendron *subgenus* Anthodendron, 1921.

EH Wilson, *Lilies of eastern Asia*: *A monograph*, 1925.

EH Wilson, *Plant hunting*, Stratford Company, Boston, 1927.

EH Wilson, *China, mother of gardens*, Stratford Company, Boston, 1929.

EH Wilson, *Aristocrats of the trees*, Stratford Company, Boston, 1930.

EH Wilson, *Smoke that thunders*, Waterstone, London, 1985. First published as *Plant hunting*, vol. 1.

ON AUSTRALIAN PLANTS

I Abbott, *Aboriginal names for plant species in South-western Australia*, Forests Department of Western Australia, technical paper no. 5, 1983.

DJ Boland, MIH Brooker, GM Chippendale, N Hall, and BPM Hyland, *Forest trees of Australia*, 5th edn, CSIRO, Collingwood, 2006.

MIH Brooker and DA Kleinig, *Field guide to Eucalypts: Volume 2, South-western and Southern Australia*, 2nd edn, Melbourne, Bloomings Books, 2001.

MIH Brooker and DA Kleinig, *Field guide to Eucalypts: Volume 1, South-eastern Australia*, 3rd edn, Melbourne, Bloomings Books, 2006.

MIH Brooker and DA Kleinig, *Field Guide to Eucalypts: Volume 3, Northern Australia*, 2nd edn, Melbourne, Bloomings Books, 2004.

R Broom, C Fahey, A Gaynor, and K Holmes, *Mallee Country: Land, people, history*, Monash University Publishing, Clayton, 2020.

GM Chippendale, *Eucalypts of the Western Australian Goldfields (and adjacent Wheatbelt)*, Australian Government Publishing Service, Canberra, 1973.

L Diels, *Die Pflanzenwelt von West-Australien sudlich des Wendekreises: Mit einer Einleitung uber die Pflanzenwelt Gesamt-Australiens in Grundzugen*, Verlag Von Wilhelm Engelmann, Leipzig, 1906.

A Gaynor and T Griffiths, 'Tearing down and building up', 2017, see https://insidestory.org.au/tearing-down-and-building-up/.

BJ Keighery and VM Longman, *Tuart (Eucalyptus gomphocephala) and Tuart communities*, Wildflower Society of Western Australia, Nedlands, 2002.

DA Keith, *Australian vegetation*, 3rd edn, Cambridge University Press, Cambridge, 2017.

SL Kessell and CA Gardner, *Key to the Eucalypts of Western Australia*, Government Printer, Perth, 1924.

ON AUSTRALIA: ENVIRONMENT, FORESTRY, CULTURE, AND HISTORY

P Bianchi, *Woodlines of Western Australia: A comprehensive history of the Goldfields woodlines,* Hesperian Press, Carlisle, 2019.

G Bolton, *Spoils and spoilers: Australians make their environment 1788–1980*, George Allen & Unwin, Sydney, 1981.

P Bridge, et al., *To the golden land: Exploration to the eastwards 1869–1896*, Hesperian Press, Carlisle, 2019.

B Bunbury, *Timber for gold: Life on the Goldfields woodlines, 1899–1965*, Fremantle Arts Centre Press, Fremantle, 1997.

B Bunbury, *Invisible Country, Southwest Australia: Understanding a landscape*, University of Western Australia, Crawley, 2015.

I Cunningham, *The trees that were nature's gift*, Cunningham, Maylands, 1998.

J Dargavel, *The zealous conservator: A life of Charles Lane Poole*, University of Western Australia Press, Crawley, 2008.

WH Douglas, *The Aboriginal languages of the south-west of Australia*, 2nd edn, Australian Aboriginal Studies Research and Regional Studies no. 9, 1976.

A Gaynor, 'Looking forward, looking back: Towards an environmental history of salinity and erosion in the eastern Wheatbelt of Western Australia', in *Country: Visions of land and people in Western Australia*, ed. A Gaynor, A Haebich, and M Trinca, WA Museum, Perth, 2002, pp. 105–24.

A Gaynor and K Bradby, 'Soil salinity in Australia: A slow motion crisis', *Australian Policy and History*, 2018.

E Giles, *Ernest Giles's explorations, 1872–76*, Friends of the State Library of South Australia, Adelaide, 2000.

T Griffiths, 'How many trees make a forest? Cultural debates about vegetation change in Australia', *Australian Journal of Botany*, 2002, 50(4):375–89.

M Hercock (ed.), *The Western Australian explorations of John Septimus Roe 1829–1849*, Hesperian Press, Carlisle, 2014.

M Hercock, S Milentis, P Bianchi (ed.), *Western Australian exploration 1836 – 1845: The letters, reports and journals of exploration and discovery in Western Australia*, Hesperian Press, Carlisle, 2011.

T Hughes-d'Aeth, *Like nothing on this Earth: A literary history of the Wheatbelt*, University of Western Australia Publishing, Perth, 2017.

B York Main, *Between Wodjil and Tor*, Jacaranda Press, 1967.

E Rolls, *A million wild acres: 200 years of man and an Australian forest*, Thomas Nelson, Melbourne, 1981.

J Shoobert, et al. (ed.), *Western Australian exploration 1826–1835, volume 1*, Hesperian Press, Carlisle, 2005.

G Seddon, *Sense of place: A response to an environment, the Swan Coastal Plain, Western Australia*, University of Western Australia Publishing, Crawley, 1972.

T Smith and L Southwell, *Incredible Fraser Island*, Australian Conservation Foundation, Melbourne, 1975.

INDEX

Note: Bold page numbers refer to illustrations.

Abies procera 146
Aboriginal languages, Kalgoorlie Goldfields 195
Aboriginal people
  at mia mia **193**
  bush medicine 83, 162–3
  bush tucker 203
  care for Country 215
  hunting tools **193**
  inherited knowledge 203–4
  natural resource use 42, 92
  place names 192–4
  records of megafauna 93
  sacred trees 40–1, 259
  see also Noongar
Aboriginal plant names 29, 30, 45, 141, 259
Acacia 53, 73, 90, 95, 210, 211, 235
  acuminata 83, **84**
  aneura 204, 205
  bark types 206
  citrinoviridis 209
  cyclops 148
  drummondii 300
  grasbyi 205
  harpophylla 263
  kempeana 203
  microbotrya 97–8
  papyrocarpa 204, 205
  pruinocarpa 205
acacia woodlands 204
Acidonia macrocarpa **179**
Adelaida 216, 233, 234, 235
Adelaide Hills 159, 235
Adenanthos 95
  obovatus **178**
afforestation, SA 235–6
Africa 52
Agathis 259
  robusta 254, 258
Agonis 148
  flexuosa 143
agricultural areas, loss of trees 252–3
agroforestry 236, 243
Albany 141, 146, 147, 314
Albany bottlebrush **170**
Allen, C. E. F. 314

Allocasuarina 73
  fraseriana **166**, 209
  huegelliana **74**
  torulosa 159, 238
America
  loss of trees 283
  rise of National Parks 289–90
  slow to protect native birds 223
American beech 216, 218
ancient flora 38, 46, 47, 53
Andersonia 148
angle of leaf repose (Australian flora) 33, 34
Angophora 144
  lanceolata var. *angustifolia* 350
Antarctic Circumpolar Current 53
Antarctic landmass 52, 53
  flora 53, 157, 246
Anthocercis
  anisantha **133**
  littorea **57**
Anzac Day 244
Aquila audax 87
Araucaria 254, 258
  bidwillii 244, **256**, 258
  cunninghamii 244, 258, **260**, **264**, 300
  excelsa 244
Araucariaceae 258–9, 261
  distribution 258, 259
"arboricidal mania" 252, 262, 263, 284, 291, 301
Argyle apple 17
arid zone 204, 215
arid-zone timbers, research on 205–8
Aristocrats of the Trees (Wilson) 103, 301
Arnold Arboretum (Hunnewell Building, Jamaica Plain) 17, 25, 313
  author stays at Roslindale during her research 216
  author's research 9–10, 13–14, 17–18, 309

  can no longer collect live plants or seeds due to US quarantine regulations 2, 3, 299, 303
  climate prevents including southern hemisphere plants into Arboretum's living collection 3–4, 303
  designed by Frederick Olmsted 17
  diaries and newspaper clippings 9, 13, 16, 17, 20
  glass-plate negatives (WA) 13–14
  Government of Western Australia provides support for Expedition 216
  Indian tree seeds 4
  lack of eastern Australian images 13–14, 20–1
  New Zealand conifers 200
  New Zealand glass-plate negatives 20
  Queensland rainforest photos 13, 20
  reasons for and funding of Wilson's collecting expeditions 1, 2, 216, 299
  remnant collection from Expedition to Australasia, India and Africa of 1920-2 13–21
  Tasmanian photos 20
  time and commitment expected of Wilson in plant collecting 299–300
  'tree' definition 75
  see also Harvard Herbaria
artisanal salt farm 83
Ashford, Mr (Minister for Lands, NSW) 251
Asia collecting expeditions 1, 2, 4–5, 18
Astrida contorta 210
Atalaya hemiglauca 196
Atriplex spp. 83
Australasian Virtual Herbarium 365

Australia, as 'deforestation hotspot' 306
Australian flora
  adaptation to low nutrients 161–2
  differences from New Zealand flora 36, 51
  differences from Northern Hemisphere flora 2, 305
  diversity 305
  evolutionary history 33, 53–4, 311
  geological and climate history effects 50–5
  great need for handbook of 286
  importance of 261
  leaf features 33–4
  mycorrhizal associations 41–2
  similarities to southern African flora 36
  species richness and unique biology 50
  see also eucalypts
Australian forests, as heterogeneous mix 50
Australian landmass, as fragment of Gondwana 52–5
Australian landscapes 55
Avon Valley 111

Baandee 79, 80, 82, 83, 195
Baandee Lakes 79, 82, 82
Babakin 30
backward rivers 79
Baeckea 235
Bailey, J. F. 314
Baladjie Rock 97
Balaustion pulcherrimum 115
balga 45
Ballarat 243
Ballarat's Avenue of Honour 243
Banks, Joseph 300
Banksia 31, 36–7, 73, 88, 148, 257, 300
  aemula 258, 277
  archaeocarpa 38
  attenuata 144
  baueri 300
  bipinnatifida 129
  caleyi 300
  coccinea 148, 176
  formosa 177
  fossils 38
  goodii 300
  grandis 37, 56, 64, 114, 144
  integrifolia 255, 275
  latifolia 278
  leaf colour contrast 144
  littoralis 148, 175
  marginata 235, 274
  menziesii 37, 38, 39, 144, 300
  petiolaris 38
  resilience to hot conditions 39
  speciation 38
  spinulosa 276
Bates, Daisy 40
Bauer, Ferdinand 300
beaked hakea 273
Beattie, John Watt 20
Beaufort, Captain Francis 194
beefwood 206
Belches, Richard 75
Benarkin State Forest 254, 262, 264
Bencubbin 30
Bennetts, Dr H. W. 196
Big Brook 153, 163
biotic homogenisation 285–6
birds *see* native birds
Birdsville indigo 196
Black Flag Lake 187–8, 191, 191, 194, 198–9
black morrel 205
black mulga 209
black oak 206
black Sally 241
black swans 30
black tooth-brushes 14
blackbutt 238, 251, 254, 257
Bleazby, Robert 100–1
blue gum 153, 238, 251
Blue Mountains 259
bluebush 83
Bolton, Geoffrey 85
boree (borree) 225
Boronia 144, 148
  molloyae 180
bottle tree 254–5
Bow River 155
box poison 86, 95, 111
Brach, Anthony 14, 15, 309, 315
Brachychiton rupestris 254
Bribie Island 233
brigalow, clearing 263
Briggs, Roy 300
Brisbane 234, 253, 254, 295
Brisbane Field Naturalists' Club 261
broad-leafed messmate 16, 241
broad-leafed red gum 17
Brooks, Mr (settler) 92
broom bush 210
Broome 201
Brown, Robert 33, 300
brown stringybark 17
Bruce Rock 76, 88, 159
brush box 251, 257
Bulga Plateau 238
  destruction 250–2
bullich 153, 154
Bunbury 144, 208, 209
bunya bunya pine 244, 256, 258, 259
  cone 259, 261
Bunya Mountains 254, 256
Bunya Mountains National Park 259
Burracoppin 30, 76, 94
Burrill Creek 238
bush banana 193, 194, 211
bush flies 200–1, 221
bush medicine 83, 162–3
bush tomato 203
bush tucker 94, 98, 203
Busselton 142, 143, 144
button-grass plains 247, 248

cabbage poison 196
cabbage tree palm 239
Cairns 262
Caley, George 300
California
  forest trees suitable for 78, 98, 104, 143
  ornamental shrubs/trees suitable for 4, 6, 28, 37–8, 111
Callistemon 148
  homalophyllus 117
  speciosus 170
Callitris 95, 187, 195, 198–9, 211, 218
  evolution and adaptation 199
  glauca (syn. *columellaris*) 187–8, 197–8
  propinque 235

robusta 80
tuberculata (syn. *preissii* and *columellaris*) 199, 200
water-stress tolerant xylem 199
Calytrix stringosa 116
camel bush 211
Canastota, SS
    eastern Australia plant specimens sent from Brisbane but lost 18–19, 20, 233, 298
    leaking benzine delays departure from Brisbane 19, 20
    leaves from Sydney buts fails to arrive in Wellington 19
    Lloyds assumes the ship was lost 19–20
    New Zealand Navy fails to find debris 19
    WA specimens not on 165
Canberra, national forestry school 242–3
Cape Town 295, 296
carbon farming 107
Carlson, Rachel, *Silent Spring* 289
Cassia eremophila 211
Casuarina 53, 73, 218
    fraseriana 148
    glauca 159, 238
    pauper 206
    torulosa see *Allocasuarina torulosa*
Casuarinaceae 61, 158
caustic bush 196
Cedrela australis 239
centipede bush 14, 80
Chamaelaucium uncinatum 14
Champion Bay poison 196
Charles Gardner Reserve 98
Chengu Plain, China 249
Chilean flora 51
China
    forest loss/tree loss 249, 252, 283
    Wilson's expeditions 1, 5, 7, 18, 234, 303–4
China, Mother of Gardens (Wilson) 301
Chippendale, GM, *Eucalypts of the Western Australian*

Goldfields (and the adjacent wheatbelt) 205
Chorkerup 30
Christmas Day, Sydney 237
Cleland's blackbutt 205
climate fluctuations 162
coast ash 17, 241
coast banksia 275
coast redwood 146
Cobb & Co. travel 78–9
cockatoos 38, 81, 143
cogla vine 193, 194, 198, 199, 203, 211
Colbung, Ken 30
collecting disasters 18–19
Colombo, Sri Lanka 6
colour charts for plants 35–6
colour of foliage
    Australian flora 33–5, 36
    New Zealand plants 36
Comboyne Plateau 250
conifers 146, 199, 200, 244
Conospermum 147
continental drift 51–2, 159
Cook, Captain James 300
Coolgardie 79, 159, 188, 192, 193, 194, 211, 213, 217
Coolgardie Cemetery 217
Coopernook State Forest 238
coral gum 223
corkwood 206
Corrigin 30
Corsica 249
Corsican pine 243
Corymbia
    calophylla 30, 73
    ficifolia 148, 153, 153–4, 168, 283
Creswick 242
Cretaceous period 52, 161
crimson-flowered gum 155
Crotalaria retusa 196
Cryptandra leucophracta 196
Cunderdin 30, 76
Cunningham, Allen 300
Cupressaceae 199
cycads 48

dairying 163, 252
Dalrymple-Hay, Richard 237, 238–9, 257
Dampier, William 300
Dampiera 87

linearis 58
Darling Range ghost gum 165
Darling Scarp (Darling Range) 26, 31, 73, 165, 314
Darwin 296
Darwinia 235
Datonea 87
Davidia involucrata 1
Daviesia 90
    euphorbioides 14, 80
de Vlamingh, Willem 30, 48
Deakin, H.G. 305
deciduous native tree 55, 245–6
Denmark 141, 146, 148, 283, 314
Derbal Yerrigan 30
desert 215
Dicksonia antarctica 246, 246
Diels, Ludwig 213
Dietrich, Scott and Michael 13
dinosaurs 93
Diplolaena 95
    dampieri 67
Diprotodon optatum 93
Dodonaea 211, 235
Doodlakine 76, 78, 79, 80, 81, 194–5
    salt lake 82
Dorrigo Plateau 250
Douglas fir 146
Dowerin 30
Drummond, James 300
Dryandra 36, 148
    bipinnatifida see *Banksia bipinnatifida*
Du Toit, Alexander, *Wandering Continents: An Hypothesis on Continental Drifting* 51–2
Duboisia hopwoodii 95, 132
Durban 295
Dvorak, Monika 202, 203

eastern Australia, Wilson's travels and collecting in 233–80
eastern Australian flora
    consigned on SS *Canastota* but lost at sea 18–20, 233, 298
    diaries cover 16–17, 20
    lack of in Harvard Herbaria or Arnold Arboretum 13–14, 16, 20–1, 303
ecosystem recovery 107–8
ectomycorrhizal plants 42

INDEX   389

Edna-May Mine, Westonia 95
Ednie-Brown, John 290
Eighty Mile Beach 201
Endicott, Mr A.L. 299
Endoxyla leucomochla 203
England 295, 296
environmental concerns 300–1, 305–6
   see also "arboricidal mania"; forests; land clearing
environmental movement 289
Epacris 148
Eremophila 95, 211, 300
   brownii 211
   drummondii 211
   oldfieldii 206, 211
   scoparia 210
Eriostemon thrytomenoides 130
erosion 104
Esperance 201
eucalypt leaves
   angle of repose 34
   colour 35
eucalypt woodland and forest 204
eucalyptol 162
eucalypts 153
   adaptation to fire 154, 155, 159–61, 160
   description 145–6
   distinctive smell 162–3
   distribution 144–5
   diversity 16, 205
   evolution 157–9, 161–2
   fossils 157, 158
   fruits 205
   high-latitude 158
   longevity 146, 161, 245
   medicinal properties 162, 163
   tall trees 15–16, 144, 146
   taxonomy 144
Eucalyptus 15, 53, 98, 144, 235
   accedens 16, 73, 146
   acmenoides 238, 254
   alba 269
   argutifolia 161
   astringens 16, 75
   caesia 98
   subsp. *caesia* 99
   calophylla see Corymbia calophylla
   calycogona var. *gracilis* 211, 220
   campaspe 211, 219
   capillosa 73
   cariacea 241
   cinerea 17
   clelandii 205
   coriacea 17
   cornuta 29
   creba 16
   crucis subsp. *crucis* 95, 96, 98
   curtisii 161
   dalrymphona 241
   diversicolor 30, 148, 149, 150–1
   dives 16
   elaeophora 17
   erythronema 90, 91
   eugenioides 17, 267
   ewartiana 211
   ficifolia see Corymbia ficifolia
   flocktoniae 205, 211, 222
   gigantea 241
   globulus 146, 153
   gomphocephala 30, 73, 141–2, 143, 144, 209
   grandis 157, 257
   griffithsii 211
   guilfoyei 153
   haemastoma 17
   jacksonii 146, 148, 149, 150, 152
   laeliae 165
   lane-poolei 27
   lehmannii 28, 29
   leptospermum 17
   lesouefii 205, 211
   longicornis 205
   longifolia 17
   loxophleba 98, 104, 105
   subsp. *loxophleba* 73
   macarthurii 17
   macrocarpa 28–9, 29
   macrorhyncha 17
   maculata 238
   maculosa 16
   marginata 26, 30, 31, 32, 73, 148, 167
   megacarpa 29, 148, 153, 154
   melanoxylon 205
   melliodora 17
   microcorys 254
   obliqua 268
   oleosa 28, 161, 205, 209, 211
   var. *longicornis* 206
   paniculata 17, 238
   pauciflora 146, 241
   pellita 257
   phenotypic plasticity 159
   phylacis 161
   pilularis 238, 254
   piperita 16
   platypus 76
   punctata 17, 238
   pyriformis 59, 94–5
   radiata 16
   ravida 214
   regnans 15, 16, 146, 233, 245
   resinifera grandiflora 17
   rubrida 241
   rudis 60, 73
   salicola 205
   saligna 238
   salmonophloia 27, 73, 77, 205, 207–8
   salubris 81, 189, 189–90, 205, 207, 211
   sargentii 75, 110
   siderophloia 237, 254
   sieberiana 17
   smithii 17
   speciation 155, 159
   stelluate 241
   stricklandii 211, 224
   tereticornis 17
   terminology 75
   torquata 211, 223
   transcontinentalis 205, 211
   as very large genus 15
   viminalis 17, 146
   wandoo 30, 32, 73, 78, 209
   subsp. *wandoo* 147
   websteriana 221
eucalyptus–banksia woodlands 144
eucalyptus oil 162
Europe, forest loss 249
evolution
   Australian flora 33, 53–4, 311
   eucalypts 157–9, 161–2
   Southern Hemisphere plant families 157–8, 159
   see also Gondwana
Expedition to Australasia, India and Africa of 1920-2
   assistance provided to

Wilson in Australia 310
  see also Gardner; Herbert; Lane-Poole; Swain; White
collection remnants, Harvard 13–21, 298, 303, 309
  cost of 299
  eastern Australian material sent towards the Expedition due to loss of collection on *Canastota* 314
  expedition seen by Wilson as a failure? 303–4
  loss of eastern Australian material 18–20, 233, 298, 303
  purpose of expedition to Australasia 2–3, 6–8, 23–5, 234–5, 236, 299
  seeds collected sent to England due to US quarantine 299
extinct animals 215

Failford 238
Farrer, William 84–6, 107
Federation wheat –685
fence posts 266
feral animals control 112
feral cats 302
Finlay, William 196
fire
  eucalypts adaptation to 154, 155, 159–61, 160
  xanthorrhoeas adaptation to 47
flies 31, 200–1, 221
Flinders, Matthew 300
flooded gum 73, 257
Florence, Ross 159
flowering shrubs, extraordinary wealth of 33
flowers in vases 28, 29
fluoroacetate 111, 112
Ford Tourer 78, 196, 197
forest conservation/preservation 248, 250–1, 261, 265, 284, 287, 288, 290, 306
  men advocating for 289–90
forest trees, suitable for California 78, 98, 104
forestry, and agriculture 236, 243
forestry school

Creswick, Vic. 242
Ludlow, WA 142
national, Canberra 242–3, 285
NSW 242
forests
  clearing, Wilson's newspaper pleas about "arboricidal mania" 252, 262, 263, 284, 291, 301
  destructiveness, New Zealand 302
  gradual cutting down, Europe 249
  habit of destructiveness, Australia 283–92
  need for public education about value of 243, 283–5
  need to balance preservation with use for timber 290
  slow cutting down, China 249, 252
  Wilson appalled by loss of/ misuse of, Australia 16, 26, 163, 164, 236, 242, 248, 249–52, 262, 266, 302
  see also land clearing; timber; trees; woodlands
forests/forestry/forest management 242–3
  NSW 237, 238, 250
  political control 248, 265
  Qld 253, 261–6
  SA 235–6
  Tas. 244–8
  Vic. 248
  WA 26, 90, 187
Formosa (Taiwan) expeditions 1, 7, 237
Forrest, John 193
fossils
  Araucariacea 259
  banksias 38
  eucalypts 157, 157, 158
Frankland River 141, 148, 153, 154
Fraser Island 233, 254, 257–8, 260, 264
Fraxinus 243
Fremantle, Wilsons's arrival in 6, 23
French, Malcolm 98, 106

Eucalypts of Western Australia's Wheatbelt 98

Gardner, Charles Austin 195, 196, 303, 305, 310, 314, 365
  as botanist and watercolour artist 195–6
  collects with Wilson 195, 196, 204, 211, 214, 314
  The Toxic Plants of Western Australia 196
Gastrolobium 110, 111
  calycinum 110
  crassifolium 86
  glaucum 196
  melanocarpum 111
  oxylobioides 196
  parviflorum 86, 95
  propinquum 196
  retusum 111
  spinosum 196
  tomwntosum 196
Gawler 235, 236
Geelvink (ship) 48
Geraldton wax 14
German botanists, plant collecting in WA 213
giant mallee 205
giant redwood 146
gidgee 205
Giles, Ernest, explorer 217–18
gimlet 77, 78, 81, 95, 104, 189, 189–90, 199, 205, 207, 207, 211, 214
Gippsland 233
glaciation 55
glass-plate negatives 13, 20
  digitisation 13
gnamma holes 90
Gnowangerup 30
gold prospectors 192
Golden Pipeline Heritage Trail 79
Goldfields 188–231
Goldfields blackbutt 205, 211
Goldfields Speciality Timber Industry Group (GSTIG) 204–8
  favourite timber 207
  research on arid-zone timbers 205, 206
Goldfields Water Pipeline 48–50, 78, 79, 189, 213

Goldfields Water Supply
   Scheme 50, 212
Gompholobium 110
Gondwana 41, 52–4, 55, 93,
   157–9, 199, 246, 258
Good, Peter 300
Government of Western
   Australia, supports Wilson's
   Arnold Arboretum's
   Expedition 216
granite wattle 203
grass-leaved hakea 226
grasstrees 45–7
Gray Herbarium 313, 315
Great American Biotic
   Interchange (GABI) 54, 55
Great Salt River 80
Great South Land 14
Great Western Express (train)
   188, 215
Great Western Woodland 165,
   204, 215
green parrots 87
Grevillea 73, 87, 211, 300
   apicoloba 14, 87–8
   bipinnatifida 66
   endlicheriana 48
   eryngioides 88
   excelsior 88, 123
   hakeoides 88
   integrifolia 88
   nematophylla 228
   paradoxa 88, 94, 126
   pritzelii 88
   pterosperma 88
   quercifolia 144
   secunda 203
   striata 206
   teretifolia 88, 125
   tridentifera 88
   uninculata 88
   wilsonii 88
grey gum 17, 238, 241
grey ironbark 17, 241
grey mulga 205
groundwater hydrology and
   salinity 100–1, 106
Group Selection Schemes 163
Guichenotia 48
gully ash 17
Gumbi Gumbi 83
Gundagai, NSW 16

Gymnoschoenus sphaerocephalus
   248

hairpin banksia 276
Hakea 87, 88, 91–2, 235
   ambigua 65
   ceratophylla 144
   commutata 88
   incrassata 88, 127
   leaves 91, 92, 93
   lignotubers 92
   meissneriana 88, 128
   multilineata 226
   platysperma 122
   recurva 91, 92–3
   rostrata 273
   suberea 206
   varia 174
handkerchief tree 1
Hardy, A. D. 314
Harvard Herbaria 13, 36
   absence of Wilson's eastern
      Australian flora specimens
      16–17, 18–20, 303
   Australian-plant specimens-
      names, species in honour of
      their collectors 300
   author's research 8–10,
      13–21, 309, 315
   digitisation of Wilson
      collection 15
   Goldfields collection 211–12
   inteegrated collections 313
   Myrtaceae collection 14–17,
      153
   remnants of Expedition
      to Australasia, India and
      Africa of 1920-2 13–21
   replacement specimens 20
   selected images of Wilson's
      herbarium specimens
      313–15
   specimens from Wilson's
      Australasian Expedition in
      349–61
   see also Arnold Arboretum
haustorium 210
Heim, Arnold, on forest
   destructiveness in Australia
   286–7
hemiparasites 41, 210

Herbert, Desmond (Government
   Botanist) 28, 42, 44, 303, 304,
   305, 310, 365
   accompanies Wilson to
      Goldfields 190, 196, 204,
      210, 211, 216
   accompanies Wilson to
      southern forests 141, 146,
      147, 154
   accompanies Wilson to
      Wheatbelt 74, 76, 88, 95,
      110, 314
   Friendly Fruits and Vegetables
      28
Hibbertia 235
Himalayas 52
Hines Hill 76
Hobart 234
Hobart Botanical Gardens 200,
   244
holly-leaved hovea 148
Holmes, Oliver, *Principles of
   Physical Geology* 52
Homalospermum firmum 171
honeypot ant 203
hoop pine 244, 258, 260, 264
Hoover, Herbert 192, 193, 201
Hopper, Stephen 159
Hovea chorizemifolia 147
Howard, Richard 25
Hunt, Charles Cooke 79
Huntington, Henry 6, 28
Hunt's wells 79, 90, 189
Hutchins, Sir David 78
Hutt River poison 196
Hynes Hill 14
hypersalinity 81
Hypocalmma angustifolium 62

Ice Age 142
iceplants 83
Ichang, China 18
Imbil State forest 253, 254, 264
India 4, 52, 237, 295, 296
Indian sandalwood 209
Indian-Pacific (train) 188–9
Indigenous names 30, 76, 192–4
   preservation in WA 194–5
   standardisation 194
Indigofera enneaphylla 196
insect pests 265
inselbergs 89, 90, 95, 97
introduced trees 34

392   PLANT COLLECTING IN ANOTHER PLANET

Irby, LG (Forest Conservator in Tasmania) 247–8
ironbark 237, 254
Isopogon 36
   divergens 121
   occidentalis 148
   sphaerocephalus 173
Isotropis 110

Jacksonia 148
Jam wattle 83, 84
Japan
   expeditions 1, 18, 237
   tree loss 283
jarnockmert 91
jarrah 26, 30, 31, 32, 48, 73–4, 148
Java 296
Jerramungup 30
Jillabenan cave 241
Joobaitch (Noongar man) 40
Journama State forest 241
jungle scrub 262, 263

Kajewski, Sethrick 314
Kalgoorlie 48, 50, 79, 165, 187–8, 192–4, 207, 213, 311
   Aboriginal peoples and languages 195, 203
   ex hat shop and rabbit story 201–2
   Goldfields 188–9, 192, 195, 196, 198, 208
   Goldfields Speciality Timber Industry Group 204–8
   Indigenous art gallery and artworks 201, 202, 203
   timber felling 212–13, 215
Kanowna 192
karlkurla 192
karri 30, 31, 148, 149, 150–1, 152, 163
kauri pine 254, 257, 258
Kellerberrin 30, 76, 78, 88, 89, 97, 98
Kenya 295, 296, 304, 305
Kerri (friend, Connecticut), author spends Christmas with 309–10
Kessell, Stephen 145, 163, 187, 195
K'gari 257–8
Kingia australis 47, 47

Kings Park, Perth 29, 38
Kojonup 30
Kondinin 30
Königin Luise (ship) 6, 23
Korea expeditions 1, 18, 247
Kulikup 30
Kulin 30
Kununoppin 30
Kunzea 90
kurti-kurti 192

lace sheoak 209
Ladiges, Pauline 157
land clearing
   as "arboricidal mania" 252, 262, 263, 284, 291, 301
   brigalow 263
   Bulga Plateau, NSW 250–2
   and destruction of trees 104–5
   Fraser Island 261
   Goldfields 212–13
   Government insists on 102, 103
   impact on salinisation 100–1, 102
   and loss of trees in agricultural areas 252–3
   New Zealand 302
   and soil loss 104
   southern forests 163
   to "improve the country" 107, 249, 251, 286, 291
   tropical rainforests 262–3
   warnings about impact of 100–2, 287–8, 288
   Wheatbelt 76–7, 81–2, 97, 98, 100–3
   Wilson appalled by 76–7, 98, 102–3, 104, 215, 250–2, 291–2, 302, 305
Lane-Poole, Charles (Conservator of Forests, WA) 23, 26, 27, 37, 43, 101, 187, 195, 211, 213, 216, 284, 298, 304–5, 310
   accompanies Wilson to southern forests 141, 142, 143, 146, 150–1, 153, 163–4
   concept of forest preservation 250, 289
   ideas about agriculture and forestry 236
   proposes national forestry school in Canberra 242–3
   sends plant material home from WA to Boston 165
   The Primer of Forestry for Western Australia 195
   travels with Wilson and Herbert 28, 37, 42, 48
Laurasia 51, 52, 54–5, 199
leaf colour
   Australian flora 33–4
   banksias 144
   eucalypts 35
leaf position, Australian flora 33, 34
Leptospermum 87, 148
Leschenault de la Tour, Jean-Baptiste 300
Leschenaultia 300
Leucadendron 36
Leucopogon 147
Leucospermum 36
lignotubers 16, 92, 161
Lilium regale 7
Lindley, John 110
Linton, James Walter 195–6
Litsea glutinosa 209
Livistona australis 239
Lophostemon confertus 257
Ludlow 141, 142
   forestry school 142
Ludlow Tuart Forest 142, 143

maalok 75–6
McHarg, Ian, *Design with Nature* 289
McLean, Forman 235
Macrozamia riedlei 48, 49
Madagascar 52
Maiden, Joseph 237–8
Main, Barbara York, *Between Wodjil and Tor* 104, 107–8
Maireana spp. 83
Malaya 296
mallee 75, 90, 91, 94, 95, 96
mallee hen 87
mallet 75, 81, 199, 205
Mangart 83
Manjimup 30
manna gum 17, 146
manna wattle 98
Manning River region 238

INDEX   393

marri 30, 31, 73
Marsdenia australis 193, **194**, 203, 211
martilgarang 29
Maryborough, Qld 257
medicinal oils 162
Mediterranean climate 84
megafauna 80–1, 93–4
Melaleuca 42, 83, 87, 90, 95, 148, **227**
   alternifolia 162–3
   cuticularis **169**
   glaberrima **118**
   glauca **170**
   lanceolata 30
   linariifolia 311
   preissiana **43**
   rhaphiophylla **44**
   sheathiana **225**
   squarrosa **270**
   thymifolia **271**
   viridiflora 297
Melbourne 233, 242, 244, 301
Menzies, Archibald 37, 300
Merredin 30, 76, 78, 79, 86, 88, 89, 94, 189, 198
merrit 205, **222**
Merrotsy, Captain A. L. 314
Mesembryanthemum spp. 83
messmate **268**
Miles, Harriette 201
minniritchi 205, 206
Mirning group of languages 195
Mitchell, Sir James 24
Mitchell's Island State Forest 238
Mizzi, Johnn 83
moa 302
Molloy, Georgiana 300
Mombasa 295, 296
Montana (ship) 295
Monterey pine 23
moodjar 40
moonah 30
'Mooranoppin' (property) 81
Moore, David Moresby 314
Moowattin, Daniel 300
Morrison, Alexander 110–11
Moss Vale 16, 241
mottlecah 29
Mount Caroline 98
Mount Crawford Forest Reserve 235

Mount Lofty Ranges 235
mountain ash 15, 146, 241, 245
   logging 245
   named trees 245
Muir, John 259, 289
Mukinbudin 30
mulga 204
Mummballup 30
Mundaring 31, 48, 78, 79, 212
mungee 40
Munro, Forester (in Kenya) 305
Munsell Soil Color System 36
Muntadgin 30
Musca vetustissima 200
mycorrhizal plants 41, 42
Myrtaceae 14–17, 144, 257

Nabiac 238
Nairobi 295
Nannup 30
Narrikup 30
Narrogin 30
narrow-leaved ironbark 16, 241
narrow-leaved messmate 17, 241
National Parks, rise of 289–90
native apricot 83
native birds 30, 38, 81, 87
   habitat 143
   for insect control 264
   protection 265–6
native fauna, unaffected by poison plants 111
native peach **202**
native pomegranate **115**
native trees
   lack of in parks and gardens 285
   and shrubs, availability in nurseries 285–6
native vegetation, replanting with 106–7
native willow 206
Natural Colour System of Sweden 36
A Naturalist in Western China with Vasculum, Camera, and Gun (Wilson) 1
New Caledonia 258
New South Wales 16–17, 234, 236–41
New Zealand
   forest clearing and biodiversity loss 302

   as part of Gondwana 52, 53
   Wilson's collecting in 234, 242
New Zealand conifers 200, 258
New Zealand flora
   ancient flora 157
   differences from Australian flora 36, 51, 54
   glass-plate negatives 20
Nicolle, Dean 160
Nijptangh (ship) 48
noble fir 146
nonmycorrhizal plants 41, 42
Noongar
   bush medicine 83
   coastal plain changes 142
   holds *Nuytsia floribunda* sacred 40–1
   names of geographical features 30
   place names 30, 76
   plant names 29, 30, 45, 141
   scar trees 146
Noongar Country 30, 81–2
Norfolk Island 258
Norfolk Island pine 244
Nornalup 30, 147, **149**
Norseman 211
North America 54, 55
   tall trees 15, 16, 146
Northam 76, 77
Northern Hemisphere 50, 54–5, 305
Nothofagaceae 158
Nothofagus 53, 54, 245–6
   gunnii 54, **245**
Nullarbor Plain 204, 215
Number 1 Rabbit Proof Fence 94, 201
Number 2 Rabbit Proof Fence 201
Nuytsia floribunda 39, 40–1, **40**
   as hemiparasite tree 41

O'Connor, Charles Yelverton 48–50
old growth forests, preservation 306
Olmsted, Frederick Law 17
Ongerup 30
orchids 110
Orites revolutus **272**
Oxylobium 110, 111

heterophyllum 196
parviflorum see *Gastrolobium parviflorum*
retusum see *Gastrolobium retusum*
spectabile 196

paleo-system 83, 191
Pangea 199
paperbarks 42–4, 44, 45
Parker, Captain R. N. 4
parks and gardens
   lack of native trees in 285
   mixture of exotic and native trees 286
Patagonia 157
Pearson, Lisa 28
Pelloe, Emily 196
peppermint tree 143
Percy-Bowers. Julia 365
Persoonia 95
   saundersiana var. *diadena* 120
Perth, Wilson's arrival 23
Perth Zoological Gardens 29
Petrified Forest National Park, Arizona 259
Petrophile
   circinata 119
   diversifolia 172
   linearis 63
Picea sitchensis 146
Picton 16, 241
pied magpie 87
Pimelea 144, 148, 235
   hispida 182
   imbricata 131
   lehmanniana 181
Pinus
   insignis 243
   nigra 243
   patula 244
Pittosporum
   angustofolium 83, 206
   phylliyreoides see *P. angustifolium*
pituri 95
Pityrodia 95
pixie bush 206
plant explorers in Australia who have had species named in their honour 300
Plant Hunting (Wilson) 1–2, 89, 233, 290, 304

plants
   colour charts 35–6
   evolution with megafauna 93
   leaf colour 33–5
plate tectonics 52
Podocarpaceae 158
poisonous plants 48, 49, 110–12, 196
Porongurup 30
Port Said, Egypt 6
powderbark wandoo 73, 146
Prasophyllum sargentii 110
prickly poison 196
Pritzel, Ernst 213
Protea 36
Proteaceae 36, 88, 144, 158, 206
proteas 301
Pseudotsuga menziesii 146
Pterostylis sargentii 110
Puccinia graminis 85
Pumping Station Number 1 48, 50
Purdie, Alexander 109–10
Purdom, William 252

Quairading 26, 76, 88
Qualeup 30
quandong 202, 203
quarantine regulations, US, prevents importing live plants or seeds 2, 3, 299, 303
Queensland 234, 253–66, 295
Queensland box 257
Queensland rainforest photos 13, 20
Queensland Waratah 262
Quercus 243

rabbits 87, 201, 202
railway track gauges 189, 215
rainforest ecosystems 254
rainforests 262
re-afforestation
   NSW 238
   Qld 264
red cedar 238, 239
red gum 17, 241
red mahogany 17, 241, 257
red mallee 209
red morrel 205
red stringybark 257
red-tailed black cockatoos 81
red tingle 146, 148, 149, 150

red-flowering gum 154
red-flowering paperbark 297
redwood (*Eucalyptus transcontinentalis*) 205
redwood (*Sequoia sempervirens*) 15, 146
regal lily 7
reptiles 87
resprouting 161, 214
revegetation practices
   Wheatbelt 106–8
   Wheatbelt towns 108–9
ribbon gum 17
ring-barking 102, 286–7
Ripper, Mr 81
river jam 209
Robinson, Robert T. 28
Roe, John Septimus 100, 110, 196
Roe's poison 196
Roslindale (Boston suburb) 216–17
Royal Botanic Gardens of Sri Lanka, Peradeniya 6
Royal Society of Queensland 261
rubber plantations 296
rusty poison 196

Salicornia australis 80
salinity/salinisation, Wheatbelt 81, 98, 100–1, 102–3, 105
salmon gum 27, 73, 77–8, 77, 81, 205, 207–8, 214
salmon white gum 27
salt-based beauty products 83
salt gum 205
salt lakes 79, 80, 82, 83, 100, 191
Salt River mallett 110
salt-tolerant plants 83
saltbush 83
samphire 80, 83
sandalwood 95, 208–9
   export trade 209
   as hemiparasite 210
   licenses 210
   slow-growing tree 209–10
Sandford Rock 95, 96
sandplains 86–7
sands (Fraser Island) 257
Santalum 148
   acuminatum 202, 203
   album 209

spicatum 95, 206, 208–9
Sarcostemma australe 196
Sargent, Charles Sprague (Director of Arnold Arboretum) 4, 18, 19, 21, 28, 302, 314
    complains about US quarantine regulations 299
    on importance of good plant collectors and photographs 4–5
    on the purpose of Wilson's Australia trip 8
    suggests Wilson extends his journey to India, Kenya and South Africa 295–6
    time and commitment taken by Wilson on plant collecting 299–300
    Wilson's letters to 200, 216, 237, 244
    Wilson's letters to on his expedition in WA 164–5, 298
    Wilson's letters to on his travels to WA 5, 6
Sargent, Oswald (plant collector in York, WA) 109–10, 111
Scaevola 211
scarlet banksia **176**
scarlet flowering gum 155, 283
scented sandalwood 206
sclerophyllous vegetation 92, 144, 161, 217
Scott, Pauline and James 89, 95, 97
scribbly gum 17, 241
'scrub' 261, 262, 301
scrub box 257
Sebald, WG 249
sedge 248
Sequoia sempervirens 15, 146
Sequoiadendron giganteum 146
sheep farming
    ecological damage to Country 101
    total-ecosystem approach 105–7
showy dryandra **177**
sighing sheoak **74**
silky pear 193, **194**
silver-topped gimlet **214**
Sitka spruce 146

sleepers 31, 32
slender poison 196
Smoke that Thunders (Wilson) 2, 8, 15, 33, 54, 55, 143, 290, 300, 301, 310
    does not name those who assisted him in Australia 305
    no mention of his environmental anger in newspaper reports 301
snow gum 17, 146, 241, **241**
sodium monofluoroacetate 111, 112
soil colour 35–6
soil health/soil management 104, 106, 107
soil nutrient status 161–2
Solander, Daniel 33
Solanum centrale 203
soldier settlement schemes 163
Sons of Gwalia Mine at Kanowna 192
South Africa 295, 296, 301
South America 51, 52, 54
South Australia 234–6
south-western flora 31
    forests 144, 146, 147–53, 314
    species richness 50
southern Africa, close botanical association with Australia 36
southern beech 53, 54
southern bluegum 146
southern cross (plant) 148
Southern Cross (town) 79
Southern Cross silver mallee 95, **96**, 98
southern forests 141–89
Southern Hemisphere 54
    flora, differences from Northern Hemisphere 2, 50, 305
    landscapes, differences from Northern Hemisphere 54–5
    plant families evolution 157–8
Southwest Australian Floristic Region 21
Sowden, Sir William 234–5, 289
speciation 38, 155, 159, 191
spike poison 196

Spongberg, Stephen, *A Reunion of Trees: The Discovery of Exotic Plants and Their Introduction into North American and European Landscapes* 299
spotted gum 238, 285
Straham 247
Stratham 208
strychnine bush 196
Strychnos lucida 196
Stylidium scandens 148
sugarcane 262
Swain, Edward (Queensland Director of Forests) 253, **254**, 257, 265, 289, 305, 310, 314
swamp karri 153
swamp paperbark **44**
Swan Coastal Plan 31, 40, 73, 141–4
    changes during last Ice Age 142
Swan River 30–1
Swan River mahogany 31
Sydney 159, 234, 237
Sydney peppermint 16, 241
Synaphea 148
Syncarpia glomulifera 257

tall boronia **180**
tall trees
    Australia 15, 16, 144, 146
    North America 15, 16, 146
tallowwood 251, 254, 257
Tambellup 30
Tammin 30, 76, 88, 104
Taree 238
Tarn Shelf 245
Tasmania 234, 244–8, 301
    use of 'waste lands' for exotic conifers 247–8
Tasmanian remnant forests 53
Tasmanian tiger 215
Tasmanian tree fern 246
Tasmanian Wilderness World Heritage Area 246, 248
tea tree 83
Tecticornia spp. 83
The Lakes 88
Thelymitra sargentii 110
thick-leaved poison 86
Thursday Island 295, 297
thylacine 215
thyme honey myrtle **271**

timber 83, 101
  arid zone, research on 205–8
  imports 252
  management, Bulga Plateau, NSW 250–2
  proper use, Goldfields region 204–8
  salmon gum 208
  warping 208
  wasted for mediocre purposes or burnt, Australia 31–2, 163, 212, 250–2, 261, 262, 266, 286, 290–1, 302
  see also forests/forestry/forest management
Toodyay 76, 194
Torst, Mandrop 48
Totadgin/Totadgin Rock 89–91, 90, 93–4, 105
total-ecosystem approach to farming 105–6
Toulon, France 5
tourism 245, 248, 261, 302
Trans-Australian railway 188
transportation disasters 18–19
tree colour 34
trees
  creation of public sentiment for 283–5
  lack of native trees in parks and gardens 285
  loss by removal or neglect 283, 305–6
  removal by ring-barking 102, 286–7
  Wilsons' enthusiasm for 283
  world census 2–3, 234, 302
  see also forests; land clearing; timber; woodlands
tropical palms 239, 240
tropical rainforests, clearing of 262–3
tuart 30, 73, 141–2, 209, 314
  genetic diversity 142
Tuart Forest National Park 142
Tuncurry prison plantation 238, 239
Turner, Neil 208, 209
turpentine bush 206, 257

US quarantine regulations impact on collecting for

introductions 2, 3, 299, 303

Valley of the Giants 149, 153
Vancouver, Commander George 300
Vavilov, Nikolai 85
Velleia discophora 196
Verticordia 95
  acerosa 61
Victoria 234, 242–4, 248
Victoria Falls 295
Victorian Forestry School, Creswick 242, 243
von Mueller, Baron Ferdinand 218

Wadi group of languages 195
Wagin 30
Wallum banksia 258, 277
Walpole 149
Walter Street Cemetery, Roslindale 217, 218
Wanderrie wattle 203
wandoo 30, 73, 78, 104, 146, 147, 209, 286
Wangkathaa (or Wongutha or Wongi) people 192–3
warping 208
water oak 238
water transport, trees 199
water trees 92
weather conditions 5, 6, 78, 239, 241
Weddellian Biogeographic Province 157
wedge-leaved rattlepod 196
wedge-tailed eagle 87
Wegener's theory of continental drift 51
Wemba Wemba language 75
Werner's Nomenclature of Colours 35
West Australian Christmas tree 39, 40
Western Australia
  Goldfields 188–231
  locations and timetable of Wilson's travels 25
  south-western forests 141–89
  "veritable botanic garden" 23–71
  Wheatbelt 73–141, 189, 201
  Wilson's arrival 6, 23–4

Western Australian Government, pushes land clearing 102, 103
Western Australian Herbarium 110, 303
  Wilson's specimen's in 365–81
western myall 204, 205
Westonia 14, 76, 86, 88, 95, 189, 211
wheat belt, Gawler 236
wheat breeding 84–6
wheat cropping 80, 83–4
wheat rust 85
Wheatbelt 73–141, 189, 201
  flowering shrubs 86
  Government encourages land clearing 102, 103
  land clearing 76–7, 81–2, 97, 98, 100–2
  revegetation practices 106–8
  salinity 102–3
  sandplains 86–7
  woodlands 76, 81, 97, 101
Wheatbelt towns, need for revegetation 108–9
wheatbelt wandoo 73
White, Cyril T. (Queensland Government Botanist) 28, 253–4, 254–5, 257, 264, 289, 305, 314
White, Captain Samuel 256, 257, 260, 289
white cypress pine 197
White Flag Lake 191
white gum 16, 211, 241, 269
white mahogany 238, 254
white stringybark 17, 241
whitewood 196
'Whollandra' (property) 81
Widgiemooltha 194, 195, 211
Willdampia formosa 300
Wilson, Ernest 32, 150, 152, 246
  absence of environmental critique and exposition on his return to Harvard 301–3
  arrival in Perth 6, 23–4
  arrives back in Harvard 296–8, 299
  collecting for Arnold Arboretum 25
  killed in motor accident 217

newspaper reports on the purposes for his expedition to Australasia 2–3, 6–8, 23–5, 234–5, 236
as President of the Horticultural Club of Boston 301
prior Asia expeditions 1, 5, 7, 18
timetable and collecting sites in WA 25
timetable and travels in eastern Australia and New Zealand 234
train from Kalgoorlie to Adelaide 215, 311
train from Perth to Kalgoorlie 188–9
travels beyond Australia 295–6
voyage from France to Australia 5–6, 23
Wilson, Nellie 212, 217
Wilson, Thomas Braidwood 88
Wilson's transcribed diaries, 1920-1, Australia 316–48
Windsor, NSW 237
Wingham Brush' 238
winter wheat 83–4
witchetty grubs 203
Woggerup 30
Wollemi pine 259
Wollemia 259
   nobilis 259
woodcutters camp 213
woodlands 76, 81, 97, 101, 204
   clearing of 102, 103, 212–13
   eucalyptus–banksia 144
   Goldfields 212–13
   Great Western 165, 204, 215
   impact of clearing 101–2
   regeneration and resprouting 213–15
   Wilson appalled by destruction of 215
woodworking 208, 209
woody pear 88, 89
Woogenellup 30
wool exports 101
woolly poison 196
woollybutt 17
World Heritage rainforest 262
Wright, Judith 263

Xanthorrhoea 46–7, 87, 157
   nana 113
   preissii 45, 46
Xanthosia 148
   rotundifolia 148
Xylomelum angustifolium 88, 89

Yalbarrin 30
yellow box 17
yellow tingle 148, 153
Yilgarn Craton 76, 95
Yilgarn River 79, 81, 83
York 76, 88, 109, 111
York gum 73, 98, 104, 105
York Road poison 110
Yoting 89, 106

zamia palm 48, 157
Zeehan 247